Geodesign by Integrating Design and Geospatial Sciences

GeoJournal Library

VOLUME 111

Managing Editor

Daniel Z. Sui, Columbus, Ohio, USA

Founding Series Editor

Wolf Tietze, Helmstadt, Germany

Editorial Board

Paul Claval, France
Yehuda Gradus, Israel
Sam Ock Park, South Korea
Herman van der Wusten, The Netherlands

GEODESIGN

Managing Editor

Daniel Z. Sui, Columbus, Ohio, USA

Editorial Board

Danbi J. Lee, The Netherlands
Eduardo Dias, The Netherlands
Henk J. Scholten, The Netherlands

More information about this series at http://www.springer.com/series/13477

Danbi J. Lee • Eduardo Dias • Henk J. Scholten
Editors

Geodesign by Integrating Design and Geospatial Sciences

Editors
Danbi J. Lee
Department of Spatial Economics, SPINlab
VU University Amsterdam
Amsterdam
The Netherlands

Henk J. Scholten
Department of Spatial Economics, SPINlab
VU University Amsterdam
Amsterdam
The Netherlands

Eduardo Dias
Department of Spatial Economics, SPINlab
VU University Amsterdam
Amsterdam
The Netherlands

Series Title   ISSN 0924-5499   ISSN 2215-0072 (electronic)
Subseries Title ISSN 2405-4828   ISSN 2405-4836 (electronic)
ISBN 978-3-319-08298-1   ISBN 978-3-319-08299-8 (eBook)
DOI 10.1007/978-3-319-08299-8
Springer Cham Heidelberg New York Dordrecht London

Library of Congress Control Number: 2014946298

© Springer International Publishing Switzerland 2014
This work is subject to copyright. All rights are reserved by the Publisher, whether the whole or part of the material is concerned, specifically the rights of translation, reprinting, reuse of illustrations, recitation, broadcasting, reproduction on microfilms or in any other physical way, and transmission or information storage and retrieval, electronic adaptation, computer software, or by similar or dissimilar methodology now known or hereafter developed. Exempted from this legal reservation are brief excerpts in connection with reviews or scholarly analysis or material supplied specifically for the purpose of being entered and executed on a computer system, for exclusive use by the purchaser of the work. Duplication of this publication or parts thereof is permitted only under the provisions of the Copyright Law of the Publisher's location, in its current version, and permission for use must always be obtained from Springer. Permissions for use may be obtained through RightsLink at the Copyright Clearance Center. Violations are liable to prosecution under the respective Copyright Law.
The use of general descriptive names, registered names, trademarks, service marks, etc. in this publication does not imply, even in the absence of a specific statement, that such names are exempt from the relevant protective laws and regulations and therefore free for general use.
While the advice and information in this book are believed to be true and accurate at the date of publication, neither the authors nor the editors nor the publisher can accept any legal responsibility for any errors or omissions that may be made. The publisher makes no warranty, express or implied, with respect to the material contained herein.

Printed on acid-free paper

Springer is part of Springer Science+Business Media (www.springer.com)

Preface

By Henk J. Scholten and Jack Dangermond

The first GIS Summer Institute was held in Amsterdam in August 1989 at the School of Architecture, Town Planning, and Landscape. The Institute's work foresaw the evolution into a digitized world and the adoption of computerized spatial analysis. 22 years later, two of the original participants from that event, Henk Scholten and Jack Dangermond, met again. At that meeting, Jack told Henk about his past professor, Carl Steinitz, who was writing a book about geodesign.

It was an inspiring new concept; a new perspective on an old and familiar problem. In the geodesign framework long-term models are coupled with short-term (impact) models. In its vision, spatial planners should use the outcomes of the long-term models to assess whether the developments occurring in a region necessitate intervention. If true, several solutions to the encountered issues are almost always available. The most promising solutions are worked out in scenarios and the effects are calculated using the impact models. This provides the possibility to test whether the planners' intended goals are attained and what negative effects might occur. If the outcomes are unsatisfactory, go back a step and try again.

In a way, this process is what we are doing today with geodesign. It is a framework for how we can design together to solve complex problems. We step back, using traditional techniques of spatial analysis and modeling, in order to step forth towards innovative technologies and collaboration frameworks.

Our world faces serious challenges, and it's clear that we need to work together to collectively create a better future. We need to leverage our very best brains, our best creative talent, our best design talent, and our best science, and use all of these combined to create a better future. To meet the geographic challenges we face, we need to grow geodesign from a concept understood by a few to a framework used by all. We need to inform the world about the value of geodesign, while at the same time making it easy to implement and use throughout organizations and across society.

Inspired by the successes of Geodesign Summits by Esri in Redlands, California, we organized the first Geodesign Summit in Europe in September of 2013. Designers, planners and geospatial scientists from around the world gathered to share ideas on how to design with spatial information in Europe. This book is testament to the momentum that geodesign is gaining both academically and professionally.

We're confident that through the continued good work of many, geodesign will in due course be widely adopted and recognized as one of the most important ideas to come out of this century.

Acknowledgments

Great thanks are owed to the speakers and participants at the 2013 Geodesign Summit Europe and especially the authors of this book, who graciously agreed to share their geodesign experiences to the world. This book was made possible only through the hard work and support of the VerDuS-programme ('Verbinding Duurzame Steden') of the Netherlands Organization for Scientific Research (NWO), the VU University Amsterdam, Geodan, Esri Europe, Esri Headquarters, and the GeoFort.

Contents

Part I Introduction

1. Introduction to Geodesign Developments in Europe 3
    Danbi J. Lee, Eduardo Dias and Henk J. Scholten

2. Which Way of Designing? 11
    Carl Steinitz

Part II Resilience and Sustainability

3. Energy Resilient Urban Planning.............................. 43
    Perry Pei-Ju Yang

4. PICO: A Framework for Sustainable Energy Design.............. 55
    Steven Fruijtier, Sanneke van Asselen, Sanne Hettinga
    and Maarten Krieckaert

5. Holistic Assessment of Spatial Policies for Sustainable
    Management: Case Study of Wroclaw Larger
    Urban Zone (Poland)... 71
    Jan Kazak, Szymon Szewranski and Pawel Decewicz

6. Recent Applications of a Land-use Change Model in Support of
    Sustainable Urban Development 87
    Eric Koomen and Bart C. Rijken

7. Using Geodesign to Develop a Spatial Adaptation Strategy for
    Friesland... 103
    Ron Janssen, Tessa Eikelboom, Jos Verhoeven and Karlijn Brouns

8. Geodesign to Support Multi-level Safety Policy
    for Flood Management 117
    Sanneke van Asselen, Henk J. Scholten and Luc Koshiek

9. The Multi-Layer Safety Approach and Geodesign: Exploring
    Exposure and Vulnerability to Flooding 133
    Mark Zandvoort and Maarten J. van der Vlist

10 **Interactive Spatial Decision Support for Agroforestry Management** 149
André Freitas, Eduardo Dias, Vasco Diogo and Willie Smits

Part III Heritage and Placemaking

11 **History Matters: The Temporal and Social Dimension of Geodesign** 173
Jan Kolen, Niels van Manen and Maurice de Kleijn

12 **Urban Landscape archaeology, geodesign and the city of rome** 183
Gert-Jan Burgers, Maurice de Kleijn and Niels van Manen

13 **GIS-based Landscape Design Research: Exploring Aspects of Visibility in Landscape Architectonic Compositions** 193
Steffen Nijhuis

14 **3D LOS Visibility Analysis Model: Incorporating Quantitative/Qualitative Aspects in Urban Environments** 219
Dafna Fisher-Gewirtzman

15 **Space Syntax in Theory and Practice** 237
Akkelies van Nes

16 **A Standard-based Framework for Real-time 3D Large-scale Geospatial Data Generation and Visualisation over the Web** 259
Massimo Rumor, Eduard Roccatello and Alessandra Scottà

17 **Crowd Sourced Public Participation of City Building** 271
Ana Sanchis, Laura Díaz, Michael Gould and Joaquín Huerta

Part IV Adopting Geodesign Thinking

18 **Geodesigning 'From the Inside Out'** 287
Kitty Currier and Helen Couclelis

19 **People Centered Geodesign: Results of an Exploration** 299
Simeon Nedkov, Eduardo Dias and Marianne Linde

20 **Enhancing Stakeholder Engagement: Understanding Organizational Change Principles for Geodesign Professionals** 315
Lisa A. McElvaney and Kelleann Foster

21 **Geodesign in Practice: What About the Urban Designers?** 331
Peter Pelzer, Marco te Brömmelstroet and Stan Geertman

22 **Open Geospending: Bridging the Gap Between Policy and the Real World** 345
Egbert Jongsma

23 **Towards Geodesign: Building New Education Programs and Audiences** 357
John P. Wilson

Contributors

Sanneke van Asselen Geodan, Amsterdam, The Netherlands

Karlijn Brouns Ecology and Biodiversity, Utrecht University, Utrecht, The Netherlands

Gert-Jan Burgers Faculty of Arts, VU University Amsterdam, De Boelelaan 1105, 1081 HV, Amsterdam, The Netherlands

Helen Couclelis Department of Geography, Santa Barbara, CA, USA

Kitty Currier Department of Geography, Santa Barbara, CA, USA

Pawel Decewicz Center of Spatial Management, Warszawa, Poland

Eduardo Dias Faculty of Economics and Business Administration, Department of Spatial Economics, SPINlab, VU University Amsterdam, Amsterdam, The Netherlands

Geodan, Amsterdam, The Netherlands

Laura Díaz Institute of New Imaging Technologies, Universitat Jaume I, Castellón, Spain

Vasco Diogo Faculty of Economics and Business Administration, Department of Spatial Economics, SPINlab, VU University Amsterdam, Amsterdam, The Netherlands

Tessa Eikelboom Institute for Environmental Studies, VU University Amsterdam, Amsterdam, The Netherlands

Dafna Fisher-Gewirtzman Faculty of Architecture and Town Planning, Technion - Israel Institute of Technology, Haifa, Israel

Kelleann Foster College of Arts and Architecture, Pennsylvania State University, University Park, PA, USA

André Freitas Faculdade de Ciências e Tecnologia, Universidade Nova de Lisboa, Caparica, Portugal

Geodan S&R, Amsterdam, The Netherlands

Steven Fruijtier Geodan, Amsterdam, The Netherlands

Stan Geertman Urban and Regional research centre Utrecht (URU), Utrecht University, Utrecht, The Netherlands

Michael Gould Institute of New Imaging Technologies, Universitat Jaume I, Castellón, Spain

Sanne Hettinga Geodan, Amsterdam, The Netherlands

Joaquín Huerta Institute of New Imaging Technologies, Universitat Jaume I, Castellón, Spain

Ron Janssen Institute for Environmental Studies, VU University Amsterdam, Amsterdam, The Netherlands

Egbert Jongsma Netherlands Court of Audit, The Hague, Netherlands

Jan Kazak Wroclaw University of Environmental and Life Sciences, Wroclaw, Poland

Maurice de Kleijn Faculty of Economics and Business Administration, Spatial Economics, SPINlab, VU University Amsterdam, De Boelelaan 1105, 1081 HV, Amsterdam, The Netherlands

Jan Kolen Faculteit Archeologie & Centre for Global Heritage and Development (CGHD), Universiteit Leiden, Leiden, The Netherlands

Eric Koomen Faculty of Economics and Business Administration, Department of Spatial Economics, SPINlab, VU University Amsterdam, Amsterdam, The Netherlands

Luc Koshiek Bevelandseweg 1, Hoogheemraadschap Hollands Noorderkwartier, Heerhugowaard, The Netherlands

Maarten Krieckaert Geodan, Amsterdam, The Netherlands

Danbi J. Lee Department of Spatial Economics, SPINlab, VU University Amsterdam, Amsterdam, The Netherlands

Marianne Linde TNO Built Environment, Delft, The Netherlands

Niels van Manen Faculty of Economics and Business Administration, Spatial Economics, SPINlab, VU University Amsterdam, De Boelelaan 1105, 1081 HV, Amsterdam, The Netherlands

Lisa A. McElvaney Business Transformation Consulting, Redlands, CA, USA

Simeon Nedkov VU University Amsterdam, Amsterdam, The Netherlands

Steffen Nijhuis Faculty of Architecture and the Built Environment, Delft University of Technology, Delft, The Netherlands

Peter Pelzer Urban and Regional research centre Utrecht (URU), Utrecht University, Utrecht, The Netherlands

Bart C. Rijken Faculty of Economics and Business Administration, Department of Spatial Economics, VU University Amsterdam, Amsterdam, The Netherlands

PBL Netherlands Environmental Assessment Agency, The Hague, The Netherlands

Eduard Roccatello 3DGIS, Padua, Italy

Massimo Rumor University of Padua, 3DGIS, Padua, Italy

Ana Sanchis Institute of New Imaging Technologies, Universitat Jaume I, Castellón, Spain

Henk J. Scholten Faculty of Economics and Business Administration, Department of Spatial Economics, SPINlab, VU University Amsterdam, Netherlands

Geodan, Amsterdam, The Netherlands

Alessandra Scottà Geodan, Amsterdam, The Netherlands

Willie Smits Institute of Technology Minaesa (ITM Tomohon), Jl. Stadion Selatan Walian, Tomohon, Sulawesi Utara, Indonesia

Carl Steinitz Harvard Graduate School of Design, Harvard University, Cambridge, MA, USA

Szymon Szewranski Wroclaw University of Environmental and Life Sciences, Wroclaw, Poland

Marco te Brömmelstroet Amsterdam Institute for Social Science Research (AISSR), University of Amsterdam, Amsterdam, The Netherlands

Sanneke van Asselen President Kennedylaan 1, Geodan, Amsterdam, The Netherlands

Maarten J. van der Vlist Wageningen UR Landscape Architecture Group, Rijkswaterstaat, Wageningen, The Netherlands

Akkelies van Nes Faculty of Architecture, Delft University of Technology, Delft, The Netherlands

Jos Verhoeven Ecology and Biodiversity, Utrecht University, Utrecht, The Netherlands

John P. Wilson Spatial Sciences Institute, University of Southern California, Los Angeles, CA, USA

Perry Pei-Ju Yang School of City and Regional Planning and School of Architecture, College of Architecture, Georgia Institute of Technology, Atlanta, USA

Mark Zandvoort Wageningen UR Landscape Architecture Group, Wageningen, The Netherlands

Part I
Introduction

Chapter 1
Introduction to Geodesign Developments in Europe

Danbi J. Lee, Eduardo Dias and Henk J. Scholten

1.1 When Geodesign Crosses over the Atlantic

In 1985 the Dutch National Planning Department was charged with the formidable task of drawing up a new masterplan describing the National Spatial Policy. This resulted in the creation of the Fourth Note on Spatial Planning (Vierde Nota over de Ruimtelijke Ordening). Naturally, the best known and most experienced spatial planners and landscape architects were involved in drafting the document. For the first time in its history a working group was tasked with the preparation of a spatial information system (Scheurwater 1984) which lead to the acquisition and implementation of the first GIS application in Europe, namely Esri's ArcGIS 3.0. Up until that moment the term 'Geographical Information System' was little known in the Dutch language.

Although the Vierde Nota has become a zenith of Dutch spatial planning, and the first use of geographical information for a large scale spatial issue is considered a breakthrough, it presented a number of challenges (Scholten and Meijer 1988). It became apparent, for instance, that the viewpoints and culture represented in scientific models did not match those of planners and designers. It was also an interesting challenge to establish a comprehensive collection of digital geographical data of The Netherlands and its neighbouring countries.

Based on the experiences gained in drafting the Vierde Nota, the decision was made to host the first GIS Summer Institute in Amsterdam in August 1989. It was held in the 'Academie van Bouwkunst', the seventeenth century home of Amsterdam's School of Architecture, Town Planning and Landscape (Scholten and

---

D. J. Lee (✉) · E. Dias · H. J. Scholten
Faculty of Economics and Business Administration, Department of Spatial Economics,
SPINlab, VU University Amsterdam, Amsterdam, The Netherlands
e-mail: danbi.lee@vu.nl

E. Dias
e-mail: ess580@vu.nl

H. J. Scholten
e-mail: h.j.scholten@vu.nl

© Springer International Publishing Switzerland 2014
D. J. Lee et al. (eds.), *Geodesign by Integrating Design and Geospatial Sciences,*
GeoJournal Library 1, DOI 10.1007/978-3-319-08299-8_1

Stillwell 1990). In his concluding remarks at the conference, Peter Nijkamp stated that he thought the term GIS would no longer be in use 20 years hence. He predicted that it would be replaced by so-called CIAs (Computerized Information Analysis) for urban and regional management. In his speech, Nijkamp touched on an issue that we still encounter today—the need for careful evaluation of alternative courses of action due to the high costs of misinformation (Nijkamp et al. 1990). In fact, several contributions from the 1990 proceedings put forward the notion that the differences between planners and analytical scientists and models are reflected by the distinctive systems they use, i.e. CAD versus GIS. Bridging the gap between these systems is seen as the main ambition of geodesign (see for example Wood 1990).

In 2013, the momentum for geodesign in Europe was earmarked by the first Geodesign Summit Europe, hosted by Esri, Geodan and the Vrije Universiteit Amsterdam at the GeoFort in the Netherlands. Researchers and planners from 28 different countries gathered over a few intensive days to discuss the way spatial issues should be approached in light of an extraordinarily digitized and technology-based world. The main difference with a quarter century ago was the growing request to connect scientific models of processes with proposed design solutions in order to understand design impact in a collaborative way between policy makers, scientists, designers, and citizens.

1.2 Geodesign as Concept, Method, and Product

Formal definitions of geodesign appear from around 2010 (e.g. Dangermond 2010; Flaxman 2010; Zwick 2010; Ervin 2011; Steinitz 2012). Brewed from these developments, we define geodesign to be an iterative design and planning method whereby an emerging solution is influenced by (scientific) geospatial knowledge derived from geospatial technologies. Whereas traditional planning and design processes separate context analysis, design, and evaluation into explicit steps, geodesign integrates the exploration of ideas with direct evaluation in the same moment, generating an advanced design solution. In other words, the design impact can be examined through geospatial technology (simulations, modeling, visualization, and communication of design impacts) and be immediately fed back into the evolution of a design. This yields a fitter, more robust and context-sensitive design solution. Geodesign enables systems-thinking, which makes it an attractive approach for today's complex, dynamic, and multi-disciplinary design challenges.

Some argue that geodesign is merely an alluring alias for design methods that have been practiced for many decades already. This is, to a degree, quite true if we dissect the components of our own definition and look at the series of events leading to this point. The academic discourse on spatial and context analysis using overlays began with the first hand drawn overlays documented by landscape designer Warren Henry Manning in the 1913 Billerica Town plan (see Steinitz et al. (1976) for an historical overview of hand drawn overlays) (Manning 1913). In the 1960s, marked by British–Canadian geographer Roger Tomlinson's development of the first geographic information system, the promise of digitized maps sparked a fantastic dream of computerized planning support. All the while, in 1965 American city planner Britton Harris proposed integrating "sketch planning" (as the drawing of alternatives) with state of the art analytical modeling to directly visualize the design implications (Harris 1965).

Scottish landscape architect McHarg's (1971) seminal book Design with Nature elaborated the idea of an empirical layering of spatial information, and the early integration of scientific and regulatory information in the design phase to filter-out unwanted options. The Harvard Computer Graphics Lab was pioneer in this movement when Fisher (1966) demonstrated SYMAP could perform automated overlays. Steinitz then applied and developed SYMAP into planning applications (Sinton and Steinitz 1969). From there, other mapping and analysis tools emerged as the GIS field matured through the 1970s and 1980s. It strayed, however, from the design realm as designers faced barriers in adopting analytical tools with unfriendly user interfaces, difficulty in collecting spatial data and low computational power. Today, the advent of intuitive user interfaces, increased processing capabilities and wider availability of base datasets allow for the emergence of geodesign as it's own field (Batty 2013; Dias et al. 2013).

The term 'geodesign' as an alias to these past efforts offers two strategic advantages. Firstly, as a moniker for a group of mutually dependent fields of research, it sets a new research agenda aiming to explore symbiotic outcomes between them (data visualization and participatory planning, for example). Geodesign is thus in its infancy relative to traditional sciences. It's youth incites innovation through debate and dialog on what geodesign really means, which we hope is apparent in this book. More importantly, the journey through which we question and struggle with something apparently 'new' allows anyone to participate in the conversation. City planners, designers, architects, hydrologists, traffic engineers, first responders, public health officials, sociologists, biologists, computer scientists, politicians and citizens all have a stake in the future of geodesign for improving the way we plan, manage, develop, protect, and pay for our quality of life.

Secondly, we observe that as a consequence of academic exploration into geodesign, multiple disciplines take ownership of advancing geodesign theory. We hope to see an alignment in the way we work and think through design solutions collaboratively. This is 'geodesign thinking', akin to design thinking as explored by many cognitive scientists since the 1960s, except it has a specific geo-spatial requirement to the design problem at hand. By adding the 'geo' to 'design', barriers into the design world previously encountered by other disciplines are eroded. Design thinking as a problem-solving process becomes accessible, and the lines of communication between technocratic analysts, qualitative scientists, skillful designers, and local citizens become incredibly short.

For our European debut of geodesign (as relatively new concept), Professor Carl Steinitz's chapter on the Redlands Experiment (Chap. 2) offers a comprehensive syllabus of his seminal work on the geodesign framework, with which many of our authors have aligned their own research. Steinitz's framework forms the central discourse for exploring all the far corners of the geodesign world. He rightfully points out that all stakeholders should participate in answering the question "What if?", which is the driver of every geodesign project. He demonstrates, that as an iterative and multi-disciplinary problem-solving method, it is fruitful for the process to result in many designs rather than 'The Design' and that by simply combining existing urban design and analysis skills with inventive spatial technologies and local citizens, we can generate a potent mix of problem-solving ability. It is the combination of ingenuity and skill with spatial technology that excites us and gives us a safe space to experiment and play with geodesign. This book offers a strategic glimpse into European geodesign innovations that embody this excitement and is supplemented by interesting case studies from the around the world.

1.3 Benchmarking Geodesign Innovations in Research and Practice

Stemming from a theoretical foundation set by Carl Steinitz, the next chapters showcase a plethora of practical applications (or aspirations) where geodesign thinking is beginning to emerge. In Part 2, the reader will explore ideas on how we measure and evaluate our efforts to build resilient and sustainable cities and regions. Coming from an urban designer's perspective, Yang (Chap. 3) gives an eye-opening introduction to how we might visualize and evaluate the energy performance of cities, embedding Planning Support System (PSS) theory as a sister of geodesign. By framing urban energy problems as design problems he purports that energy solutions can be explored by altering urban block forms and observing changes in energy fluxes. In doing so, the designer introduces the much needed 'science' into their traditional design process. Fruijtier et al. (Chap. 4) discusses energy on a regional level in the Netherlands from a data management perspective, giving a clear overview of the system architecture and digital tools to be utilized in evaluating the feasibility of different regional energy scenarios, towards consensus-driven energy planning.

Regional sustainability can also be measured by evaluating the impact of different urban growth scenarios. In Kazak et al. (Chap. 5), the CommunityViz extension of Esri's ArcGIS is assessed as a tool for evaluating the impact of spatial policies in Wroclaw, Poland (with and without densification), and found that even without a complex integration of sustainability indicators, policy-makers were better informed. Koomen and Rijken (Chap. 6) draw the same conclusion when undertaking a similar exercise using an interactive touch table with local stakeholders, using Friesland, the Netherlands as a case-study.

Regional resilience can be built by improving adaptability to natural disasters such as overland floods. Janssen et al. (Chap. 7), van Asselen et al. (Chap. 8) and Zandvoort and van der Vlist (Chap. 9) argue that the geodesign framework aligns neatly with a multi-layer safety approach to flood risk assessment and adaptation planning in the Netherlands. Vulnerability assessed by executing 3D flood models with different planning scenarios enables governing bodies to quickly assess new threats and react preemptively. This ability to evaluate and asses across aspects of economics, social and environmental impacts to guide decision-making is the same research agenda adopted by Freitas et al. (Chap. 10) who focus on sustainable agro-forestry (mixed-species) management practices in Borneo, Indonesia. They developed spatial 3D visualizations of the operational trade-offs of different agro-forestry practices, which they suggest will facilitate collaborative decision-making.

In Part 3 of the book, readers will enter the realm of placemaking, where culture, history, and the socio-economic matters of design problems collide. Kolen et al. (Chap. 11) and Burgers et al. (Chap. 12) introduce geodesign from a historian and archaeologist's perspective. They note that landscape history and archaeological heritage have more to contribute to planning and design processes than conventional practices allow. A geodesign framework facilitates the consistent connection of these aspects by first developing Spatial Data Infrastructures (SDIs) that add the much needed quantification and digital documentation of heritage that planners and designers are thirsty for. Heritage digitization becomes the intellectual arena where planners, urban designers, and historians exchange ideas.

Visibility of culturally sacred places (past and present) is of recent interest in geodesign research. It is one of the more obvious and practical measures in evaluating places. Visible landmarks and landscapes that are well composed influence the sense of history (and thus sense of belonging), wayfinding ability and even place satisfaction. Nijhuis (Chap. 13) gives a sharply curated tour of how GIS can be used to explore aspects of visibility in landscape compositions, and comes as a welcome introduction for information technologists and developers to a landscape designer's needs. Visibility is analyzed from the interior in Fisher-Gewirtzman's (Chap. 14) research on visibility impacts on place satisfaction. Here she correlates volume of visible outdoor space (voxel calculations) to a survey of several students from their apartments and notes the influence of individual cultural on place 'level of satisfaction'.

On the other hand, van Nes's (Chap. 15) chapter on the Space Syntax method outlines a very objective and quantifiable tool to analyze street interconnectedness in order to predict economic and cultural hubs in European cities, but notes that the cultural element of where economic centers are placed are not singularly tied to street arrangements. It can be the effort of local associations or concerned stakeholders that generates activity. It is within this public power that Rumor et al. (Chap. 16) and Sanchis et al. (Chap. 17) place their faith to test urban designs. By using 3D geospatial data and (web) visualization tools, they purport that valuable crowd-sourced public opinion can be collected to evaluate and even redesign public spaces.

In Part 4, authors focus on how to adopt geodesign thinking as a common language for exploring design solutions. Currier and Couclelis (Chap. 19) eloquently relate Steinitz's definition of 'people of the place' to the soft aspects of the design problem (what do stakeholders value, establishing sense of place), and argue that before determining a hard outcome (the design interventions), a soft outcome must first be achieved. They propose a methodology called 'perspectives mapping', a process that gives a spatial dimension to 'soft' design criterion that lead to 'hard' design decisions. This builds ownership and understanding of design decisions since stakeholders are empowered with spatial information. The citizens become a source of intelligence, which is a novel notion supported by Nedkov et al. (Chap. 20). By streaming spatial information (relief, air pollution and heat distribution etc.) to citizens during a design exercise, they found that situational awareness was improved and discussions concretized. However they note the lingering organizational challenge of building consensus in large groups, and the scant inclusion of financial information tied to spatial decisions. This is an ongoing challenge for design solutions generally.

McElvaney and Foster (Chap. 21) continue this argument by underscoring the relevance of understanding organizational change principles in the process of dismantling barriers to collaboration among stakeholder groups (and even trained professionals) towards adopting geodesign thinking. This helps the reader in understanding some of the barriers to geospatial technology adoption by urban designers and planners in practice as also discussed by Pelzer et al. (Chap. 22). Both papers distill the challenge down to change resistance due to perceived professional barriers. One of the newer uses of spatial information science is in the field of auditing. Jongsma (Chap. 23) describes the new mandate for the Dutch Court of Auditors to use geo-spatial technology and spatial thinking to improve the way policy is formed, by tracking the impact of government investments.

However, the real challenge, and the backbone, to adopting geodesign thinking is in training and education, as Wilson (Chap. 24) explains, and it is expected that this book will mark a clear starting point for discussion and dissemination of geodesign in Europe in practice as well as the classroom—as a framing concept for collaboration, creative problem solving, and connecting people together through geospatial science and technology. The way in which we collaborate is the most important element that differentiates geodesign from concepts like DSS, PSS, impact assessments and change models, and should be obvious after studying this book. Starting with a concise overview of the history behind geodesign in Europe, to an overview of Steinitz's geodesign framework, we have hand-picked case-studies and technological advancements that represent the most relevant geodesign applications and developments from various disciplines. The book is meant to inspire research, conversation, and analysis into geodesign theory so that we may continue to define it together and in doing so, adopt geodesign thinking.

References

Batty, M. (2013a). Defining geodesign (=GIS+design?). *Environment and Planning B: Planning and Design, 40*(1), 1–2.

Dangermond, J. (2010). GeoDesign and GIS: Designing our future. In: Buhmann E, Pietsch M, Kretzler E (Eds.), Proceedings of Digital Landscape Architecture (pp 502–514.). Wichmann.

Dias, E., Linde, M., Rafiee, A., Koomen, E., Scholten, H. J. (2013). Beauty and brains: Integrating easy spatial design and advanced urban sustainability models. In S. Geertman et al. (Eds.), *Planning support systems for sustainable urban development, Lecture notes in geoinformation and cartography* (vol. 195, pp. 469–484). Berlin: Springer. doi:10.1007/978-3-642-37533-0_27.

Ervin, S. (2011). *A system for geodesign*. Germany: Anhalt University of Applied Science.
Fisher, H. T. (1966). *SYMAP—Selected projects: 1966–1970*. Cambridge: Laboratory for Computer Graphics and Spatial Analysis, Harvard Graduate School of Design.
Flaxman, M. (2010). *Fundamentals of geodesign* (pp. 28–41). Berlin: Anhalt University of Applied Science.
Harris, B. (1965). New tools for planning. *Journal of the American Institute of Planners, 31*, 90–95.
Manning, W. (1913) The Billerica town plan. *Landscape Architecture, 3*(5), 108–118.
McHarg, I. L. (1971). *Design with nature*. New York: Doubleday/American Museum of Natural History.
Nijkamp P., Rietveld, P., & Voogd, H. (1990). *Multicriteria evaluation in physical planning*. Amsterdam: North Holland.
Scheurwater, J (1984). Toward a spatial demographic information system. In Henk ter Heide & Frans J. Willekens (Eds.), *Demographic research and spatial policy: The Dutch experience* (pp. 69–93). Orlando: Academic Press, (Studies in Population).
Scholten, H. J., & Meijer, E. (1988). From GIS to RIA. Paper presented at the URSA-NET conference, Patras, Greece.
Scholten, H. J., & Stillwell J. C. H. (1990). *Geographical information systems for urban and regional planning*. Delft: Kluwer.
Sinton, D. F., & Steinitz, C. F. (1969). *GRID: A user's manual. Laboratory for computer graphics and spatial analysis*. Cambridge: Harvard Graduate School of Design.
Steinitz, C. (2012). *A framework for geodesign: Changing geography by design*. Redlands: Esri Press.
Steinitz, C., Parker, P., & Jordan, L. (1976). Hand-drawn overlays: Their history and prospective uses. *Landscape Architecture, 66*(5), 444–455.
Wood, S. J. (1990). Geographic information system development in Tacoma, geographical information systems for urban and regional planning, *The Geodesign, 17,* 77–92.
Zwick, P. (2010). The world beyond GIS. *Planning, 76,* 20–23.

Chapter 2
Which Way of Designing?

Carl Steinitz

2.1 Introduction

> Everyone designs who devises courses of action aimed at changing existing situations into preferred ones. (Herbert Simon, The Sciences of the Artificial, 1969)

I have organized many collaborative, multidisciplinary studies of major landscape change over more than 40 years at Harvard and in collaboration with other universities. The framework within which I organize most of my work and teaching strategies has been published in my recent book A Framework for Geodesign (Steinitz 2012). In this paper I will focus on one of the most significant decisions which the geodesign team must make when organizing the methods for its study: Which of the change models—which of the many ways of designing—shall we use? The change model which is selected may be the most important part of any professional or academic project because if the methods are unsatisfactory, then the products are also likely to be unsatisfactory.

2.2 The Framework for Geodesign

The framework for geodesign consists of six questions that are asked (explicitly or implicitly) at least three times during the course of any geodesign study. They all have sub-questions that are modified as needed by the geodesign team. The answers to those questions are models, and their content and levels of abstraction are particular to the individual case study. Some modeling approaches can be general, but data and model parameters are local to the people, place, and time of the study.

C. Steinitz (✉)
Harvard Graduate School of Design,
Harvard University, Cambridge, MA, USA
e-mail: csteinitz@gsd.harvard.edu

These six key questions are the following:

1. How should the study area be described in content, space, and time? This question is answered by *representation models*, the data upon which the study relies.
2. How does the study area operate? What are the functional and structural relationships among its elements? This question is answered by *process models*, which provide knowledge about the study context.
3. Is the current study area working well? This question is answered by *evaluation models*, which are dependent upon the values of the decision-making stakeholders.
4. How might the study area be altered? By what policies and actions, where and when? This question is answered by *change models*, which will be developed and compared in the geodesign study. These generate data that will be used to represent future conditions.
5. What difference might the changes cause? This question is answered by *impact models*, which are knowledge produced by the process models under changed conditions.
6. How should the study area be changed? This question is answered by *decision models*, which, like the evaluation models, are dependent upon the values of the responsible decision makers.

Questions 1–3 refer mainly to the past and the existing conditions of the study's particular geographic context. They focus on assessment. Questions 4–6 of the framework concern the future rather than the past and present. They focus on intervention.

Over the course of a geodesign study, each of these six primary questions and their subsidiary questions are asked at least three times (Fig. 2.1). In the first iteration the questions are asked beginning with question 1 as we define the context and scope of the work. In this first iteration we treat these as WHY questions for the project. The aim of the second iteration is to choose and clearly define the methods of the study, the HOW questions. In this stage, the framework is used in reverse order, working from question 6 to question 1. This reversal of the regular sequence of conducting a study is crucial to designing a set of potentially useful methods. In this way, geodesign becomes decision-driven rather than data-driven. The third iteration carries out the methodology designed by the geodesign team in the second iteration. During this round we ask the WHAT, WHERE, and WHEN questions as we implement the study and provide results. In this third stage, the framework is again from questions 1–6, through models of representation, process, evaluation, change, impact, and decision. Once a geodesign team has worked its way through the three iterations of the framework questions, there can be three possible decisions as an outcome: "No," "Maybe," or "Yes".

Reaching a "No" implies that the study result does not satisfy the geodesign team and is not likely to meet the requirements of the decision makers. Then any or all of the six steps are subject to feedback and alteration. This makes geodesign particularly nonlinear in its application. If the team's decision is a "Maybe" or perhaps a contingent "Yes" decision, it may also trigger a change in the scale, size, or time frame of the study. Shifting the scale of the project may lead to either larger or

2 Which Way of Designing?

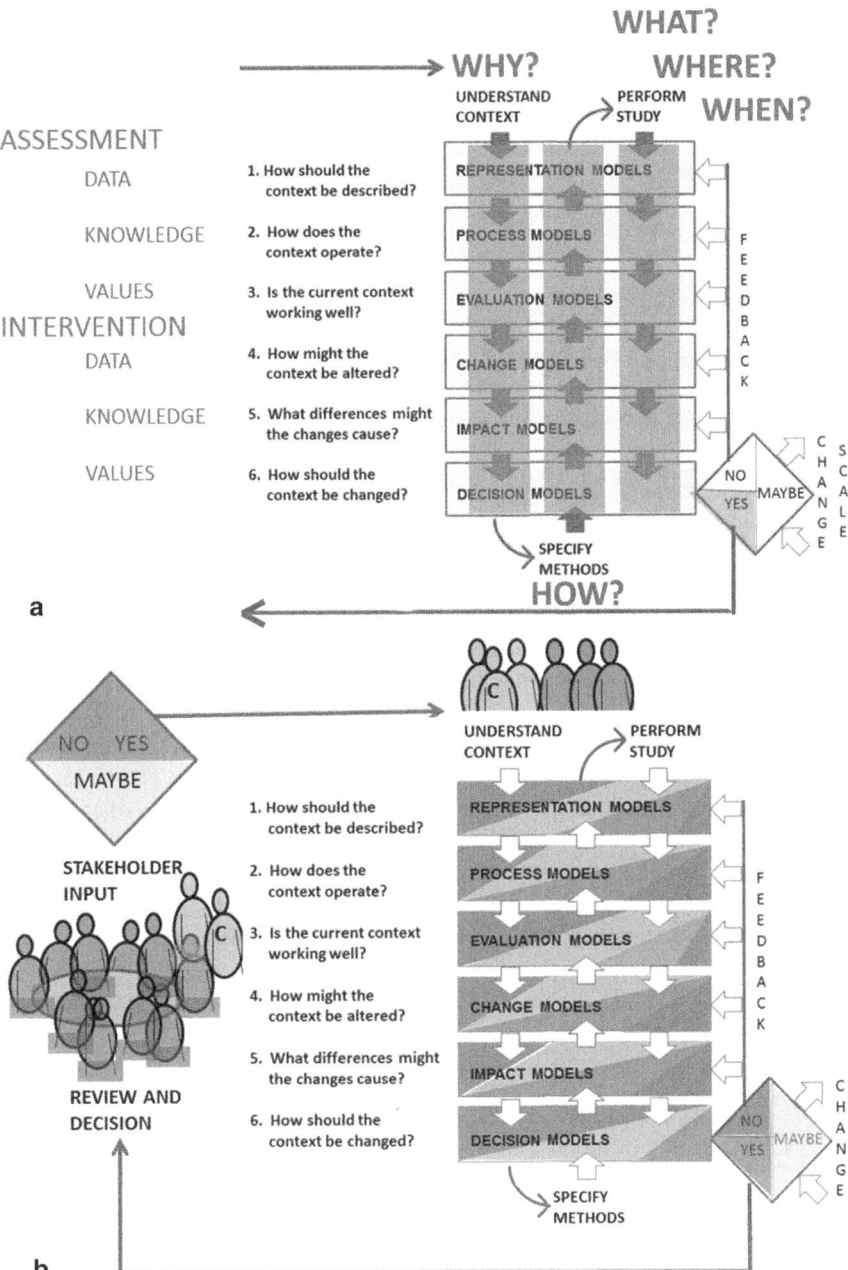

Fig. 2.1 The framework for geodesign

smaller geodesign activities, and the structure and content of several model types may require modifications. Nevertheless, the study will again proceed through the six questions of the framework and continue until the geodesign team achieves a positive ("Yes") decision.

If a "Yes" decision is reached by the geodesign team, the resulting study or proposed project is poised for presentation to the stakeholders for their review towards implementation. The decision makers (and there may be many layers of decision making) also have the choices of "No," "Maybe," or "Yes". A "No" may trigger the end of the study. A "Maybe" will likely be treated like feedback and require changes in the geodesign methods or their results. A "Yes" decision implies implementation and updating for future representation models.

Implementation of agreed-upon designs is not necessarily automatic or immediate, especially for larger and longer-term projects. In whatever ways the geography changes, there will be forward-in-time changes to new representation models. Future generations are likely to seek changes in their geography and see the implemented consequences of the geodesign team's study as part of their data, and so the cycle continues for generations of people of that place. All geography, designed or otherwise, is always in a state of change.

At first glance, the framework may appear to be excessively linear. Yet while the framework's questions and models are purposely presented in an orderly and sequential manner, the framework is normally not linear in its application, and the route through any study is not straight forward. There will always be unanticipated issues, false starts, dead ends, and serendipitous discoveries along the way.

When repeated and linked over scale and time, the questions of the framework may be the organizing basis of a very complex and ongoing study. The result may be a 2-, 3-, or 4-dimensional study, and at a range of scales. Regardless of complexity, the same questions are repeated in any applications of geodesign. However, the answers, models, methods, and results, and the ways by which they were developed and applied will vary according to the case under study.

It is important to emphasize that geodesign (indeed, any design) is not just proposing changes, as question 4 alone might suggest. Whether explicitly or implicitly, all six questions must be satisfied throughout all three iterations of the framework for a geodesign study to be complete.

2.3 Change Models: Ways of Designing

> Many devices which succeed on a small scale do not work on a large scale. (Galileo Galilei, Discorsi e dimostrazioni matematiche, intorno à due nuove scienze, 1638.)

The basic problem of geodesign can be stated as, "How do we get from the present state of this geographical study area to the best possible future?" In the framework we answer the question: "How might the landscape be altered?" with change models, the ways of designing and achieving the products of the geodesign study. The relative influences of the methodological choices made in the second iteration of the

2 Which Way of Designing?

framework will not be equal, and change models are a particularly important element within the geodesign framework. There are multiple strategies for approaching change models. In the book I describe eight different ways of designing for change and a ninth mixed example.

All change models combine decisions related to allocation, organization, and expression, and all require visualization and communication (Fig. 2.2). Allocation refers to where changes are located, such as the placement of new housing in the landscape, the conversion of forest to agriculture, or the protection of a rare animal's habitat, and so on. Organization refers to the interrelationships among the elements of the design, such as how the school, the shopping area, the park, the bus system, and both low- and high-density housing all fit together in the design of the new community. Expression refers to the way in which the design is perceived. For example, is it seen as a residential community, or as a friendly place, a beautiful or an expensive one, etc.

These three characteristics of allocation, organization, and expression are rarely applied with equal emphasis in change models. As a general rule, the larger the size of the design study, the more emphasis is placed on allocation. By contrast, the smaller the project, the more emphasis can be placed on expression. This change of emphasis is characteristic of the differences between landscape planning and garden design, or regional planning and architectural design, or demography and being a parent.

I think that the extremes of size and scale are relatively well served. Design professionals such as architects, landscape architects, urban planners and civil engineers are generally capable at serving client needs at the scales which are symbolized in Fig. 2.2 by the house and the urban design. They increasingly work

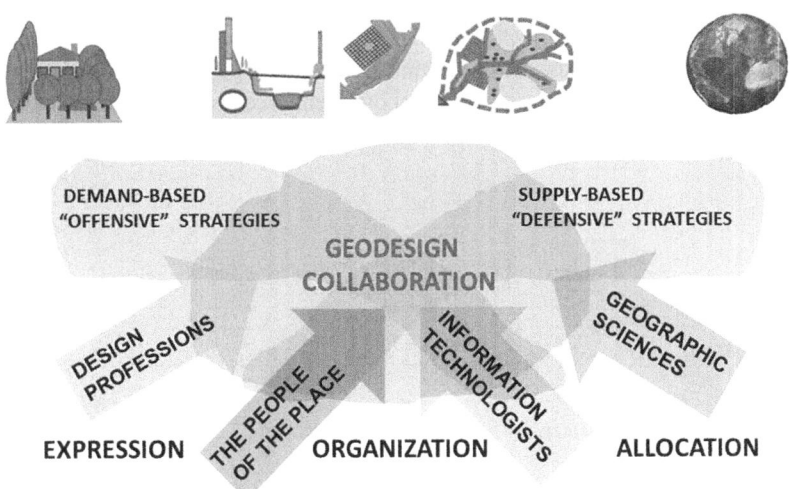

Fig. 2.2 Influences of size and scale

with the people of the place, and on rare occasion with geographic scientists. Similarly geographically oriented scientists are generally capable of understanding the needs of the environment at the scales symbolized from the globe to the large region. They increasingly work with information technologies, and on rare occasion with design professionals. I believe that collaboration in geodesign can be most significant and effective where the extremes overlap, between the large project and the large region. This requires the participation of all four groups: design professionals, geographic scientists, people of the place and information technologies. This range of geodesign "problems" is where the decisions that can really shape the world's environments (plural) for the better are and should be made.

Regardless of size or scale, every geodesign study has four groups of influences which should be considered: the history of the place and its past designs and proposals, the "facts" of the area which are not likely to be changed during the period of study, the "constants" which should be incorporated into any proposed alternative, and the requirements of the project. Yet, while all change models are different they share the same overarching template (Fig. 2.3). The parallelograms can best be understood as map layers of spatial representations needed for the geodesign study, such as drawn diagrams or data layers within a GIS. The arrows are the links in the cumulative process of making the design.

The first influence is history. Knowing the history of the geographical context within which the geodesign study will occur is essential, particularly the history of any previous designs for that area. In my long experience, I have never worked in a region that didn't already have past designs, and the people who made them were not fools.

Next are facts. Facts are aspects of the geography that are assumed not to change over the life of the design. These can be aspects or results of the study's representation, process or evaluation models. We might be working toward a point in time 20 or 30 years in the future, and such things as subsurface geology or a major river pattern or the evaluation of an historic palace are not likely to change within that time frame.

Then there are constants, the things that are certain to occur in the time-frame of the geodesign study. You must find out about them, because if you don't, none of your alternatives will be implemented. An example of a constant could be a high-

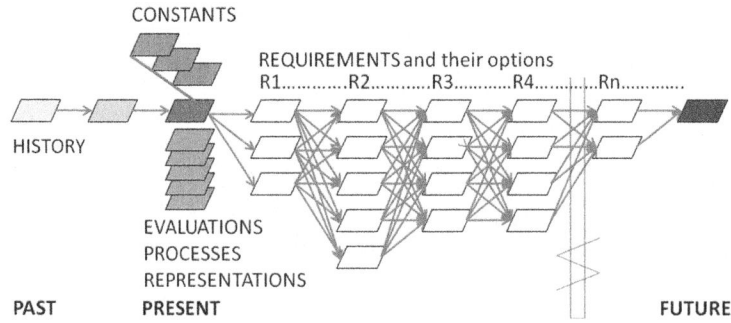

Fig. 2.3 The template for change models

2 Which Way of Designing? 17

way or sewage treatment system in the study area which has already been proposed, approved, designed, funded, and though not yet constructed, is contracted to begin within the next year or two.

Lastly, there are the requirements and their options, the things that should and could happen. Capturing the major, strategic, and generating requirements and their alternative choices is key here. The most important assumptions must be part of the beginning of the sequence of change-decisions, since if you make the first steps wrongly, you will certainly end up wrong. Then again, if you make the right first steps you still may end up wrong, but you have a better chance of success. Spatial analysis frequently plays its most critical role in the assessment of these initial and strategic alternative choices. You have to be able to say: "either here, here, or here," or "in one or more of these several ways."

Each of the eight change models and one mixed example shown in Fig. 2.4 represents a different strategy for approaching and organizing the design and/or simulation of change. (Each is described in Part III of my 2012 book with a case study of major landscape change as applied within the framework for geodesign.) The names of each of these change model strategies reflect their primary approach or characteristic: anticipatory, participatory, sequential, constraining, combinatorial, rule-based, optimized, and agent-based. All eight support the use of scenarios, recognizing that there are an infinite number of future options. At the same time, all of them eventually reduce the possible number of alternatives from the infinite to a manageable number. In the end, the change models must include the most important issues and produce an appropriate range of policy and design choices. Although nearly all designs are the result of combinations of these eight ways, during a given geodesign project one of these eight is likely to dominate. The way that the change model is organized and started is crucial and should be preplanned in the second iteration of the framework for geodesign.

The change models can be considered in three different groups. The anticipatory, participatory and sequential change models assume that the designer or the geodesign team is confident in the ability to directly develop the design for the future state of the study area. The constraining and combinatorial ways assume that the geodesign team is not certain of the crucial initial decisions and must first assess the major requirement-variables before developing the rest of the design. The rule-based, optimized, and agent-based approaches assume that the geodesign team is assumed to understand the rules that guide the processes of change, but also is obligated to test the variability of the main requirements in order to develop the most beneficial design solution.

Well before writing the book I had often argued that there is no such singular thing as "THE Design Method" or "THE Planning Method" (and I consider a plan to be a design). Rather, there are many methods and they must be chosen in the second iteration of the framework and adapted to issues and questions raised by the problem at hand. This raised the important but difficult question: Which way of designing should be chosen? My hypothesis was that the larger and more complex geodesign studies would be best served by the more complex change models (Fig. 2.5).

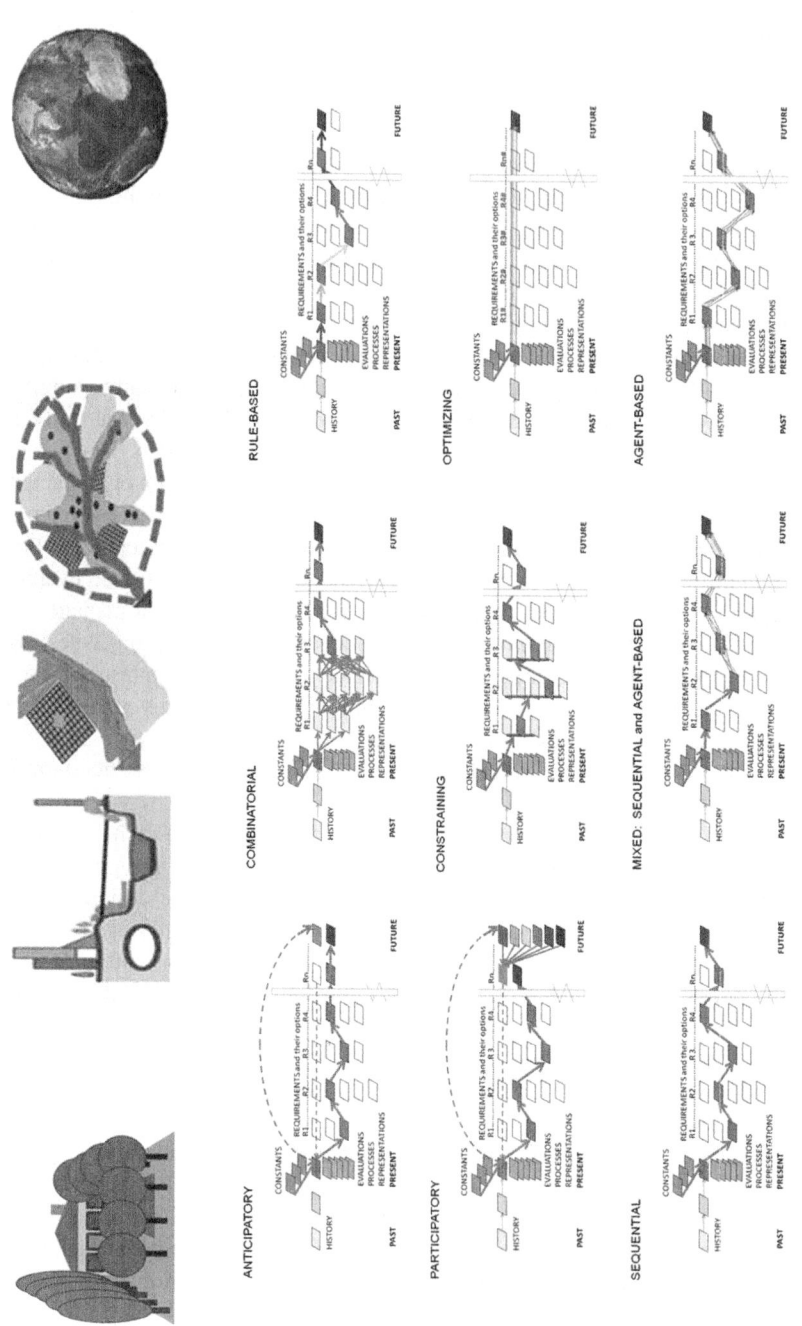

Fig. 2.4 Change models: ways of designing

2 Which Way of Designing?

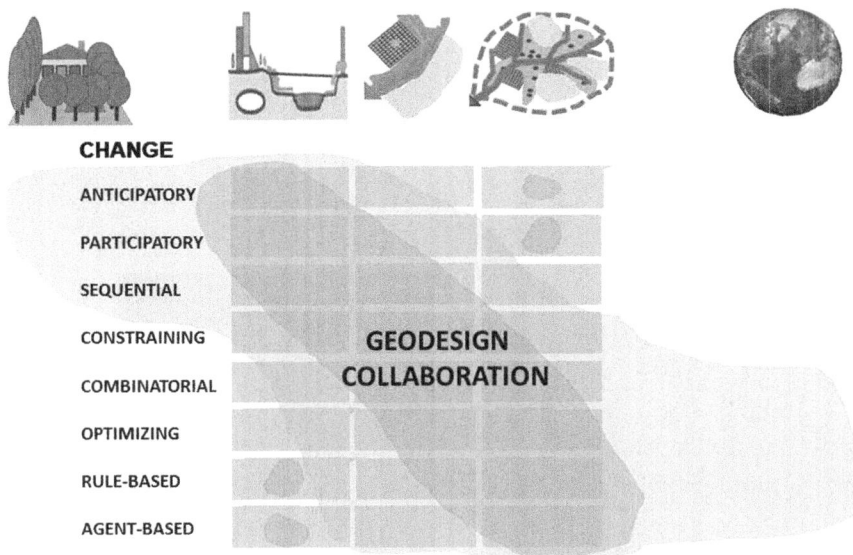

Fig. 2.5 An hypothesis regarding the link between geodesign size and scale, and the efficacy of change models

2.4 The Redlands Experiment

At the first Geodesign Summit in Redlands California in 2010, I proposed an initial experiment to test and compare the efficacy of the nine change models described in my book. This idea was taken up by Jack Dangermond and several other persons, some of whom I had previously worked with. With significant support from Jack and ESRI, an agreement was made with the City of Redlands and the University of Redlands to conduct an experimental workshop that would be of help to the city regarding two prominent issues facing it. The first was the preparation of a landscape plan for the city (which Redlands referred to as an open space plan). This was seen as 2-dimensional design. The other was the 3-dimensional design for a transit-oriented development near ESRI's corporate campus (Fig. 2.6).

In this agreement, the Redlands University and its Redlands Institute would host the workshop and have several faculty and some students as participants, the City of Redlands would organize the data and its representation, process and evaluation models, and establish all the requirements for the two geodesign studies. Its planning staff and several residents who are active in city affairs would be participants. ESRI would contribute several information-technology staff, and allow the workshop to test several geodesign management tools which were in early development. I would help recruit persons who were familiar with the framework and who would organize the work of the geodesign teams, and I would manage the workshop (with a lot of help).

The core team and several persons from Redlands met in October 2010 to plan the workshop and its information flows. There were three significant constraints: the work-

Fig. 2.6 The study areas

shop had to be conducted within five consecutive days, it had to be based on existing and pre-available public data, and we needed to have as much of a controlled experiment as possible. We recognized and discussed many other potential limitations: short preparation time, active-time for the workshop and for software pretesting, the workshop's time schedule, the varied skills among participants, the possibility of wrong or too narrowly or broadly defined models (and especially for qualitative aspects), and that the two geodesign problems were not really "real", for either Redlands' Open Space or the anticipated Transit Oriented Development (TOD).

The workshop was held from January 10 through 14, 2011. There were about 50 people as full time participants and several observers. The schedule for the five-day workshop was basically as follows: Monday was for orientation lectures which were prepared by the city of Redlands, and a site visit throughout the city and to the area of the proposed transit oriented development. It also included an overview lecture on the framework and the several change models, and the organization of the participants into the nine multidisciplinary geodesign teams. Each team would have at least one person with geodesign experience, one with information-technology skills, one resident of Redlands, and one with relevant knowledge from the arts and/or sciences. Tuesday, Wednesday and Thursday were to be devoted to preparing the two designs. Each team was encouraged to make its initial open space design first. Tasks would be organized as each team decided so long as it was as possible within the change model to which it was assigned. Friday morning was for preparing a public presentation of the team's way(s) of designing, and the products of its geodesign activities. On Friday afternoon a public presentation was made in the largest auditorium at ESRI to a capacity audience.

The flow of information (Fig. 2.7) was designed so that each geodesign team would have maximum flexibility in using the representation, process and evaluation models which had been prepared by the City of Redlands. In addition, each team would have access to three assessment dashboards. The first provided immediate feedback on whether any new element of the design was moving toward or away from the program requirements established by the city. The second provided immediate feedback on whether any new element of a design was improving or diminishing vis-à-vis the criteria of the city's decision model for each of the two designs. A third information management dashboard allowed all participants to compare the performance across the nine teams for each of the two geodesign problems.

The City of Redlands planning office and several resident—participants had organized the program for each of the two designs, and all aspects of the models needed for the assessment phase (Fig. 2.8). These were to be assumed as correct and sufficient by all the workshop's participants

On the first day of the workshop and after an orientation to the open space plan and the TOD, those participants who were residents of Redlands participated in a two-stage Delphi experiment in which they were asked to weight the several criteria established by the City of Redlands for each of the two geodesign problems. This experiment was aimed at identifying the decision models which would guide the final comparison of the designs and the assessment of the efficacy of the various change models.

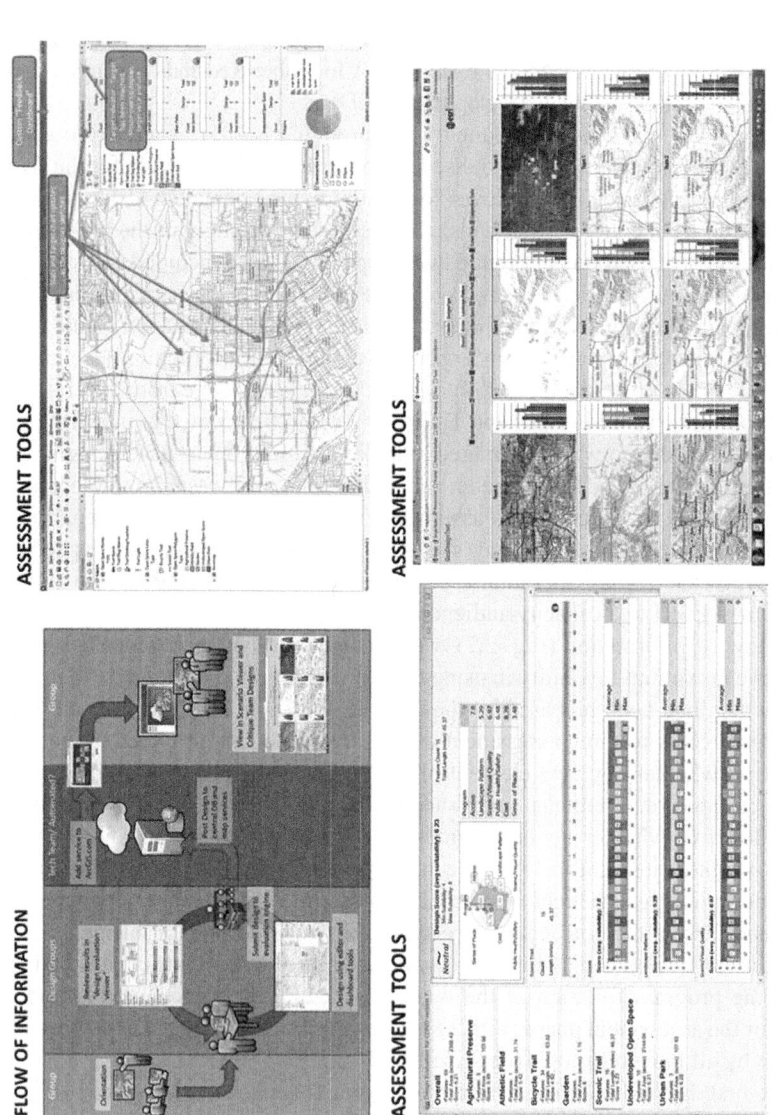

Fig. 2.7 The flow of information and the assessment tools

2 Which Way of Designing?

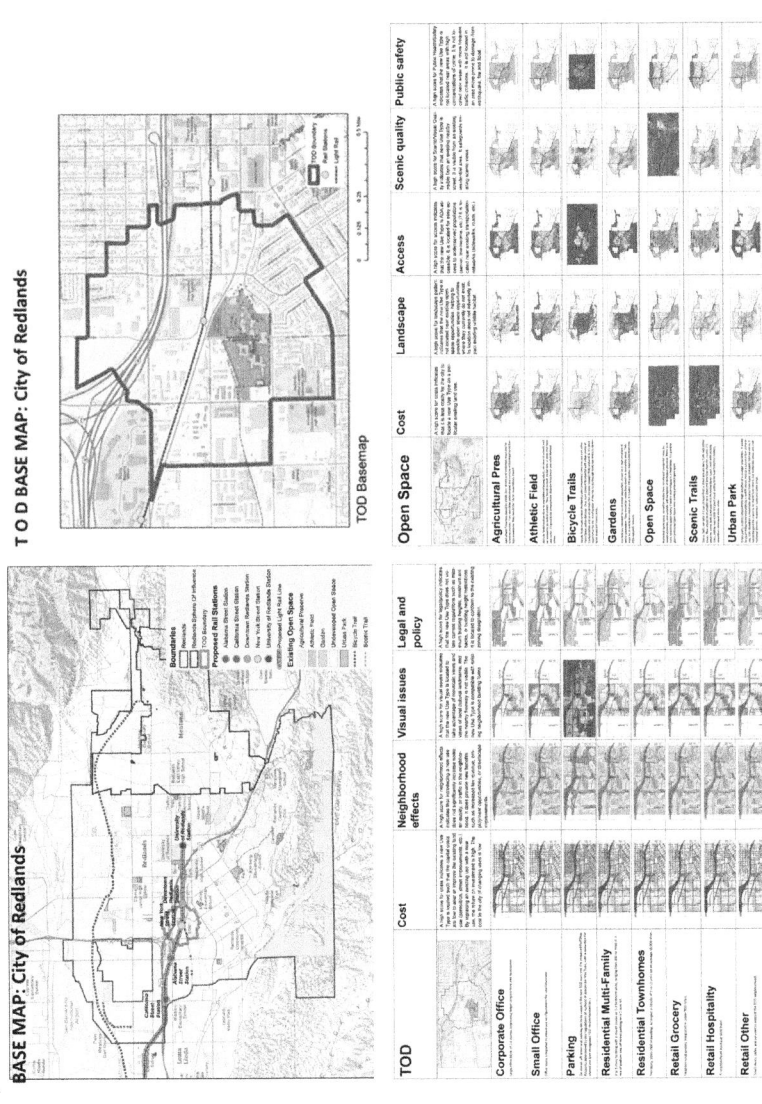

Fig. 2.8 Representation, process and evaluation models, and their resulting impact models

2.5 Anticipatory

The anticipatory approach is based on the premise that the designer's confidence and experience can provide the "great leap forward" to a concept of what might be the basis for a good design (Fig. 2.9). This necessarily assumes that the designer has a sufficient and adequate amount of experience from which to draw upon. He or she will then need to go back to existing conditions, and through deductive logic move forward through the many requirements and their options to try to achieve the preconceived design.

The anticipatory team had two participants who were very experienced professionals in design at the scales and types of the two workshop cases. The team decided that each of these persons would work as an individual designer, with the rest

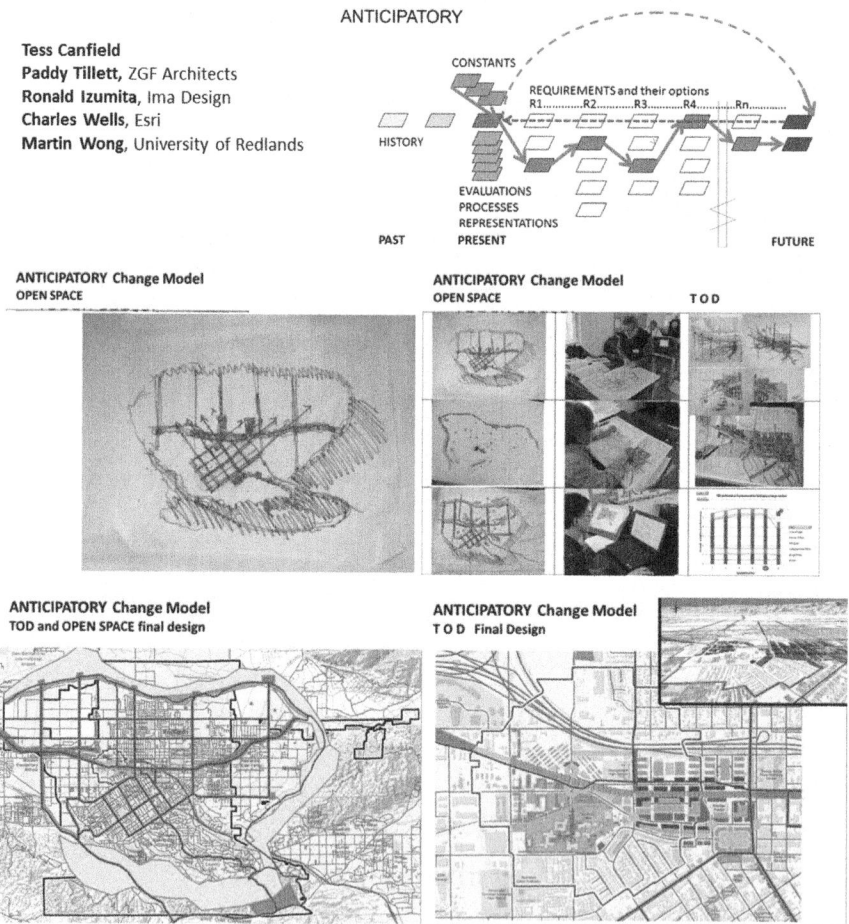

Fig. 2.9 The anticipatory change model

of the team in support of their needs as they developed. Both designers chose to work by hand-drawing on paper through several iterations of their initial conceptual diagrams, while other members of the team prepared staged digital versions for the several available impact assessments. The better-performing designs of one of the two designers is shown in Fig. 2.9.

2.6 Participatory

The participatory design approach assumes that there is more than one participating designer, and that each has a concept about what the future design should be (Fig. 2.10). This premise expects the designers to have a sufficient

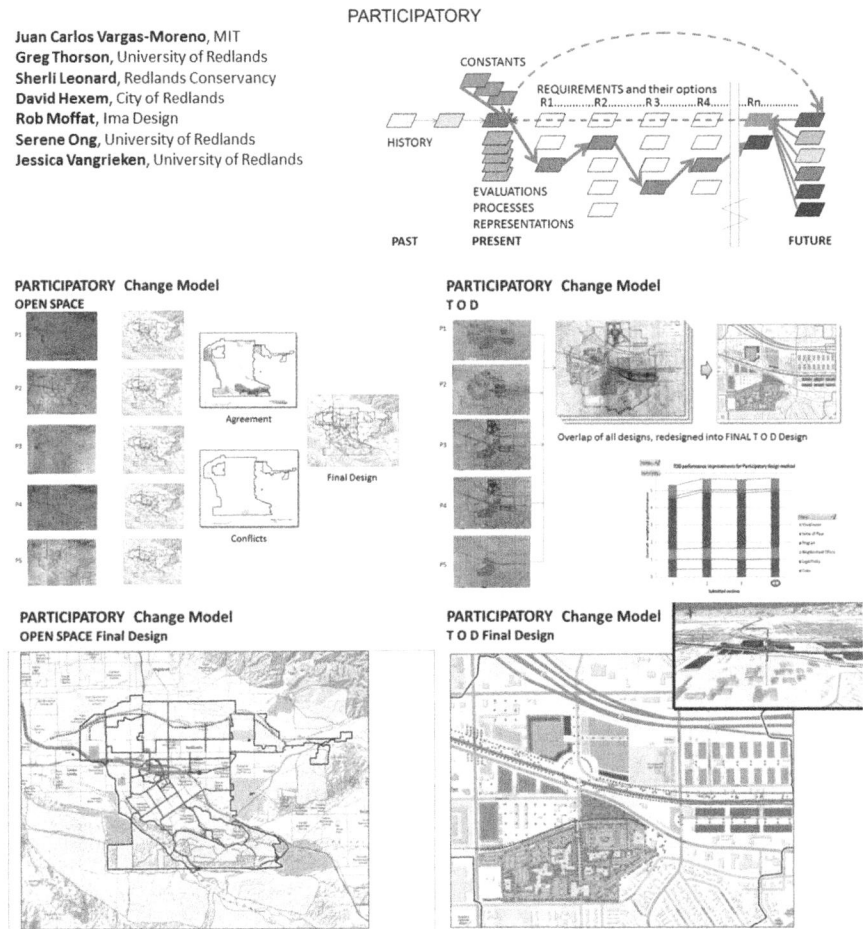

Fig. 2.10 The participatory change model

sense of place and time to provide a future-oriented design, while still recognizing that their designs are different and must be aggregated into one consensus design.

Five members of this team were residents of Redlands and they were chosen as the direct participants. Each person drew a design for the open space system. These were digitized and algorithmically compared for their spatial agreement. The final plan is the result of a series of majority—decision rules combined with a modest amount of interpersonal negotiation based on expert local knowledge. A similar strategy was applied for the TOD, but in this case the five initial drawn designs were negotiated into a first consensus. This design was then digitized and improved through a series of feedback loops to what was the final design.

2.7 Sequential

Under the sequential approach, the designer makes a series of confident choices that systematically develop into the future design (Fig. 2.11). This approach begins with present conditions and uses abductive logic as it moves with certainty directly through a single set of choices for each requirement.

This geodesign team consisted of several residents of Redlands, one of whom was experienced in design at these scales. After a group discussion in which the sequence of decisions was established and various options were given priority, the constraints for the open space design were identified digitally. The design was then rapidly generated manually and its final version made digital. A similar process was followed for the TOD, but in this case an early version was digitized and improved through feedback to its final state.

2.8 Constraining

The constraining method (Fig. 2.12) is useful when the client and/or the geodesign team are not sure of the decision models, or when the relative importance of the study's objectives or requirements approximate Zipf's Law but where there are also many options for each requirement. A strategy of making decisions by comparing and selecting options in the sequence rank order of decision importance is then followed.

This team recognized that the two decision models provided by the Redlands residents generally followed a Zipf distribution. They therefore adopted a strategy of making decisions in the rank order of importance within each decision model. Knowing the sequence made discussions highly efficient and several digital feedback iterations were conducted. The team to follow a similar process for the TOD but in this case the entire design was conducted in a series of hand drawings due to perceived time limitations.

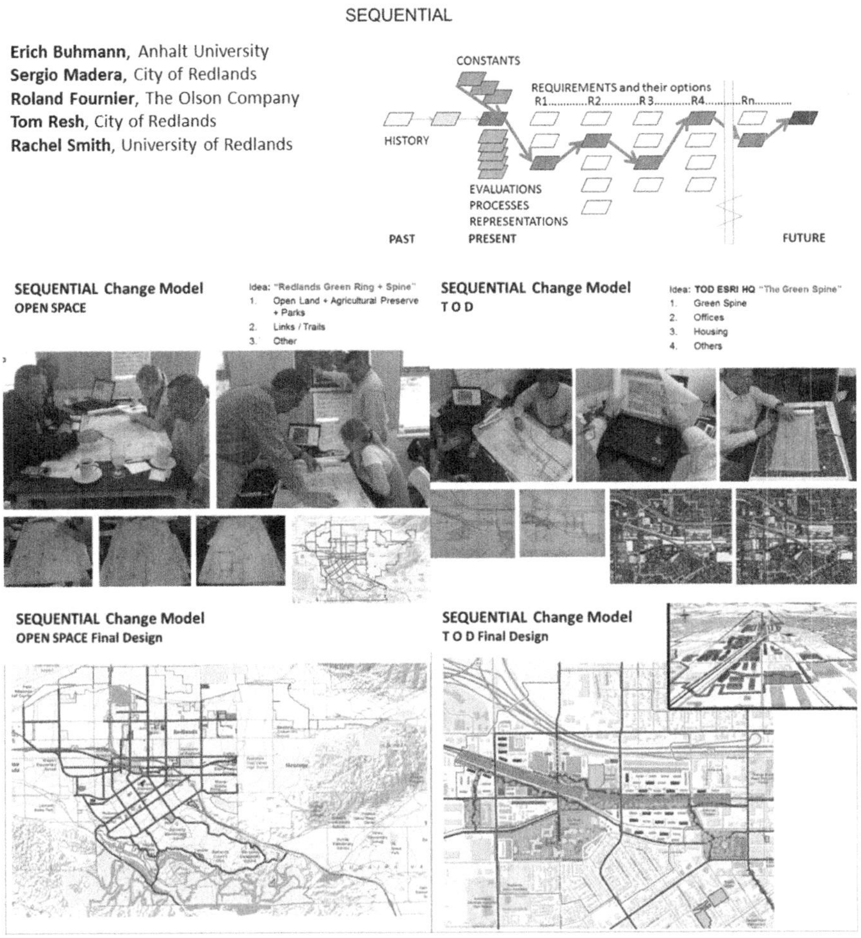

2 Which Way of Designing?

Erich Buhmann, Anhalt University
Sergio Madera, City of Redlands
Roland Fournier, The Olson Company
Tom Resh, City of Redlands
Rachel Smith, University of Redlands

Fig. 2.11 The sequential change model

2.9 Combinatorial

When the designer or the client is not sure of the appropriate choices in the sequence of decisions to create the design, the combinatorial approach is useful (Fig. 2.13). This strategy is commonly applied to investigate alternative scenarios for the future. It is especially appropriate when the few main objectives are of similar importance and a combination of the key requirements must be resolved before continuing with less important ones

This team realized that if they identified the best combination of options for the three most important requirements for each design, and spent less time and energy on less significant requirements, that their final designs had a high likelihood of

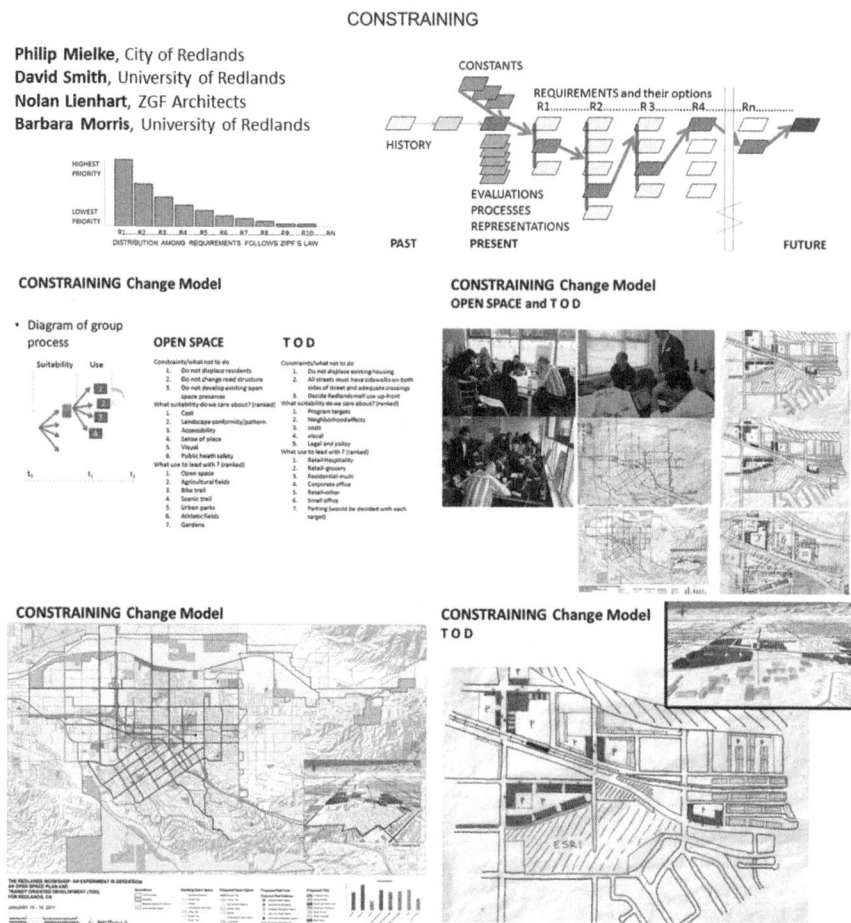

Fig. 2.12 The constraining change model

being good ones. Their first steps were to diagram and name a substantial number of options for the three most significant open space requirements. In a very efficient way, the members of the team sorted and rejected most of the combinations. A decision was made from a short list of feasible sets, digitized and improved through feedback. The same process was then applied to the TOD.

2.10 Rule-based

The rule-based approach assumes that the geodesign team is knowledgeable and confident enough to specify a set of formal rules for developing the design (Fig. 2.14). Such approaches are normally organized as a set of computer algorithms, but they

2 Which Way of Designing?

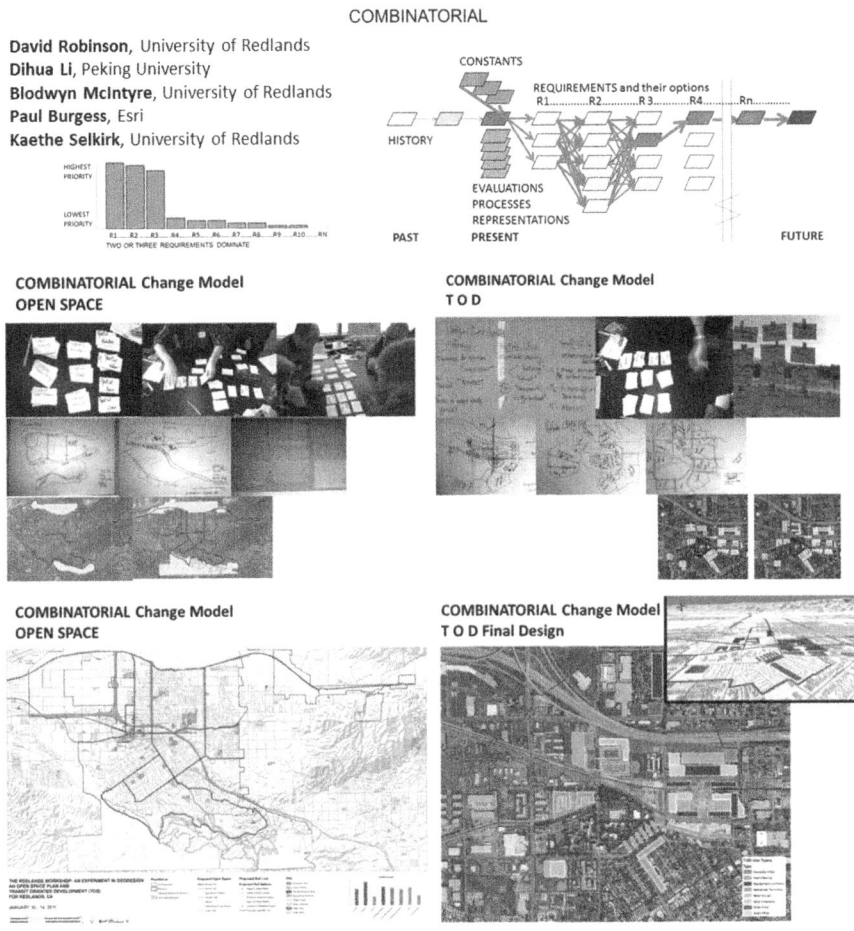

Fig. 2.13 The combinatorial change model

can also be expressed as mental steps which can be followed manually. The rules for each requirement are combined in a sequence of design decisions comparable to that in a sequential approach.

This team decided to build a rule-based model which would allocate the requirements for the open space plan in a sequence of decisions reflecting perceived importance. They were able to do this in an efficient manner, but calibration of distance related criteria required trial and error feedback within the process and evaluation models. Preparing a rule-based model for the 3-dimensional TOD was seen as requiring more time than available, so the anticipatory team's TOD design was borrowed and adapted to the open space plan.

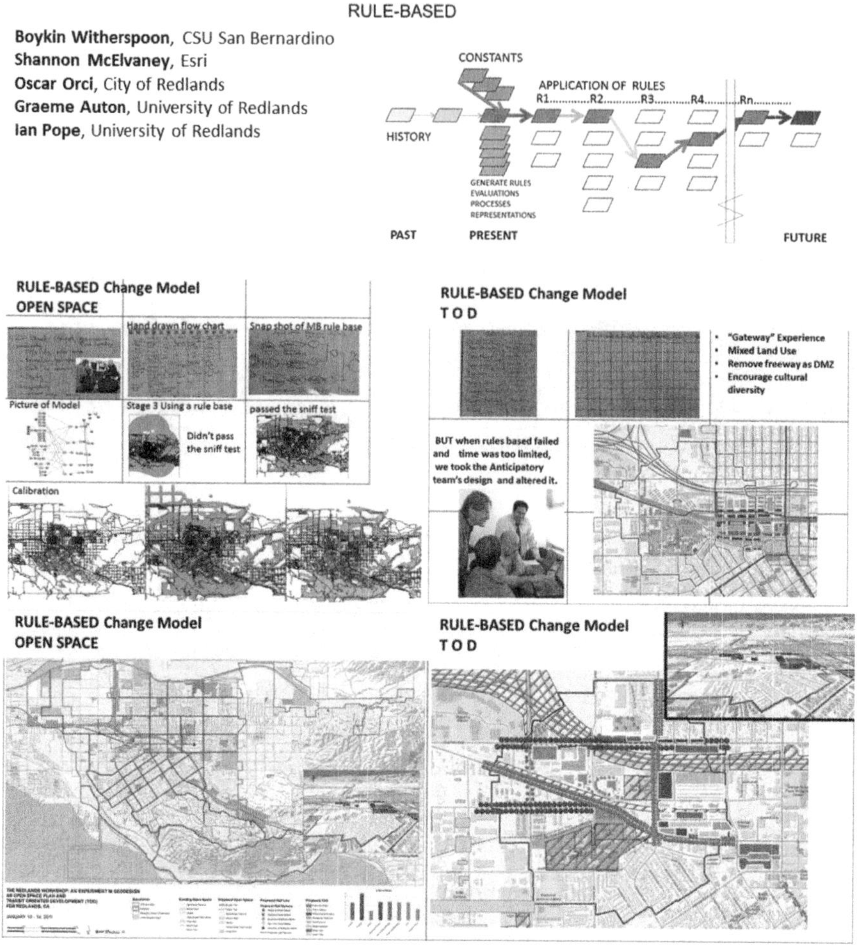

Fig. 2.14 The rule-based change model, with each colored arrow in the diagram representing a different requirement, in this case a different land use

2.11 Optimized

The optimized requires that the client and the geodesign team understand a-priori the relative importance of each of the desired requirements and also its decision criteria. The optimizing decision model is based on the values of the decision makers and reflects their goals, the values by which they make judgments about the design, and the relative importance they attach to these goals and values. This approach needs these criteria to be identified and comparable in a single metric, such as a financial rate-of-return or potential votes, etc., in order to be able to declare a design "optimal" in the end (Fig. 2.15).

2 Which Way of Designing?

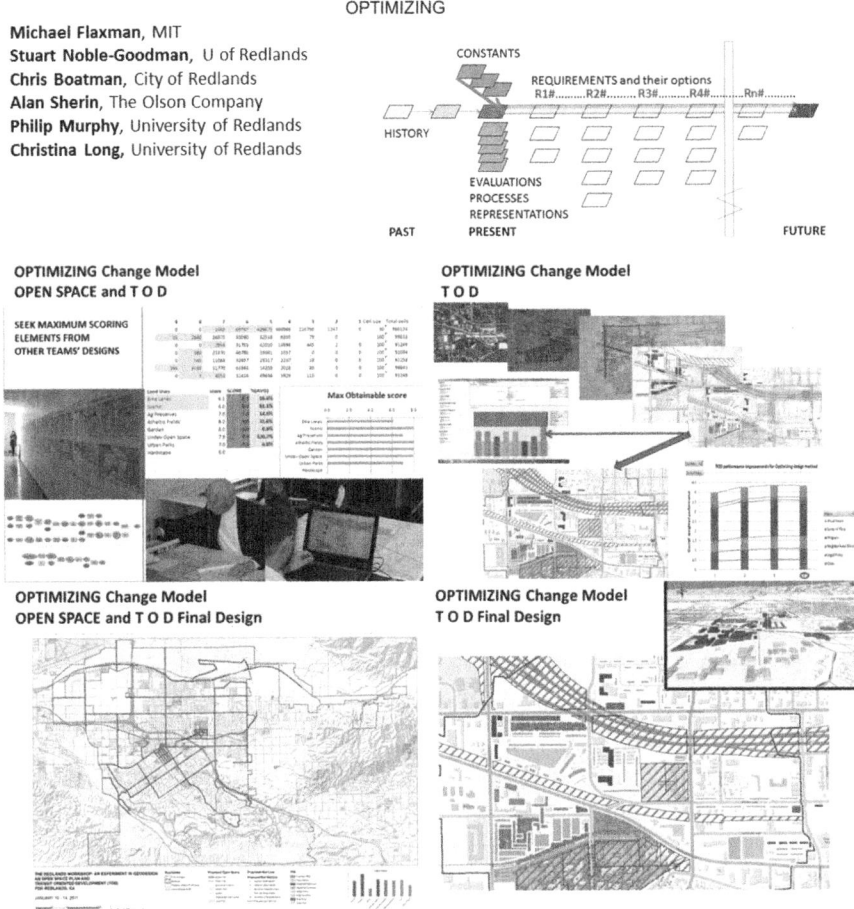

Fig. 2.15 The optimized change model, with each colored line in the diagram representing a different requirement, frequently for a different land use

While the optimizing team was aware that digital methods exist to generate optimal spatial designs, they realized that available time did not allow for their designs to be generated algorithmically. They therefore adopted the equivalent of a crowd-sourcing strategy. Since the information flow of the workshop enabled the designs of all the teams to be visible, the optimizing team developed an algorithmic approach which enabled them to identify the best elements from all other designs. These were combined first into the open space plan and then into the TOD. Several digital feedback iterations improved both designs to their final states.

2.12 Agent-based

In the agent-based approach, the future state of the study area is the result of interactions among policy and design decisions that direct, attract, or constrain the independent but rule-based actions of independent "agents" (Fig. 2.16). Agents can be stakeholders, or decision makers, or people of the place, or be defined as home seekers or developers or conservationists, for example. They can also be land uses, as demonstrated in this experimental workshop. For each type of agent, there are different "rules" for where they can be in the study area "landscape" and how they will interact with others in their own group and among other groups. These rules are embedded into a computer model, and the changes occur simultaneously and adjust in reaction to the sequence of requirements for the design.

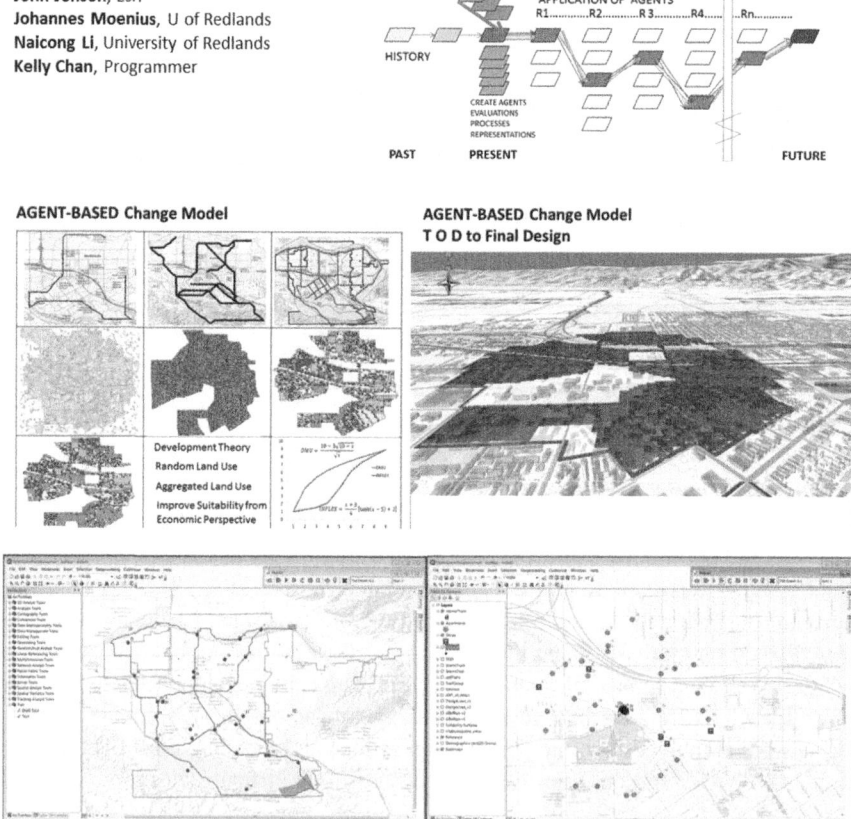

Fig. 2.16 The agent-based model, with each colored arrow in the diagram representing a different sub-model for a land use, all acting simultaneously

The members of the agent-based team were chosen in large part because of their knowledge of modeling methods and willingness to experiment with agent-based methods as change models. An early decision was made divide into two sub-teams for initial experiments, and then to focus on the 3-D TOD design. One group first borrowed the Anticipatory team's open space plan. It then located 30 "people" and specified criteria for a scenic walk from home to preferred views. The agents were then "walked" along public streets to their nearest scenic trails, and the model assessed the relative access of their home areas to trails with preferred views. The agent-based model in this application was an animated impact model which could be used as feedback.

The TOD application used agent-based modeling as a change model, to create the design in the first place. The required amounts of all land use program elements were divided into $10 \text{ m} \times 10 \text{ m}$ grid-agents. They were algorithmically allocated randomly in a first design. This pattern was then aggregated into units of appropriate size and made more economically efficient in a series of feedback re-allocations. When the design was deemed stable, a second agent-based impact assessment was made. Thirty "people" were assigned homes. They all arrived at the transit rail station and had to walk home but had to stop briefly to shop en route, thus assessing the shop locations for their pedestrian traffic. To the best of our knowledge, this TOD experiment was the first use of an agent-based model as a change model in 3-dimensions.

2.13 Mixed

In a mixed approach, several different ways of designing are combined, in whole or in part. The number of possible combinations of change models is almost infinite (Fig. 2.17).

This geodesign team had the greatest freedom to choose how it would work. It decided that mixing the anticipatory and sequential methods would be both the easiest and most efficient way of proceeding, largely because of the instinctive familiarity of these ways of designing. In addition they decided to closely observe the designs of the other teams and to borrow as needed. They organize themselves as a team with individual responsibility but frequent coordination. One consequence of this was that they were able to achieve a greater level of detail than all other teams within the limited time available to the workshop. This including alternatives for the three highway underpasses which link the two major parts of Redlands, made by a Redlands University undergraduate student.

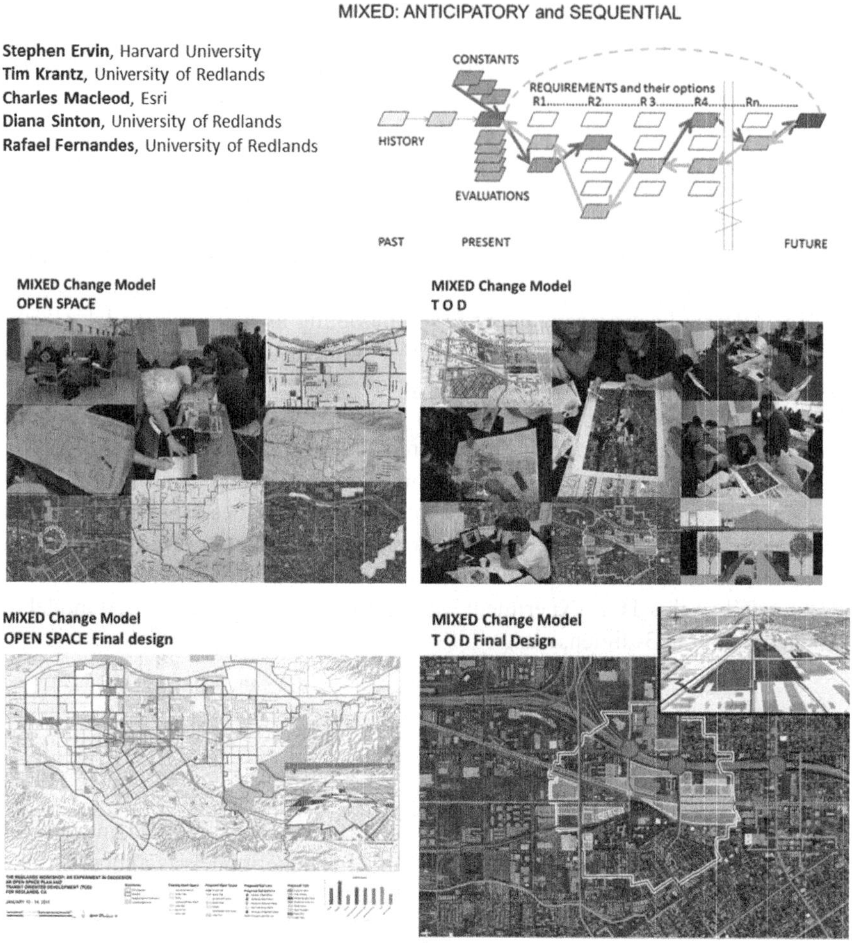

Stephen Ervin, Harvard University
Tim Krantz, University of Redlands
Charles Macleod, Esri
Diana Sinton, University of Redlands
Rafael Fernandes, University of Redlands

Fig. 2.17 Mixed ways, e.g., anticipatory and sequential

2.14 Selecting Among Change Models

On the morning of the last day of the workshop, a subgroup of the participants made comparisons among the nine open space plans and the nine TOD proposals. Using the algorithm prepared by the participatory team, and analysis was made of the majority land-use agreement among all the designs. This 10th design could be seen as the product of considering the entire workshop as participants toward a single design.

It was informally judged to be a good design by the participating Redlands residents (Fig. 2.18).

2 Which Way of Designing?

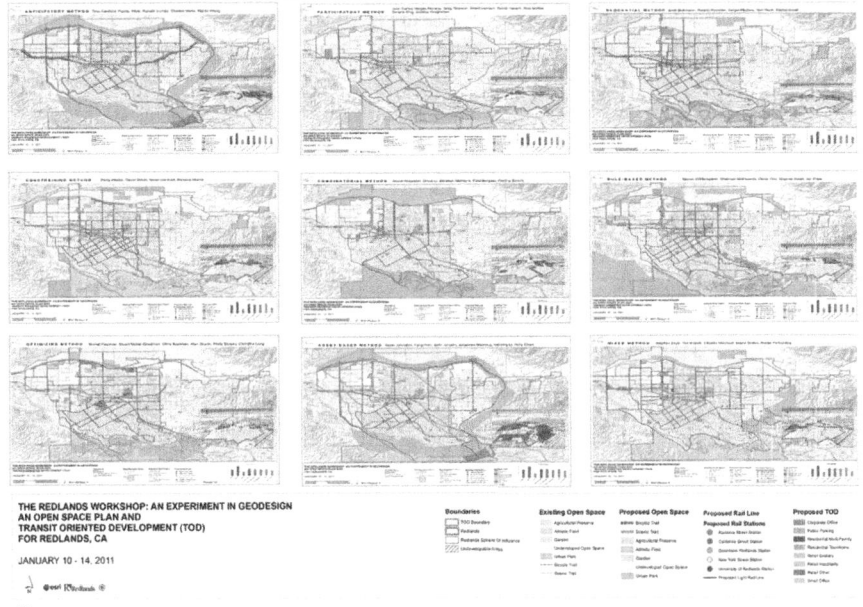

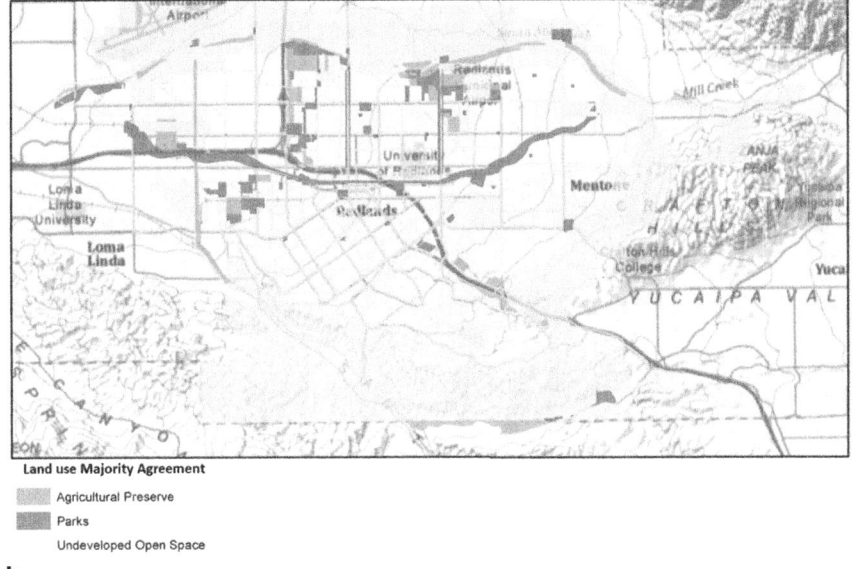

Fig. 2.18 Comparison of the open space designs

The assessment tools made available to the workshop enabled the designs to be compared for their component contributions toward matching the decision model which had previously been proposed by the Redlands residents (Fig. 2.19). The resulting graphs show a striking similarity of performance regardless of the change

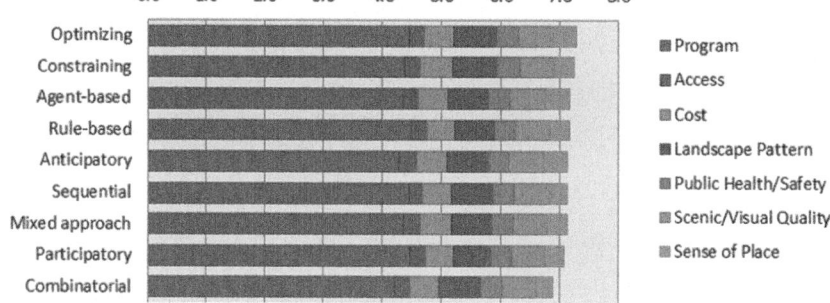

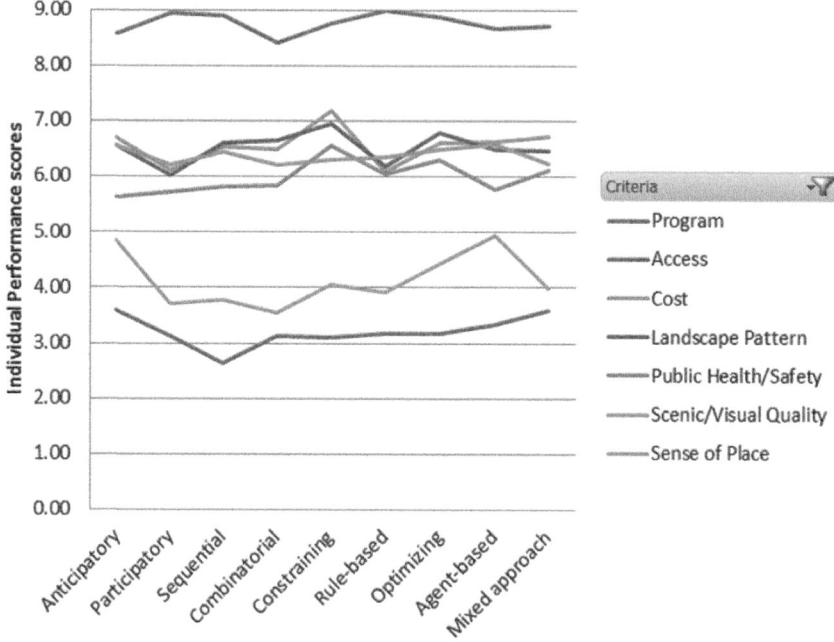

Fig. 2.19 Comparing the open space designs

model within which the designs were created. There are two reasons for this. First, all the teams began with an assessment of constraints—of what *not* to do—as indicated by the process/evaluation models prepared by the city of Redlands. Second, all the designs went through at least three rounds of feedback improvement vis-à-vis those same models. Clearly, having pre-specified assessment models and an efficient feedback process mitigated against making the wrong choice of a change model.

When the nine designs for the TOD were compared there was considerably more variance in performance among the geodesign teams (Fig. 2.20). Combining constraints is a fast early step and frequently a starting point in a 2-dimensional planimetric design, especially if known a-priori, while meeting complex program requirements in 3D is harder and needs many more feedback loops. Although time was very limited, almost all the teams made several improving iterations of their TOD designs, and these results could be monitored and reported. It is interesting to note that after a few iterations the designs at that scale did not improve greatly. Indeed, in one case further changes resulted in a distinctly less successful design which was suppressed in favor of a prior version.

When comparing the performance of the TOD designs toward their decision model criteria (Fig. 2.21), there was considerable variation. It is clear that application of feedback was again very important toward leveling the impact results related to the process/evaluation models provided by the City of Redlands. The striking exception is in the ability of the geodesign teams to meet the several programmatic requirements. Not only is there a greater variance, but there seems to be an inverse relationship between "quantity" as represented by the land-use program and "quality" as represented by the other of Redlands' decision criteria. This is not a surprising occurrence to an experienced designer, but it does raise the question of how to better integrate qualitative assessments into digital methods.

The anticipatory TOD succeeded because the designer was a very capable and experienced professional whose team took full advantage of the feedback capability. The combinatorial team also did well, as this is a fine way to begin a complex design if one has understood the decision model and its requirements and options. Finally we should note that the agent-based TOD performed very well. This is because it began by excluding constraints and then fulfilling the program requirements, albeit in a spatially random manner. It then applied many feedback changes to incrementally improve and regularize the design. This may portend a very promising change model-strategy for geodesign.

The performance of the change model teams across both geodesign problems was also assessed (Fig. 2.22). If meeting the program requirements is *excluded* from the overall calculation, the change models' performance over the two scales are essentially uncorrelated ($r = 0.05$). Clearly, the combination of an early use of the process/evaluation models to identify constraints on the design and the implementation of efficient feedback can mitigate against a badly chosen change model.

However if meeting the program requirements is *included* in the overall decision model calculation—and it almost always is—the differences in the change models performance over the two scales are significantly and negatively related ($r = -0.66$).

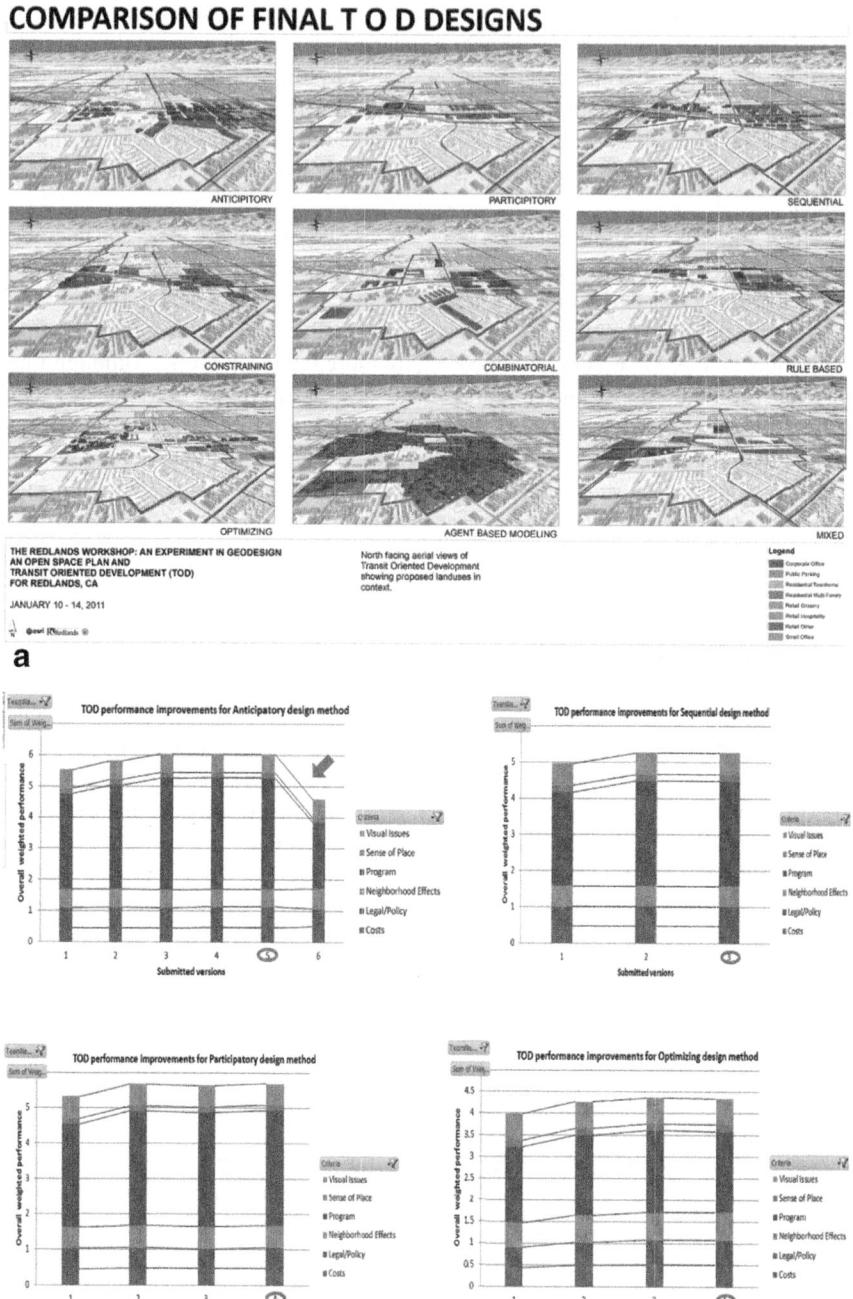

Fig. 2.20 Comparison of the TOD designs

2 Which Way of Designing?

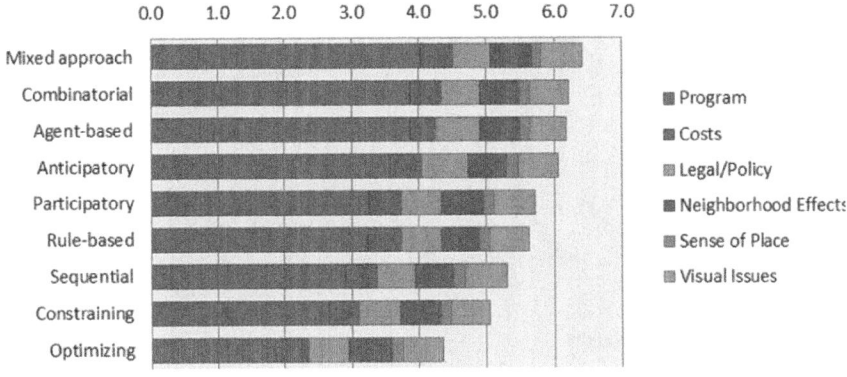

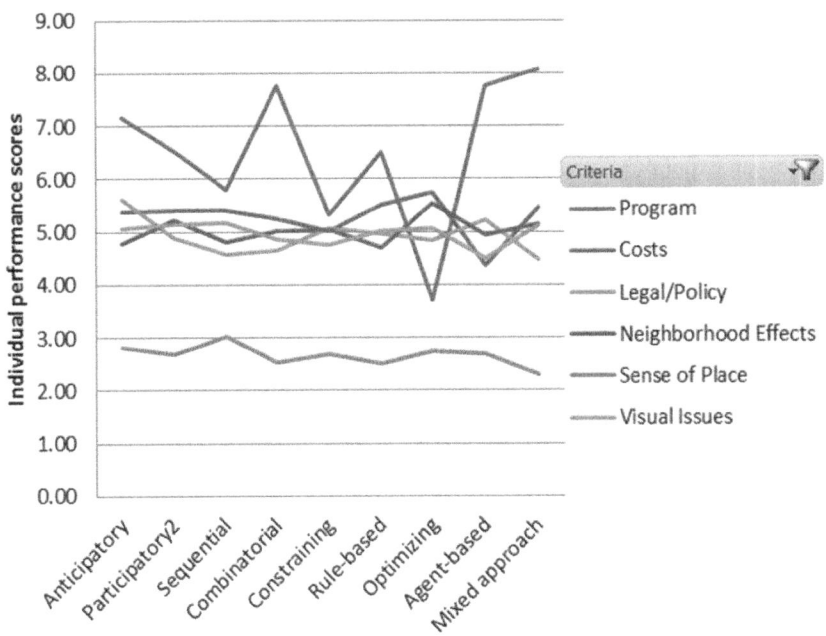

Fig. 2.21 Comparing the TOD designs

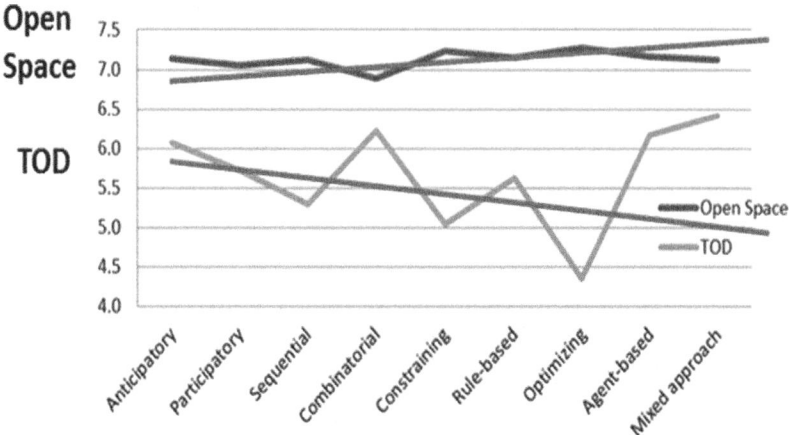

Fig. 2.22 Comparing all designs

Size and scale, and the selection of an appropriate change model do matter. Having one way of designing does *not* best-fit all geodesign conditions.

Regarding the Redlands Workshop, we all need to be very much aware that this workshop was an initial and limited experiment which compared design methods in a digital environment. It was a very interesting event and it added to our experience, but it does not "prove" anything. The Redlands workshop was first and foremost an experiment, and we must always keep in mind—"An example is not a theory". We hope it will lead to many more experiments. Nonetheless, we believe that the Redlands Workshop has been an important step forward in the development of geodesign.

Acknowledgements We thank all who have participated in and contributed to this event, especially the teams from Esri and the University of Redlands, and Jack Dangermond.

Reference

Steinitz, C. (2012). *A framework for geodesign*. Redlands: Esri Press.

Part II
Resilience and Sustainability

Chapter 3
Energy Resilient Urban Planning

Perry Pei-Ju Yang

3.1 Introduction: On Geodesign Agenda

The paper introduces a GIS-based energy and carbon performance assessment and its urban design, a forward-looking modeling approach to "designing for change" for an urban environment. It investigates how design objectives help to organize a process of geographic-based information modeling in cities, and demonstrates how design intervention drives future urban change to achieve better performance in energy and carbon emission.

Recent discussions on Geodesign draw attentions of Geographic Information System (GIS) and Planning Support System (PSS) professionals to an emerging model that is more design-oriented. How far has Geodesign moved away from the current practice of GIS and traditional model of PSS? The attempt of turning GIS technology and PSS methods to a geographic-based design brings in the concept of "designing for change" (Steinitz 2012).

However, GIS is not so explicit on addressing the design question as Geodesign (Batty 2013). The problems of lacking design dimension and adaptability in the traditional GIS and PSS models were addressed in the emerging discourse of Geodesign, an initiative that focused more on the design-oriented modeling method (Batty 2013). According to the statement of Steinitz who quoted Michael Faxman and Stephen Ervin from the Geodesign Summit 2012, Geodesign is "a design and planning method which tightly couples the creation of design proposals with impact simulations influenced by geographic contexts, system thinking, and digital technology" (Steinitz 2012). From this definition, an interventional approach through design is essential. The design intervention turns the concept of Geodesign into a matter of both science and art. Geodesign is a form of art that emphasizes creativity, individual experience, self-awareness, interpretation and expression (Steinitz 2012). It is also argued as a science-based design or "a technology of

---

P. P.-J. Yang (✉)
School of City and Regional Planning and School of Architecture, College of Architecture,
Georgia Institute of Technology, Atlanta, USA
e-mail: perry.yang@coa.gatech.edu

design" that emphasizes knowledge of how urban, environmental and social systems operate. The idea of Geodesign as a "science in design" is interventionist in contrast to the more detached and dispassionate nature of pure science (Goodchild 2010). It manipulates form and intervenes in the systems based on normative questions.

What is challenging is not only to deal with "science in design", but also to treat "design in science" (Batty 2013). The ability to raise "design questions" that connect to scientific analysis, in other words, a design-driven urban modeling, will be crucial to the shaping of future urban systems. It continues the discourse of the rational planning model in PSS where the analysis comes with purpose and objective, but it goes deeper. Geodesign aims to develop a design-oriented urban modeling method, in which its application needs to be more purpose driven and adaptable to situational change.

3.2 Research Design: Benchmarking Energy Performance of Global Downtown Settings

3.2.1 The Selection of Test Cases

In this paper, a series of test cases was conducted through the *Ecological Urbanism Studio* in 2011, and again in the *Site Planning and Urban Eco Simulation* Workshop in 2013 in the School of City and Regional Planning and the School of Architecture at the Georgia Institute of Technology. Eleven central urban districts from seven global cities were chosen from downtown urban settings in North American and East Asian cities, including Atlanta, Chicago, Macau, Manhattan, Shanghai, Tokyo and Vancouver. The test cases were used for measuring the attributes of urban form and their corresponding performance in energy and carbon emissions.

There are three steps that coincide with three criteria, centrality, urban grid pattern and variation, that determine the way we select sample urban districts:

The first step is to define the territory of the downtown urban environment. We use the term "downtown" to represent an urban district that is recognized as a city district of centrality or a relatively high-density environment, in which they are not necessarily named "Downtown" in specific contexts. For example, we also selected some districts called "Midtown" in the case of Atlanta and Manhattan or the "Loop" in Chicago.

The second step is to determine the spatial scale of an urban district. A one square kilometer urban grid system was defined as the spatial range of each sample district, in which the urban grid pattern is seen as the generator of modern cities (Martin and March 1972). There exists a strong relationship between density, spatial typology and environmental performance that operates on predetermined urban grid patterns.

The third step is to pick sample blocks to represent variation. To understand better how different density and urban typology perform in energy-carbon efficiency, high variation among sample districts would provide evidences that are easily observable. We selected urban districts with different building densities ranging from a floor area ratio (FAR) of 1.6 in Atlanta's Midtown to FAR of 15.0 in Manhattan's Midtown, urban block sizes ranging from 4,016 m^2 in Macau's Yun Han District to 59,900 m^2 in Shanghai's Pudong Financial Center, and the number of street intersections ranging from 13 in Shanghai Pudong Financial Center to 153 in Vancouver West End and 172 in Manhattan Downtown.

These measures provide quantitative parameters for describing internal organization of cities at the meso-scale territory, including the building density, building heights, block sizes, number of street junctions, land use diversity or programmatic composition, that would create different potential effects on energy and carbon performance.

3.2.2 A Framework of Geodesign Method

A Geodesign method is applied to organize a framework of representational, performative and change modeling. To respond to the definition of Geodesign from Steinitz based on the statement of Faxman and Ervin (Steinitz 2012), the proposed method couples the creation of design with performance impact assessment. It is influenced by integrating geographic contexts in an urban environment, a system thinking of spatial hierarchy from a city to a district to a block structure and digital technology including GIS and other energy simulation tools.

The method begins with the representational dimension by operating a multi-scale mapping and three-dimensional (3D) modeling of urban block structure, covering urban territories at large (L), medium (M) and small (S) scales. It focuses on the M-scale territory, in which a form-based mapping integrates quantitative attributes, such as density, diversity and street connectivity, of the urban block structure.

A performative model is proposed as a basis using GIS and other energy-related assessment tools. The objective of achieving a low energy and low carbon urban system is clearly set as the criteria for the research design. The performance of urban visibility, solar availability, energy-related analysis and carbon footprints is measured and benchmarked based on the eleven km-wide downtown urban settings. The benchmarking and comparative performance analysis among the eleven urban form structures provides results to guide future change to achieve better performance.

The final stage is to develop a change model. A test case of designing, altering and reconfiguring an urban block structure in the Chicago Loop is conducted. Alternative ecological urban block designs are designed and their performances in carbon reduction are tested. The designed urban block and their performance criteria could lead to a possible urban change. Unlike other GIS or PSS-based modeling, the design-oriented urban modeling projects future urban block structures that are grounded on a performance analysis in energy consumption, renewable energy production and carbon reduction. It involves a creative process of form making, and demonstrates how urban design can be taken as a tool for synthesizing complex scientific issues to move design decision-making to meet low carbon standards.

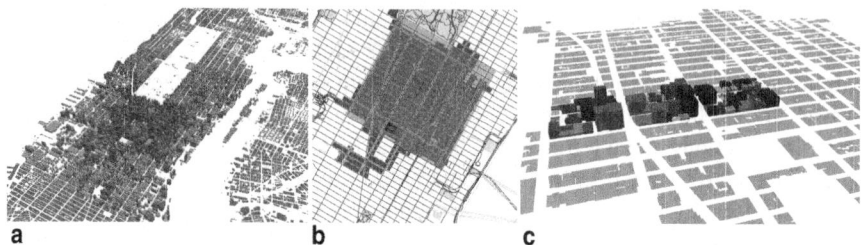

Fig. 3.1 Representational modeling for **a** Large-scale. **b** Medium-scale. **c** Small-scale urban territories, an example of Manhattan, images produced in Site Planning and Urban Eco Simulation Workshop 2013, Georgia Institute of Technology

3.3 From Representational Model to Performative Model

3.3.1 Representational Models

The modeling process starts from a description of the urban environment. Multi-scale representational models provide mappings and 3D visualization across different spatial scales ranging from the city-regional level territory to the downtown urban core (1 km by 1 km) and then to a selection of sample blocks (3 blocks by 3 blocks). The attributes of urban territories at different scales are defined based on observable environmental features and analytical issues related to the particular scales:

1. L-scale: The city-region refers to environmental features including settlement patterns, water bodies and vegetated green patches or corridors. It is defined as the L-level mapping that provides "big numbers" and major information about the cities. It deals with an administrative territory and basic parameters of cities, such as population density, the major land cover and land use, total energy consumption, total water consumption or total carbon emission, and their per capita consumption rates. To better understand each city's greater context, the project mapped each city's urban spatial structure, landscape pattern and transportation network within each large 10 km by 10 km study area. At this level of analysis, patterns of land use and density begin to emerge (Fig. 3.1a).
2. M-scale: The downtown core area is defined based on a 1 km by 1 km territory that is representative of the urban form in the densest urban area or the central city district. Examples include territories covering Lower Manhattan, Midtown Manhattan, Chicago Loop, Atlanta Downtown and Atlanta Midtown. The M-scale 1 km by 1 km study area describes a street block system. At this level, a focal scale of this particular Geodesign model, a set of parameters are developed for describing "the block structure" of downtown urban core, such as overall building coverage ratio, building density, mix of land use, street connectivity, spatial configurations of green space and building typologies (Fig. 3.1b).
3. S-scale: At the S-scale, a typological study of buildings is conducted. Building archetypes are sorted and classified based on building height, area and form.

Sixty building typologies are established and categorized based on height and massing. These typologies are then compared based on characteristics such as height, massing, FAR, carbon emissions, total solar availability and carbon offset potential. At this level, the spatial representation provides truly 3D information of building form and urban geometry that can be turned to attributes and managed in GIS for describing spatial property. For each city, a sample block structure is selected from a typical 3 blocks by 3 blocks area to form a nine blocks matrix. With eight surrounding blocks, the central block would be contextualized for measuring its performance analysis with a consideration of neighborhood effects such as shading (Fig. 3.1c).

The construction of representational layers from L, M and S-scale territories are driven by the definition of system boundaries, a conceptual boundary in a physical or functional space that encloses all components for the urban systems at different levels (Grubler and Fisk 2013). For the purpose of measuring energy and carbon performance, the system boundaries of urban territories are to be defined for figuring out the extent of energy use and carbon emissions. It is essential to select the right scale and appropriate system boundary for proposing smart strategies to promote energy performance or carbon reduction for implementation.

How do we define the system boundary for urban carbon emission measure and its reduction strategy? For an open system like cities, it is always a problematic matter (Dhakal and Shrestha 2010; Brown et al. 2009). The geographic definitions of a city or an urban area are controversial. They go from the limitation of mass transit to the entire commuter shed to political boundary by jurisdiction. Because cities are open systems, significant proportions of urban carbon emissions are normally produced or consumed elsewhere (Sovacool and Brown 2010).

Different from most social, economic and census data that are grounded on traditional administrative boundaries, energy flows cover physical and functional spaces and transcend administrative and territorial boundaries of cities. Compared with a building system or a national accounting system, the system boundary of energy flows in cities is relatively poorly defined. The literature on urban land use is of limited use in energy-related assessment despite its richness in availability of data (Grubler and Fisk 2013). For the kilometer-wide urban spatial structure, data of urban energy flows simply rarely exist at this level.

The paper relies on building energy performance measure to construct its own energy data. Using the GIS platform, a bottom-up approach is applied by aggregating the results of individual building performance to scale up to the particular M-scale urban territory according to different typologies. The system boundary of energy and carbon accounting is limited to the building sector, without considering other sectors such as transportation, water infrastructure, waste treatment or their sectorial nexus. Although the complex interaction among the components is simplified, there is an attempt to construct the relationship between urban form and energy performance at the M-scale territory and to address the question of how the internal structure of cities performs in energy and carbon efficiency.

Table 3.1 Carbon emission factors. (U.S. Department of Energy 2005)

Type of land use	CO_2 (tons/1000 sq. ft.)
Residential	7.13
Commercial	13.64
Education	10.17
Government use	12.33
Mixed use	10.91
Industrial, tourism	18.04
Construction, vacant	1.77

3.3.2 Approaches to Performance-Based Models

This study chooses two approaches for measuring energy consumption and its conversion to carbon emissions of cities. The first is to use land use categories to estimate carbon emissions based on energy intensity for each land use type (Table 3.1) (U.S. Department of Energy 2005). The office use tends to produce a higher rate of carbon dioxide (CO_2), the residential tends to be lower and the mixed use is in normally in between both.

The second approach is the building simulation method that normally produces more accurate results. The test case chooses eQuest, a building energy simulation tool from the U.S. Department of Energy DOE2.2 engine, a method that considers building envelop, system efficiency and its function. The simulation method is first applied to selected buildings based on the sixty defined building typologies. GIS is used as a platform for classifying individual building typology and for aggregating results and turning them into mapping at a broader-scale territory. The results of energy performance are converted to the carbon emissions. Other parameters such as the age of building and the energy label are not considered.

Two solar analysis approaches are taken for different purposes. The solar radiation tool in ArcGIS of ESRI is used for the L-scale analysis. Because the analysis is based on the digital elevation model (DEM) raster surface, the vertical surfaces, such as building facades, are not computed for this scale. The Ecotect of Autodesk is taken as the alternative tool for the building-level analysis. Its calculation performs incident solar radiation across a whole year, in which the output can be generated based on the annually cumulated radiation value on each surface. To study the impact of different building typologies on solar availability, all sample typologies are simulated in Ecotect for its annual surface average solar gain. The results are then applied to a GIS platform for scaling up the results from the building to urban level by aggregating all buildings in the same typology category with similar orientation in proportion to different building density values.

These simulated results provide ideal models for exploring the relationships between different urban environments and their potential carbon footprints, the availability of solar radiation and the capacity for converting it to renewable energy through solar panel deployment (Fig. 3.2). The performance model at the kilometer-wide spatial range shows that with the higher building density of the urban block structure, greater carbon offset could be achieved. We also observe from the model that the finer the grain in the block structure, the carbon emission tends to be less

Fig. 3.2 Carbon footprints and renewable energy capacity mapping for kilometer-wide urban environments from Atlanta, Chicago, Macau and Manhattan (from left to right), images produced by the Ecological Urbanism Studio 2011, Georgia Institute of Technology

(Table 3.2). These findings imply that high density, compact city form and small block development tend to perform better in total energy and carbon efficiency.

The benchmarking among various downtown urban settings in global cities is an attempt to use an urban physical model and its simulation to describe patterns of energy flow. The simple urban physical modeling approach addresses the issue of how we "manage complexity and reduce uncertainty" in cities (Grimm 2005). Urban physical form and its spatial configuration certainly matter in energy and carbon performance.

3.4 From Performative Model to Change Model

Through the mapping of eleven urban settings of seven global cities, a set of urban form parameters, including block size, building density, surface volume ratio and street connectivity, are compared to the performance results of carbon emission and potential carbon offset by solar energy. The benchmarking provides a performative

Table 3.2 Benchmarking performance of energy and carbon among the eleven km-wide urban block structures, data produced by the Ecological Urbanism Studio 2011, Georgia Institute of Technology

	Average block size (sq. m)	Building density (Gross F.A.R.)	Surface area/ building volume	Street connectivity (number of intersections)	CO_2 Emissions based on land use (tons)	Total carbon offset by solar (tons)
Atlanta/Midtown	8300	1.6	10%	27	196,667	32,991
Atlanta/Downtown	6000	3.7	12%	82	519,293	37,678
Macau/You Han	4016	13	25%	113	337,007	27,847
Macau/Xin Kou An	5858	4	22%	110	556,936	29,214
Chicago	9500	3.8	10%	61	1,853,772	50,599
Manhattan/Midtown	10,175	15	11%	68	1,815,579	58,252
Manhattan/Downtown	11,979	11.4	12%	172	1,227,094	36,709
Shanghai	59,900	6.5	8%	13	756,756	20,607
Tokyo	8000	1.9	17%	152	169,467	18,045
Vancouver/Historic Center	11,443	5.9	21%	147	165,700	8.529
Vancouver/West End	8800	10.1	21%	153	187,951	9.473

model for understanding how existing internal organizations of cities perform and function in energy and carbon efficiency.

However, the question of how to achieve a low carbon and low energy urban block system remains to be answered. It requires a "change model" to design for the future. Instead of looking at how the existing urban block system performs and functions, the design question addresses how the urban block system should be changed to achieve the low carbon objective. The future urban block system is to be designed for change and is to be transformed over time, in which its performance is to be tracked to meet the standard.

A test case was conducted to take Chicago's Loop as an urban laboratory for improving energy performance through an assimilation of renewable energy. An existing urban block structure was reconstructed, and its energy performance was modelled. A series of hypothetical urban block models were designed or speculated based on the following principles:

- To develop a green street planning strategy (Kloster et al. 2002);
- To apply green area ratio (GAR), multiple green surfaces from the ground to the façade and to the roof (Ong 2003);
- To use the "solar envelope" concept for developing high performance typologies;
- To enforce directional building setbacks to enhance daylighting for saving energy; and
- To create wind tunnels within buildings to take advantage of wind power and enhance ventilation.

In the case of Chicago, those principles target on maximizing solar gain, improving daylighting, enhancing air ventilation and increasing permeable landscape surfaces. The optimization of "solar envelope" or "solar skin" provides an enormous urban surface for solar and sunlight. The proper spacing in between buildings encourages the breeze to go through the high-density environment. It allows quality sunlight and wind to penetrate into the deep center of the block both horizontally and vertically. In other words, it provides a design strategy of the mat building system, an approach to design a condensed and open structure for the urban block, and to control building mass and void below and above ground with high porosity (Sarkis et al. 2002). It is an extension of the green street concept as a pervasive ecological landscape surface as well as an expansion of the public domain from external systems, such as streets, roads and parks, to the internal territory of the urban block. The potential programs for the urban block redevelopment would include a vegetated solar roof covered by photovoltaic, roof top garden, playground, sport field, sunken plaza, street front café above and below ground, etc.

In reality, the energy performance of urban form is a complex, uncertain and dynamic matter that is affected by local climate conditions and changing over time during diurnal cycle and seasonal change. Instead of pursuing the system optimization through a comprehensive modeling approach, the research applies design-oriented urban modeling, a specific type of Geodesign method, to project alternative "designed" future models for benchmarking. Urban design is seen as an instrument to synthesize complex and dynamic factors for deriving sustainable urban future. The hypothetical principles such as "solar envelop" are tested by alternative urban design models through reconfiguring the current urban block structure to meet lower energy and lower carbon performance criteria. The hypothetical urban models are proposed based on the same building density, however, with alternative spatial configurations that would perform differently in terms of energy consumption, potential solar energy production and the total carbon reduction.

In the case of the Chicago Loop, the forward-looking modeling reconfigures the current block structure to an open structure for maximizing "solar envelop" along the street front. The model illustrates how the existing urban block structure (Fig. 3.3a) can be reorganized to new forms that achieve 63.7% (Fig. 3.3b) and then up to 69.2% (Fig. 3.3c) carbon reduction through enhancing its solar energy capacity, and reducing carbon emission by adjusting physical layout of the urban block. The following parameters were derived from the third block model in the Fig. 3.3c that projected a 69.2% performance change in carbon reduction:

- Total building volume = 3,707,385 m^3
- Building surface area = 1,038,032 m^2
- Total floor area = 1,200,708 m^2
- FAR = 9.8
- Carbon offset by vegetation on the ground = 259 tons/year (vegetation on the ground = 17,286 m^2)
- Carbon offset by vegetation on vertical and roof surfaces = 863.1 tons/year (vegetation on vertical and roof surfaces = 57,539 m^2)
- Carbon offset by solar energy gain = 2,707.2 tons/year (Available area for solar photovoltaic = 141,578 m^2; Projected solar energy gain [18% conversion rate of solar photovoltaic] = 4,417,233 kWh/year)

Fig. 3.3 Design-oriented urban modeling in the area of Chicago Loop; models created by Nathaniel Willy, Ralph Raymond and Michael Cullen in the Ecological Urbanism Studio 2011, Georgia Institute of Technology

- Carbon reduction rate = 69.2 % (by benchmarking the carbon emission 6,457 tons/year based on the model of the current urban block structure as in Fig. 3.3a)

The design-oriented urban modeling method addresses a broader question of "science in design" on how a scientific analysis, such as an energy and carbon performance assessment, can be taken as organizational principles to inform design for future urban form. It adds a performative dimension to urban design practices that are sometimes intuitive or predetermined in form making. It underlines the fact that energy and carbon efficiency of cities is related to how urban forms are organized and configured spatially. The internal structure of urban form matters in energy performance.

It also addresses the question of "design in science" and sees how design engages performance-based analysis or how design intervention is seen as an essential variable to define energy performance-based research for future urban systems. Instead of operating sophisticated and complex system modeling, the urban design-driven model is taken as a tool or instrument for synthesizing complex factors in a process of simulating or speculating future transformation of urban spaces. Designing for change is a key to enhancing the performance of the urban future. It is important to couple urban designers' forward-looking design model and GIS-based performance assessment to meet criteria of low carbon development.

3.5 Conclusion

Why has design suddenly become central again to GIS (Batty 2013)? Instead of focusing on the representational dimension of mapping, spatial statistics and scenario making based on the "What if" question in PSS (Klosterman 2008), the proposed Geodesign model responded to the question of "design for change" (Steinitz 2012).

The paper introduces a GIS-based framework of organizing information from a representational model, performative model to change model that is a forward-looking modeling approach to address the issue of how scientific-based performance analysis articulates design. It is a specific kind of Geodesign model to deal with problems of energy performance and carbon emissions assessment in cities, focusing on a kilometer-wide or meso-scale urban block structure and its internal organization.

The objectives of achieving lower energy and lower carbon performance standards were set at the beginning of framing the Geodesign model. The research benchmarks carbon footprints of downtown areas of seven global cities and their eleven central city districts by mapping the urban physical structure, energy performance, carbon emissions and solar availability of the built environment. The benchmarking of various urban spatial structures and their simulated results on energy and carbon efficiency provide a reference for exploring the relationship between urban form and performance.

Based on a test case in the Chicago Loop, an interventional design approach was taken for showing that urban form and physical organization matter and affect the energy and carbon efficiency. Two questions of "science in design" and "design in science" were addressed and tested. First, how does energy performance enhancement and carbon reduction inform urban design or change urban form? GIS and its data analysis are more than generating "predictive" future-based results regarding the "what if?" questions. They help shape the "designed" future systems. Second, how is design integrated in the Geodesign method? In this paper, urban design is used as an instrument for making the change model. Design is turned into a variable of the forward-looking, performance-based urban modeling. It is in fact part of the informational processing and should be incorporated in the technology and tools of performance assessment.

Finally, if a city is aiming for achieving low energy and low carbon objectives for shaping future ecologically sound urban systems, the Geodesign model provides a method of "designing for change". It projects future urban choices for decisions that would facilitate a set of design principles for guiding a process of urban redevelopment incrementally over time.

References

Batty, M. (2013). Defining geodesign (= GIS plus design?). *In Environment and planning b-planning & design,40,* 1–2.

Brown, M. A., Southworth, F., & Sarzynski, A. (2009). The geography of metropolitan carbon footprints. *Policy and Society,27,* 285–304.

Dhakal, S., & Shrestha, R. M. (2010). Bridging the research gaps for carbon emissions and their management in cities. *Energy Policy,38,* 4752–4755.

Goodchild, M. F. (2010). Towards geodesign: Repurposing cartography and GIS? *Cartographic Perspectives,66,* 7–22.

Grubler, A., & Fisk, D. (2013). Introduction and overview. In A. Grubler & D. Fisk (Eds.), *Energizing sustainable cities: Assessing urban energy*. London: Earthscan.

Grimm, V., et al. (2005). Pattern-oriented modeling of agent-based complex systems: lessons from ecology. *Science, 310,* 987.

Kloster, T., Leybold, T., & Wilson, C. (2002). *Green Streets: Innovative Solutions for Stormwater and Stream Crossings, in Urban Drainage*. Redistribution subject to ASCE license or copyright; . http://www.ascelibrary.org. Accessed 3 Sept. 2013.

Klosterman, R. E. (2008). A new tool for a new planning: The what if?TM planning support System. In R. K. Brail (Ed.), *Planning support systems for cities and regions*. Hampshire: Puritan.

Martin, L. & March, L. (1972). *Urban space and structures*. London: Cambridge University Press.

Ong, B. L. (2003). Green plot ratio: An ecological measure for architecture and urban planning. In *Landscape and Urban Planning,63*, 197–211.

Sarkis, H., Allard, P., & Hyde, T. (2002). Case: Le Corbusier's Venice Hospital and the Mat Building Revival, Harvard Graduate School of Design.

Site Planning and Urban Eco Simulation Workshop (2013). Unpublished., Georgia Institute of Technology.

Sovacool, B. K, & Brown, M. A. (2010). Twelve metropolitan carbon footprints: A preliminary comparative global assessment. *Energy Policy,38,* 4856–4869.

Steinitz, C. (2012). *A framework for geodesign: Changing Geography by design*. Redlands: ESRI.

U.S. Department of Energy (2005). Household, buildings, industry and vehicles end use energy consumption and analysis, Technical Report.

Chapter 4
PICO: A Framework for Sustainable Energy Design

Steven Fruijtier, Sanneke van Asselen, Sanne Hettinga and Maarten Krieckaert

4.1 Introduction

In The Netherlands, the built environment is responsible for 35 % of the total energy consumption and 30 % of the CO_2 emissions (Topteam Energie 2012). Transforming buildings can therefore substantially reduce fossil fuel consumption. This is in line with the 2013 Energy Agreement of the Dutch government, which aims to reduce energy consumption and increase renewable energy generation (SER 2013). Besides environmental objectives, these measures also stimulate investments and employment and consequently strengthen the economy.

However, Dutch evaluations show the transition towards a more efficient energy system to be difficult, and results are so far not satisfying (Rooijers et al. 2010, EC 2011). According to the EC (2011), the main reasons for this are insufficient policy coordination, investment uncertainties and a lack of incentives and awareness among consumers. Furthermore, the energy sector has become more complex. Instead of a predominantly top-down approach, at present local initiatives supported and stimulated by the local government and private companies become increasingly important. This induces an incoherent policy.

In addition, the awareness of the multiple options for energy saving and renewable energy generation at a specific location is often insufficient. Moreover, the costs and benefits of energy interventions are frequently unclear or unknown. To facilitate an energy transition, it is needed to get a clear overview of the energy saving and renewable energy options at a specific location, at different spatial and temporal scales, as well as a survey of the related costs and benefits, for example in terms of GHG emission. Information about the local geographical context, technological innovations, social considerations and financial aspects should all be collectively analysed to find the optimal solution, meeting the desires and needs of all stakeholders involved.

S. Fruijtier (✉) · S. van Asselen · S. Hettinga · M. Krieckaert
Geodan, President Kennedylaan 1, 1079 MB, Amsterdam, The Netherlands
e-mail: steven.fruijtier@geodan.nl

© Springer International Publishing Switzerland 2014
D. J. Lee et al. (eds.), *Geodesign by Integrating Design and Geospatial Sciences*,
GeoJournal Library 1, DOI 10.1007/978-3-319-08299-8_4

Currently, contrary to many other sciences (van Manen 2009), in energy sciences it is not customary to consider location for optimisation. This is understandable: so far the energy system had only a few suppliers transporting energy in a convenient state to consumers. However, with the introduction of decentralized energy generation (PV, windmills, etc.) the generation location, distribution and transportation of energy will become more complex. Horner et al. (2011) emphasize that the cooperation between GIS and energy science will be instrumental to a smoother energy transition. Furthermore, the development of smart grids asks for a good overview of where supply and demand are geographically located and what energy is supplied or demanded in what quantity at what time. Otherwise the electricity grid will not be able to deal with the fluctuations (Moslehi and Kumar 2010).

Optimal energy solutions can be designed by an integrated approach, optimizing a range of different criteria such as the local energy generation potential, the grid potential, the economic potential, concurrent on both local and national scale. An obvious example is the determination of the future location of a wind mill. Not only should be considered where the energy output can be optimized (location with most wind), but also the most cost-effective location, the location most optimal for the energy grid, and the location where the local community is least bothered by the mill.

In The Netherlands local (mainly governmental) organizations have developed or are developing digital energy information services based on spatial data, aiming to facilitate and stimulate the energy transition. Examples are the Energy Atlas of Rotterdam[1], the Energy Atlas of Gelderland[2], the Solar Energy Atlas[3] and the Heating Atlas[4]. These initiatives give an overview of energy-related information in a geographical context solely at a regional scale, or regarding only one specific renewable energy option such as solar panels. Also, these initiatives are not meant to involve stakeholders directly, or to be used as a decision support system. Although stakeholders can browse maps, they cannot discuss problems or options, nor can they have any insight in the consequences of any decision.

The interdisciplinary PICO consortium (Project Interactive Communication and Design) is developing an energy transition support system that extends beyond these initiatives. This system offers an extensive up-to-date and transparent data collection (from different disciplines), which can be visualized in its geographical context on different platforms. These data are used as input for both energy models and financial models that can be used to calculate investment potentials and potentials for energy saving and renewable energy generation for any region in The Netherlands. Furthermore, impacts of energy-reducing measures can be modelled, using for example cost-benefit analyses and CO_2 emission models. The different aspects of the PICO support system are streamlined by the geodesign framework (Steinitz 2012).

---

[1] www.rotterdamclimateinitiative.nl/nl/energieatlas

[2] www.gelderland.nl/energieatlas/

[3] www.zonatlas.nl

[4] www.warmteatlas.nl

The use of spatial data enables the employment of large datasets describing an entire neighbourhood without requiring many different data inputs or estimated averages. Several existing tools to model energy generation and savings opportunities, require attributes to be added individually (e.g. attributes describing individual buildings), resulting in a protracted and tedious process. Using spatial data, all data regarding the selected location are instantaneously available, enabling an efficient modelling approach, saving both time and effort. The geodesign framework has already been proven to efficiently work for different spatial design projects, and has great potential for energy transition projects (McElvaney 2012, Steinitz 2012).

In the next sections, we describe the preliminary results of the PICO project. The first objective was to accomplish an overview of information requirements of stakeholders in an energy transition process. This was done by analysing three use-cases, mainly based on discussion workshops (Sect. 2). The results of these analyses are used to set up the PICO system that is embedded in the geodesign framework (Sect. 3). The architecture of the PICO system is based on the concept of interoperability and modularity using standards and specifications from the OpenGeoSpatial Consortium (Sect. 4). Each geodesign process step and stakeholder has different needs, requiring different system interactions. Therefore different tools will be developed, including a web-based energy atlas of The Netherlands and an application for touch tables facilitating collaboration and discussion between different stakeholders (Sect. 5). We conclude the PICO system will enable multi-criteria decision making to support the energy transition of towns, neighbourhoods and communities (Sect. 6).

4.2 Revealing Information Requirements for Energy Transition Processes

The PICO project aims to facilitate energy transitions in a variety of local conditions and to support all stakeholders involved. Using the PICO system and tools (user interfaces and visualization tools), different impacts of different scenarios can be reviewed directly and collaboratively. Stakeholders of three different use-cases were invited to a workshop where they could provide an overview of the main requirements and aspects that should be addressed by the PICO project. The workshops were set up around a particular use-case where a group of stakeholders was faced with an information issue that needed solving such as "how are we going to invest our money to make our neighbourhood more sustainable?" The goal of the workshop was not to answer the question at hand, but rather find out what information, tools, or ways of collaborating could help answer it. In the workshops the principle of the "five Ws and H" has been applied (see Fig. 4.1) to orient on the current situation and to explore potential improvements.

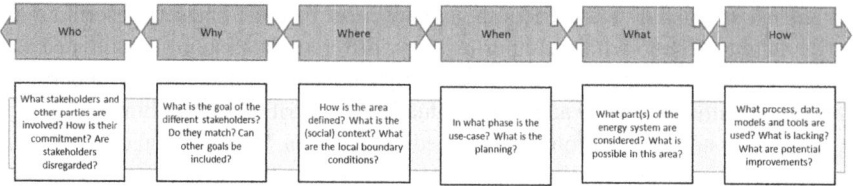

Fig. 4.1 Typical questions asked during the workshops according to the principle of the "five Ws and H"

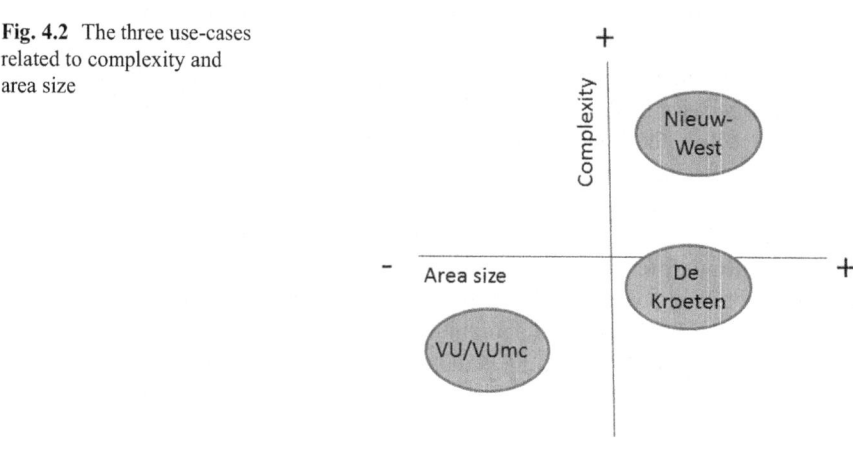

Fig. 4.2 The three use-cases related to complexity and area size

4.2.1 Three Different Use-Cases

To survey the information requirement of stakeholders, three use-cases are analysed: a residential area *De Kroeten* in Breda, a residential area *Nieuw-West* in Amsterdam, and the *VU/VUmc* university and academic hospital area in Amsterdam. Each use-case had distinctive boundary conditions and objectives. The use-cases were selected to capture a large range of situations where the PICO framework might be applied: a homogenous suburb represented by a foundation, a utility complex with a single owner and a diverse neighbourhood with many separate initiatives. Each of these use-cases has its own specific problems and needs that could be addressed by the PICO framework. Fig. 4.2 shows the positioning of the use-cases in relation to the complexity of the project and the area size. The complexity is among others determined by the number of stakeholders, ambition levels and stakeholder commitment.

In *De Kroeten*, residents collectively invested in a wind mill. All residents are a member of a sustainability foundation, making use of a revolving fund. The investment in the wind mill will pay out in the next year, providing financing to invest in new sustainable (energy) measures to improve the neighbourhood. To make sound investment decisions and plan future investments, the option for energy-reducing and generating measures and related costs need to be mapped out, evaluated, and presented understandably to the neighbourhood residents.

In *Nieuw-West* ambitions to reduce energy consumption are high. These ambitions are reflected in multiple initiatives from the municipality of Amsterdam for innovative technologies and sustainable energy solutions and also by many grassroots initiatives started by house owners. These initiatives however are not coordinated. This leads to missed chances for energy and cost efficiency and impeding the road to an integral sustainability design for the entire area because people who just invested in their house are not eager to participate in a large scale renovation. There is a lack of (financial) commitment from a larger party such as the municipality or a housing association. No larger organisation takes a leading role and implements a sustainable renovation plan for the whole area. Now only a small percentage of the residents apply sustainability measures themselves. This case is additionally complicated due to the multitude of stakeholders with very different interests and plans, who all have to agree on larger energy related investments. In this use-case, an integrated approach, where multiple stakeholders can jointly discuss different scenarios, can strongly benefit decision-making.

At *VU/VUmc*, fewer stakeholders are involved, but the energy demand is complex. The aim is to produce a large fraction of the needed energy from renewable energy sources. However, the VU/VUmc have a high constant electricity demand, as well as a heat and cold demand, which has to be 100% secure and affordable at all times. Therefore all possible energy sources and storages have to be mapped, providing that the energy supply is secure, and a clear image of the financial feasibility has to be outlined.

4.2.2 Main Insights Derived from Use-Cases

The PICO use-cases showed many different stakeholders to be involved in the energy transition process: residents, local authorities, power companies, grid operators, energy producers, and other private companies and local initiatives. Commonly, these stakeholders have different objectives and needs, whereas often the organisation responsible for the overall process is not clear or does not take the initiative. This usually results in postponed action. Frequently, communication between the different stakeholders is limited. As a consequence, stakeholders are not always aware of all possibilities at the location of interest, which may lead to a lack of commitment. Hence, there is a great need for better communication between stakeholders as improved communication will most likely stimulate public support.

Another important conclusion derived from the workshops, was that data and information needed for the energy transition project was often incomplete, or unavailable. There are different reasons for this deficiency. Firstly, most projects focus on a single energy technique (e.g. solar panels) or a single criterion (e.g. CO_2 reduction). Data required for a broader perspective is not present. Secondly, when taking into account this broader perspective a lot of data is needed from different sources and in different formats. There is no single entry to access the required data making it difficult to gather all data needed. And thirdly some of the required data does not exist but could be created using (spatial) analysis or models. The knowledge and experience for this (spatial) modelling often is not present.

The lack of data hinders the establishment of a complete overview of all energy saving and renewable energy options at a location, and precludes a complete impact analysis.

Visualizing data in its geographical context is considered highly valuable. All stakeholders, specifically citizens, want to understand the information offered to them and the reliability of this information. This applies to basic maps but also to more complex information such as results from impact models. It is important that the information is available in an understandable way, to ensure trustworthy data. This includes information about the reliability or quality of the data and models, e.g.: the creator, the actuality of the data, and how it has been derived.

Different stakeholders find different factors important for decision making. For citizens and companies the financial consequence of an energy transition measure is most important. On the other hand governmental organisations such as municipalities focus more on CO_2 reduction. Other criteria of lesser importance are the autonomy of the energy system and energy reduction. Because the financial consequence is so important a high reliability of the financial models is crucial to support the comparison of the (financial) consequences of different energy solutions.

4.2.3 A Multiple Criteria Decision Making Process

The creation of a new sustainable energy system is found to be a complex process, which can be regarded as a multi-criteria decision making process (Pohekar and Ramachandran 2004). Decisions are based on multiple objectives (e.g., CO_2 and cost reduction), a selection of multiple quantifiable options (e.g., CO_2 reduction in tons/year) and non-quantifiable variables (like feelings). The final decision will be a compromise between the interests of the different stakeholders involved in the process.

When designing a sustainable energy system, the stakeholders have different preferences for the criteria. They will bring in different points-of-view and demands, acting at different spatial and temporal scales. A framework is needed to reach mutual understanding and to establish a widely supported compromise taking into account all criteria and interests of the stakeholders. It is expected that the geodesign framework can very well be used to guide this process (Steinitz 2012).

4.2.4 Addressing many Different Stakeholders

The PICO use-cases showed that in an energy transition process, commonly many different stakeholders are involved. Steinitz (2012) subdivides the people involved in the design process into four groups: design professions, geographical sciences, information technologies and people of the place. These four groups should collaborate and form the geodesign team. For an energy system similar groups of stakeholders can be identified:

- People of the place: residents, housing cooperatives, authorities and local companies. This group delivers requirements, design ideas and boundary conditions. In the end they are responsible for making a widely supported decision.
- Organisations and companies involved in energy technology: network managers, energy companies and technology providers. They supply the technology and knowledge.
- Energy professionals including energy service companies and in some cases governmental spatial planning organizations. This group mainly has a focus on individual buildings, both in knowledge and skills.
- Geographical sciences. At the moment, these are not typically involved in energy transition processes. However, cooperation between geographical and energy science is expected to be instrumental to a smoother energy transition (Horner et al. 2011). One of the objectives of PICO is to facilitate and stimulate the involvement of geographical sciences.

4.3 PICO Embedded in the Geodesign Framework

The geodesign framework facilitates a change in a geographical context, essentially by iteratively tackling different steps in the design process in each of which stakeholders are involved and can give feedback (Steinitz 2012). Involvement of stakeholders from start to finish of the design process results not only in the best solutions, satisfying most stakeholders, but also generates their commitment.

As has been concluded from the PICO use-cases, design problems and options are not always completely understood, due to a lack of data and information sharing. By going through all six geodesign steps iteratively it is assured that most important data, processes and options are collected and analysed. Moreover, the lessons-learned from previous steps or iterations can be included to finally come up with the best solution or better said, with the best compromise.

The next sections describe how the PICO framework can be embedded in the geodesign framework. The sections are divided according to the six steps from the geodesign framework and respond to the 'why', how' and 'what' questions for the PICO framework. Fig. 4.3 shows the framework steps with questions asked during the process.

The PICO framework does not predefine a fixed process; it only provides a basis for making decisions about energy systems by offering data and tools. It is expected that during the design process new questions will arise and additional data and tools will be needed and included. In this way PICO will provide a framework for sustainable energy design which can be used 'stand-alone' but can also be embedded

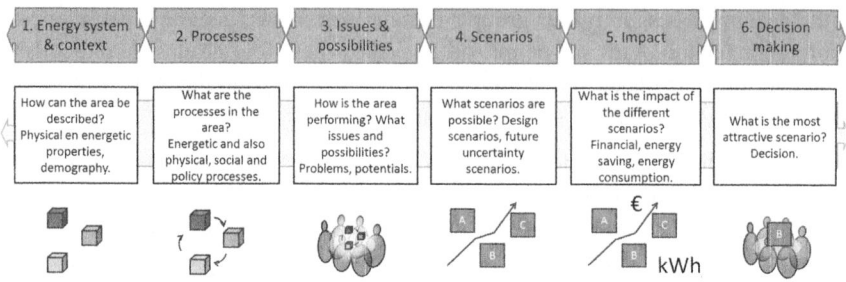

Fig. 4.3 Geodesign used for designing a new sustainable energy system.

in other processes implementing the geodesign framework. These processes initially focussing on other issues can conveniently include the energy system as well. For example social or demographic problems can be the initial starting point for a geodesign process but if needed the tools and data from the PICO framework can be included.

The application of the framework usually does not follow a linear process. It is conceivable and likely that during the process e.g. new ideas, wrong assumptions and unexpected intermediate results are discovered. These discoveries may require the process to go back one or more steps leading to non-linear iteration of the geodesign steps. Some examples are given in the description of the steps below.

4.3.1 Step One: The Energy System and Context

The goal of the first step for the PICO framework is to describe the current energy system and context of an area. This is used to gain insight in an area, to assess the initial opportunities, to identify stakeholders and to inform stakeholders. The data needed concern not only data describing the energy network itself but also data describing the built environment as a whole, including the social and demographic context of the area. A large amount of datasets is available in The Netherlands, describing the physical environment. For example the key register of buildings and addresses (Ellenkamp and Maessen 2009) describes the footprint, building year, area and other relevant attributes of all buildings.

Other important data concern the energy-efficiency of buildings (expressed by 'energy labels' (Van Hal 2007), the presence of renewable energy initiatives, technological information of renewable energy solutions, and information about regulations and subsidies. Much energy data are available as open data from the Dutch government (Fig. 4.4). Other datasets need to be created or complemented.

In case of energy labels, only 30 % of the houses have an energy label—most of which are rental houses (WoOn 2013). Based on open data and typical data for example house types (AgentschapNL 2011), the expected energy labels from the other buildings can be modelled.

4 PICO: A Framework for Sustainable Energy Design

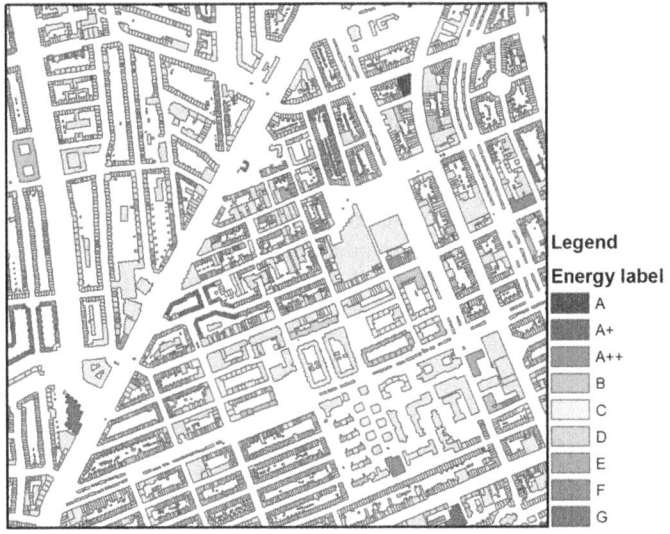

Fig. 4.4 Example of accumulated open datasets: buildings and energy labels for buildings.

All data and information from steps 1 to 3 are made available using open standards to any tool supporting these open standards. The data is visualized in easy to use visualisation tools offered by PICO giving a clear overview of the energy system. Stakeholders can use this overview to identify missing data and information.

4.3.2 Step Two: The Energy Processes and other Processes

Step two inventories the energy processes. What is the energy consumption and energy efficiency? How do these processes relate to each other and possibly to other processes? In this step dynamic data are collected to describe processes in the energy system. For instance energy consumption and generation, but also CO_2 emissions, which have an important impact on climate change. In The Netherlands, national scale data on the average yearly electricity and gas consumption per neighbourhood is available as open data at the CBS Statistics Netherlands. At more regional scales, energy grid operators like Liander provide open data on energy consumption at zip code level[5]. Furthermore, average energy consumption profiles have been estimated for 'example house types' (AgentschapNL 2011).

---

[5] http://www.liander.nl/liander/campagne/open_data.htm

4.3.3 Step Three: Issues and Possibilities Determination

The goal of this step is to identify and evaluate issues and possibilities regarding the energy system. Issues can be a mismatch between energy efficiency and energy consumption, like areas using more energy than is expected looking at the energy efficiency and social context. Possibilities are the potentials for energy savings and energy production in the specific area. These potentials can be estimated using GIS analysis or modules of energy models such as the Quick Urban Energy Scan (QUES; Celie et al. 2011) and Vesta (Folkert and Wijngaard 2012). Furthermore exploring ideas of the people of the place can lead to new possibilities and issues eventually leading to boundary conditions used for the scenarios.

4.3.4 Step Four: Scenario Definition

In this step it is decided how the energy system can be altered. Based on the energy potential maps and boundary conditions set by the involved people and organisations in the previous steps different scenarios for an energy transition in an area are inventoried and selected. Two different types of scenarios are distinguished: design scenarios and future uncertainty scenarios. The design scenarios are the geographic interventions that can be taken in the area. Examples are the designated areas for solar panels or windmills and the design of energy efficient morphologies (buildings or neighbourhoods). The future uncertainty scenarios are possible external developments like increasing energy prices or demographic developments. The determination of the aspects which are uncertain and important to model is carried out in this phase with the participation of stakeholders. The scenarios will be stored and can be used as input when going back to the previous steps.

4.3.5 Step Five: Modelling the Impacts of Different Scenarios

The impacts of the selected scenarios are subsequently calculated in step five of the framework. Energy simulation models are used in this step. Models like the aforementioned QUES and the Vesta models assess impacts of energy-reducing and generating measures in buildings (e.g., isolation, PV) and at a regional scale (e.g., geothermal energy), in terms of for example CO_2 emission, energy consumption, and investments. The results of this step can be used for decision making (step 6) but also for going back to the scenario definition if the results do not meet the expectations. They can also be included in step 1 for a new iteration of the design process. Tools supplied by the PICO framework will include user interfaces to easily control the models and to visualize the results in maps and diagrams. The results can also be included in other tools such as Microsoft Excel allowing people to use the results in their own models.

4.3.6 Step Six: Evaluation of Different Impact Scenarios by Multiple Stakeholders

The impact analyses are carried out preferably at the location, with the stakeholders present, and are immediately available for analysis and evaluation with the information used before. Depending on the implementation of models the results will be available real-time or within a few minutes. The results of the impact analyses are discussed with all involved organisations and people. Not only the results, but the complete process from inventory to scenario definition will be used for discussion to ultimately reach a mutual understanding of the interventions needed. The end result of this step will be a decision. This is not necessarily a linear step. It is conceivable the process will step back to for example the scenario definition or issue and possibilities definition. Also the decision made will often not be a final decision but input and starting point for a next iteration of the process.

4.4 Architecture of the PICO System

As stated by Ervin (2011) no single software product or approach will suffice to support a geodesign process. The geodesign system requires a combination of tools and techniques. The system cannot be implemented by perfecting one particular software product. It is needed to design a system that supports interoperability and modularity to support the required flexibility. Infrastructures in the geospatial domain, commonly known as SDI's or Spatial Data Infrastructures (Williamson 2004) are implemented based on this approach.

4.4.1 Conceptual Architecture

Figure. 4.5 shows the conceptual architecture of the PICO system. The infrastructure of the PICO system is implemented using concepts and standards of SDI's; like most current SDI's the PICO system is based on service oriented architecture. The interfaces and formats of these services are implementing the open standards and specifications from the OpenGeoSpatial Consortium OGC (Percivall 2003).

4.4.2 Services

The tools (or clients) will be using services supplied by the PICO system. These services supply the functionality to the tools giving standardized access to the data and models. The data (maps and features) and models can be accessed using the appropriate specifications from the OGC, like Web Mapping Service (WMS) for map

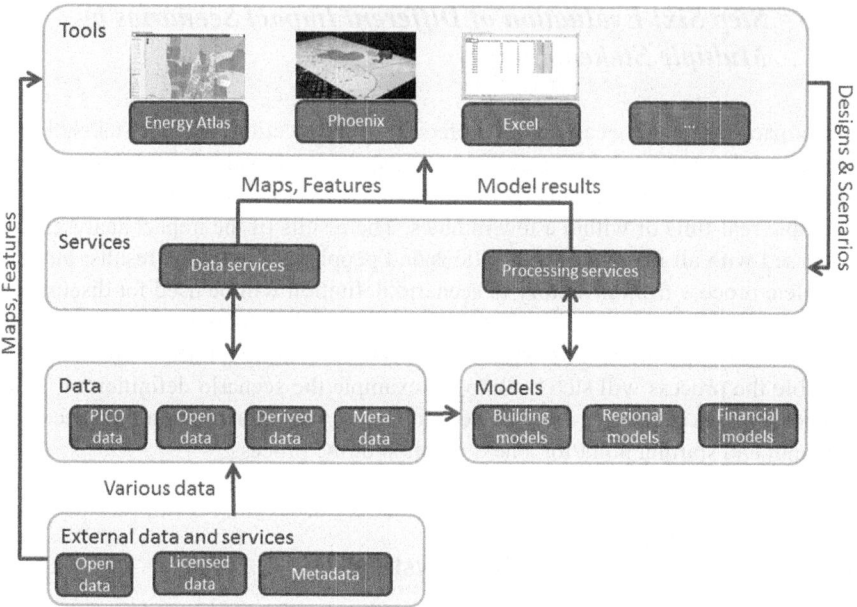

Fig. 4.5 Conceptual architecture of the PICO system.

portrayal, Web Feature Service (WFS) for feature querying, retrieval and storage and Web Processing Service (WPS) for handling model requests and results. The data and functionality are thus not only available to tools developed specifically for PICO; also applications from other parties can benefit from the information and functionality offered by the PICO framework. This interoperability ensures these and new tools can be incorporated within processes supported by the PICO platform.

4.4.3 Data

The PICO system will offer a large amount of needed datasets. These datasets come from a wide range of external sources and other datasets are created within the PICO framework. Much of the data is available as open data, while some data can have a restricting license. The open data needed is offered in many different formats (e.g. CSV, plain text and even PDFs) and often need to be transformed in order to make them available to the services using open standards. Other processes implemented by the PICO system are the creation of derived datasets out of other datasets. An example is the determination of the type of building (terraced building, semi-detached house, etc.) based on the geometry of buildings. The result is subsequently used for the analysis described before to model the energy label of

4 PICO: A Framework for Sustainable Energy Design 67

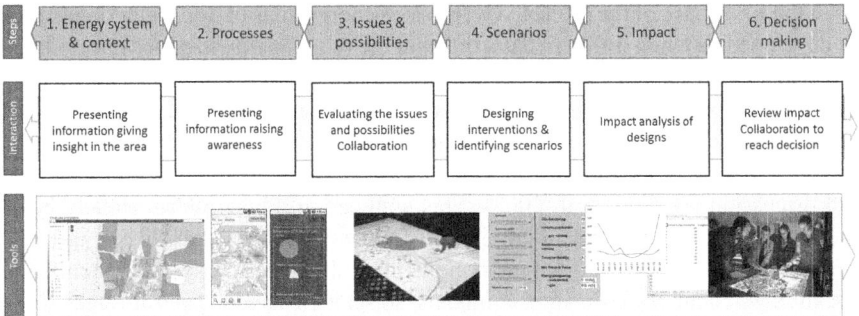

Fig. 4.6 System interactions in the six steps and accompanying tools for different steps and stakeholders.

a house. Metadata is needed to describe the lineage and quality of the data, which is especially important for describing the transformations performed by the PICO system. The metadata can also be used to search for specific datasets. Moreover, during the workshops it became apparent transparency about the data and models is essential for stakeholders to trust the data and models, and ultimately, to trust the entire process of designing a sustainable energy system.

4.4.4 Models

The created designs and defined scenarios can be used as input for the impact models which can be stored in the PICO system. Models use the boundary conditions and scenarios to calculate the impact of the possible choices or scenarios. An example is calculating the energy and CO_2 reduction when selected energy designs are implemented in a certain area. Simulation models are run ad-hoc with boundary conditions and parameters set by the geodesign team. These models can be run at different scales and for calculating different criteria. Building models calculate for example the potential solar production for individual buildings while regional models calculate the potential geothermal production for an area. Next to energy technology models, financial models can be used in the process depending on criteria set by the stakeholders.

4.5 Developing the PICO Tools

Each geodesign process step and stakeholder has different needs, requiring different system interactions. Therefore, different tools will be designed and developed to facilitate the process and stakeholders (Fig. 4.6).

In the first phase of the project, a web-based energy atlas of The Netherlands will be developed showing different data, offering the functionality to query information and statistics of specific locations. This allows obtaining for example the mean age, total energy consumption, mean energy consumption per household, per m^2 and per person. In order to enable collaboration and discussion between the different stakeholders, an application for touch tables will be used to stimulate this collaboration and discussion (Hofstra et al. 2008). This application, Phoenix, has already been used for spatial planning discussion in different settings (Dias et al. 2013), and will be extended to include energy transition planning. In a later phase, the Phoenix application will connect to the impact models, so scenarios and ideas from stakeholders can be computed and visualised directly during discussions.

During the execution of the design process, it is expected additional tools may be needed during the different phases. Tools supplied by the PICO framework will include user interfaces to easily control the models and to visualize the results in maps and diagrams. As not all stakeholders need a map-based application during the decision-making process, also an Excel based application is developed allowing the stakeholder to use their own business models with the data and models from PICO. The existing and envisioned tools, serving different stakeholders and phases, will be developed and validated based on the existing and on new use-cases. It is expected this will almost continuously lead to new and improved tools.

4.6 Conclusions

Energy transition in The Netherlands, aimed at a reduction of energy consumption and an increase of renewable energy, is a complex process for which a framework is needed to enable optimal decision making supported by multiple stakeholders. PICO meets this need by providing a system based on the geodesign framework that facilitates multi-criteria decision making processes for energy transitions.

By going through all six geodesign steps it is assured that most important data, processes and options are collected and analysed. The geodesign framework promotes stakeholder involvement from the start of the design process. This results not only in the best solutions, satisfying most stakeholders, but also generates their commitment to achieve the energy transition aimed for.

The PICO framework provides up-to-date and transparent digital data services for The Netherlands (1), supports the definition of ambitions, potentials and scenario's (2), performs impact analyses (3) and serves as a communication platform (4). Each function is associated with different tools or a set of tools. Data and model output can be communicated as for example digital maps and Excel files, on different devices, depending on stakeholders needs.

The architecture of the PICO system is based on open standards. Different parties can connect to the PICO platform to retrieve data that is otherwise difficult to find and maintain. By testing the results in existing and new use-cases, the tools will be validated for the different steps in the geodesign framework. PICO will provide a framework for sustainable energy design which can be embedded in other processes implementing the geodesign framework.

Acknowledgements The PICO project is co-financed by RVO (AgentschapNL) within the framework of the Dutch Topsector Energy, "TKI EnerGO". The PICO consortium consists of six partners: Geodan, TNO, Waifer, Ecofys, Alliander and Esri Nederland

References

Agentschap, N. L. (2011). Voorbeeldwoningen 2011. publicatie-nr. 2KPW1034, Agentschap NL. http://www.rvo.nl/sites/default/files/bijlagen/4.%20Brochure%20Voorbeeldwoningen%20 2011 %20bestaande%20bouw.pdf. Accessed 11 Jan 2014

Celie, N., van Luijpen M. R., & Sweeb, M. (2011). Stap voor stap naar een energiezuinige wijk. In: TWL Magazine Duurzaam. 04–2011.

Dias, E., Linde, M., Rafiee, A., Koomen, E., & Scholten H. (2013). Beauty and Brains: Integrating Easy Spatial Design and Advanced Urban Sustainability Models. In S. Geertman, et al. (Eds.), *Planning Support Systems for Sustainable Urban Development* (pp. 469–484). Heidelberg: Springer.

Ellenkamp, Y., & Maessen, B. (2009). Napoleon's registration principles in present times: The Dutch System of Key Registers. Paper presented at the 11th International Conference for Spatial Data Infrastructure. Rotterdam, The Netherlands 15–19 June 2009. https://gsdi.org/gsdiconf/gsdi11/papers/pdf/101.pdf.

Ervin, S. (2011). A system for GeoDesign. In: E. Buhmann, et al. (Eds.), Peer Reviewed Proceedings of Digital Landscape Architecture 2011 at Anhalt University of Applied Sciences (pp. 145–154). Berlin/Offenbach: Wichmann.

European Commission. (2011). *Impact assessment, commission staff working paper. Accompanying the document Directive of the European Parliament and of the council on energy efficiency and amending and subsequently repealing Directives 2004/8/EC and 2006/32/EC, SEC (2011) 779 final*. Brussels: European Commission.

Folkert, R., & van den Wijngaart, R.A. (2012).*Vesta ruimtelijk energiemodel voor de gebouwde omgeving. Data en methoden*. Den Haag: Planbureau voor de Leefomgeving.

Hofstra, H., Scholten, H., Zlatanova, S., & Scotta, A. (2008). Multi-user tangible interfaces for effective decision-making in disaster management. In: S. Nayak, et al. (Eds.), *Remote Sensing and GIS Technologies for Monitoring and Prediction of Disasters* (pp. 243–266). Heidelberg: Springer.

Horner, W., Zhao, T., & Chapin, T.S. (2011). Toward an integrated GIScience and energy research agenda. *Annals of the Association of American Geographers,101*(4), 764–774.

McElvaney, S. (2012). *Geodesign: Case studies in regional and urban planning*. Redlands: Esri.

Moslehi, K., & Kumar, R. (2010). A reliability perspective of the smart grid. *Smart Grid,1,* 57–64.

Percivall, G., Reed, C., Leinenweber, L., Tucker, C., & Cary, T. (2003). OGC Reference Model. Open Geospatial Consortium. http://portal.opengeospatial.org/files/?artifact_id = 3836. Accessed 12 Dec 2013.

Pohekar, S.D., & Ramachandran, M. (2004). Application of multi-criteria decision making to sustainable energy planning—A review. *Renewable and Sustainable Energy Reviews,8,* 365–381.

Rooijers, F.J., Leguijt, C., & Groot, M.I. (2010). *Halvering CO2-emissie in de gebouwde omgeving, een beoordeling van negen instrumenten*. Delft: CE Delft.

Scholten, H. J., van Manen, N., van de Velde, R. (Eds.), (2009). *Geospatial technology and the role of location in science*. Dordrecht: Springer.

SER. (2013). Energieakkoord voor duurzame groei. 6 Sep 2013. Netherlands: Den Haag.

Steinitz, C. (2012). *A framework for Geodesign: Changing geography by design*. Redlands: Esri.

Topteam Energie. (2012). Innovatiecontract—TKI Energiebesparing in de gebouwde omgeving. 15 Feb 2012, Den Haag

Van Hal, J.D.M. (2007). A labeling system as stepping stone for incentives related to the profitability of sustainable housing. *Journal of Housing and the Built Environment, 22*(4), 393–408.

Williamson, I.P., Rajabifard, A., & Feeney, M.E.F. (Eds.), (2004). Developing *spatial data infrastructures: From concept to reality*. London: Taylor & Francis.

WoOn. (2013). Energie. WoonOnderzoek (WoOn) Nederland 2012. http://www.rijksoverheid.nl/bestanden/documenten-en-publicaties/rapporten/2013/04/11/energie/energie.pdf. Accessed 14 Jan 2013.

Chapter 5
Holistic Assessment of Spatial Policies for Sustainable Management: Case Study of Wroclaw Larger Urban Zone (Poland)

Jan Kazak, Szymon Szewranski and Pawel Decewicz

5.1 Introduction: Changes in Spatial Planning

Changes in urban space, occurring both in its center and its outskirts, in recent years, tend to accelerate more and more. People are migrating from rural to urban areas, which seems to be the most important development of the twenty-first century. This process creates sprawling arrival cities, or bedroom communities without any government intervention. Suburban and lower density urban areas seem to be less efficient for local environments and are operationally more expensive than compact cities (Arbury 2005; Saunders 2010; Thompson 2013). Those changes have an impact not only on spatial development but also on social transformation (Duany et al. 2001). In post-socialist countries, the phenomenon of sprawling cities occurred later than in Western Europe. The causes and effects of urban sprawl seem to be quite similar but not completely the same (Pichler-Milanovic et al. 2007). It is also hard to define benchmarks for future development according to the desirable density because of strong implications on the cultural context (Jenks and Burges 2004). A major problem in this regard is the knowledge of the mechanism behind urban expansion in order to redirect future development towards more sustainable forms. There has been a debate in Poland on necessity of monitoring the changes in the environment. The debate shows that one of the obstacles in assessing risks in land development is a lack of consistent and comprehensive way of monitoring the development processes and space transformation at the local level. Thus, the authorities cannot efficiently predict or assess the effects of their decisions. As one

---

J. Kazak (✉) · S. Szewranski
Wroclaw University of Environmental and Life Sciences, Grunwaldzka 53, Wroclaw, Poland
e-mail: jan.kazak@up.wroc.pl

S. Szewranski
e-mail: szymon.szewranski@up.wroc.pl

P. Decewicz
Center of Spatial Management, Wiejska 18/9, Warszawa, Poland
e-mail: decev@geoportal.pl

© Springer International Publishing Switzerland 2014
D. J. Lee et al. (eds.), *Geodesign by Integrating Design and Geospatial Sciences*,
GeoJournal Library 1, DOI 10.1007/978-3-319-08299-8_5

of the conclusions arising from the scientific discussions on the phenomenon of urbanization shows there is an urgent need to create efficient techniques of assessing total impact of urban sprawl (Gutry-Korycka 2005; IGSO 2005). The experts of The European Union described it as a "priority task of the Member States of the Union" (Kozlowski 2006).

In order to mitigate negative global changes and processes efficiently it is necessary to act in a coordinated way. That is why it is crucial to organize a meeting among policy-makers to share opinions about specific spatial problems and to set common goals to achieve. Results of such decisions can have an implication for global as well as national or regional policy (Thiaw and Munang 2012). This is why in 2012, following the United Nations General Assembly Resolution, the United Nations Conference on Sustainable Development Rio+20 (UNCSD) was organized. One of its aims was to mark the 20th anniversary of the 1992 United Nations Conference on Environment and Development (UNCED), in Rio de Janeiro, and the 10th anniversary of the 2002 World Summit on Sustainable Development (WSSD) in Johannesburg. This meeting gave an opportunity to review directions of efforts and actions which are taken. Referring to sustainable cities and human settlements final report of UNCSD says that there is a *need for a holistic approach to urban development and human settlements*. Moreover, participants committed themselves *to promote an integrated approach to planning and building sustainable cities and urban settlements, (...) and enhancing participation (...) in decision making*. Regarding to technology they recognized *the importance of (...) reliable geospatial information for sustainable development policy-making* (United Nations 2012). According to those conclusions it is relevant to create a system which will assist spatial planning by monitoring and assessing urban development and at the same time support stakeholders in decision-making by giving them reliable and accurate information.

Frameworks for spatial policies impact the real estate market. Master plans give the parameters and conditions on how each plot can be developed. All details which are included in spatial planning documents (e.g. houses volume, soil sealed and biologically active ratio, rain water management system) have an impact on design and adaptation to the local natural conditions by available patterns (Alexander et al. 1977). Those patterns can have an impact on landscape integrity (Nassauer 2012), which refers to a complete and undamaged system (Morrison 2007; Su et al. 2014). Regional planning bodies and all planning authorities should propose provisions for new development, its spatial distribution, location and design to limit negative impacts. New development should be planned to make an opportunity for decentralized and renewable energy. It should be also planned to minimize future vulnerability to a changing climate (Wilson and Piper 2010). According to Steinitz's opinion (2012) those actions (here: spatial policies) can solve large, complicated and significant problems ranging from a neighborhood to a city, landscape or region. The management in landscape is not an easy task because of socio-economic implications beyond the environmental impact of all actions (Farina 2000). Combining design according to environmental issues with regional planning is highlighted as an important aspect of geodesign (Paradis et al. 2013). Treating geodesign as a tool to support decision-making highlighted the need for further research about construction of those systems, which is much more than just a single optimization analysis (Goodchild 2010). Additionally, they can combine different types of data which makes this topic more complex (Keranen and Kolvoord 2014).

5.2 Decision Support Systems in Spatial Planning

The role of decision support systems in spatial planning is still increasing. Local economic systems and communes are transforming from government to governance. This system depends less on political decisions and a much bigger role is assigned to other stakeholders from NGOs or well-functioning public sector (Vries 2013). Those stakeholders have to know the whole context of decisions to be made. It is necessary to give them appropriate data and to present it in a clear and understandable way. Public participation is still improving, especially by using virtual tools (Tsai and Pai 2014; LeGates 2005). That is why it is necessary to define first the construction of a decision support system used in spatial planning processes, to facilitate stakeholder involvement.

Efficient monitoring system should be based on indicators characterized by: simplicity, relation and synthesis (reflecting relationship with other elements and wider background of the phenomena) as well as clear context (showing relations between different areas or variants of phenomena). This point of view was noted by Czochanski (2009). All those metrics support building models of spatial development and environmental changes which are very helpful in defining problems more precisely and concepts more clearly (Turner et al. 2001; Kane et al. 2014; Fan and Myint 2014). Finally, stakeholders should have appropriate data to know which decision will lead to a healthier environment and help to improve the quality of human life (LaGro 2001; Brown et al. 2014; Tian et al. 2014). There are already many indicators to monitor environmental components according to different kinds of spatial development (Bhatta 2010; Morales et al. 2013; Lerman et al. 2014; Middel et al. 2014; Hardt et al. 2014; Li et al. 2014) but still there are few good examples of combining those elements to create one integrated model (Valenzuela and Mataran 2008). The advantage is that most of those indicators can be applied at the higher strategic level, such as in the European Union, to support decision-making on a regional and international scale (Rinne et al. 2013, Sebastien and Bauler 2013).

The examples of integrated systems are for instance European Common Indicators for Urban Environment, Sustainability A-Test, Complete Community Indicators for U.S Towns and Cities (Visvaldis et al. 2013), Key Indicators for Territorial Cohesion and Spatial Planning (2013)) or Reference Framework for European Sustainable Cities (2008). The last example (RFSC) is a framework which is now the subject of this research and is currently being developed. It is based on sustainable development as a fundamental principle shared by the European Union to constant improvement in the quality of life and wellbeing of present and future generations. Due to RFSC, promoting sustainable urban development is a key element of the European Cohesion Policy[1]. This framework can help to assess spatial as well as

---

[1] http://www.rfsc.eu/.

non-spatial components of development. As EU Commissioner for Regional Policy Johannes Hahn stated that *the RFSC tool is a vital part of our effort to ensure that cities of all sizes can play their full role in the achievement of a smart, sustainable, inclusive Europe.* However, so far there has not been found any published study about availability of consistent data. But still, potential future analysis drawn by this framework seems to be proper and interesting.

One of the options to create a decision support system is by using ontologies, which is an innovative approach in the field of spatial decision support systems. In this case, meeting the established decision-making criteria is performed by an inference, based on the formalism of description logic (DL). Reasoning includes the classification of ontology elements (instances, classes), which meet specified conditions. It is worth noting that research in this area is undertaken, mainly due to the dynamically developing the Semantic Web. However, this solution is not very suitable because of its complexity and it leads to many difficulties in the implementation by public administration (Lukowicz et al. 2012). The approach based on ontologies is not considered in the article, although the authors believe that these works can be a promising direction in spatial decision support systems. Tools based on classical GIS systems seem to be more effective because of easier navigation for people who are not specialists in informatics. The discussed system is CommunityViz, an extension of ArcGIS Desktop. It has two main components: Scenario 360 and Scenario3D and it is to support stakeholders to make decisions in the planning process. It may be helpful in many procedures like: evaluating future traits defining the area and factors affecting local community, carrying out experiments with hypothetical scenarios, performing parametric assessments, modifying assumptions of spatial calculations, presenting visual effects of suggested actions, making decisions based on extensive information and finally, connecting your work with three-dimensional pictures. As U.S. experience shows, provided that the system is operated by skillful staff, CommunityViz can be an efficient tool supporting decision-making (Walker and Daniels 2011). Secondly, it has been shown that the software is simple, thus it can be successfully used not only by specialists[2]. This research shows the possibility of applying a decision support system to the Polish spatial planning system in accordance with available data.

5.3 Methodology and Research Area

Research presented in this chapter contains the final result of the project about environmental impact assessment of unsustainable development of large cities, which was prepared for the Ministry of Science and Higher Education (grant no. NN305384838). Preliminary results were already presented before (Kazak et al. 2013a; Kazak et al. 2013b). This chapter contains analysis of the final results as well as experiences based on the conducted workshops with the use of the model.

---

[2] http://placeways.com/communityviz/gallery/casestudies/pdf.

The research has been taken in Wroclaw and surrounding rural municipalities located around the city. Wroclaw is the historical capital of Silesia and the largest city in western Poland. Today, Wroclaw is an important city in Poland with over 650,000 citizens, second after Warsaw. The total number of citizens in Wroclaw Larger Urban Zone is about 780,000. Research was taken on local spatial development policies for City of Wroclaw and municipalities located in the suburbia of Wroclaw: Kostomloty, Miekinia, Oborniki Slaskie, Wisznia Mala, Czernica, Dlugoleka, Katy Wroclawskie, Kobierzyce, Siechnice, Zorawina and Kostomloty.

Spatial planning documents from all the communes were converted to digital versions in vector format (Fig. 5.1). This data together with statistical assumptions became a source for calculations presented in this chapter. All formulas for holistic assessment were created in a decision support system—CommunityViz. To make the assessment suitable for the needs of shaping sustainable development the indicators are describing demographical, economic and environmental impact of development.

The main objective of this research was to evaluate the impact of different growth scenarios by indicator-based assessment. Discussion on strategies of Wroclaw LUZ development were based on a diversity of spatial and non-spatial data including cadastral landuse, demographic changes, consumption of natural resources, energy and waste consumption, approximate load of the road network, as well as statistical data, including latest census results from National Statistical Office. Eleven municipalities provided their land-use plans differing in terms of level of abstraction and the meaning of definitions of landuse designations. The plans created the unified landuse zoning plan for the region and it became the foundation of the project. The dataset presented horizontally includes 5748 planned landuse polygons (UAZ). CommunityViz Version 4.1 and ArcInfo 9.3.1 interpreted the polygons into a common landuse model.

Unlike the UrbanSim[3] (software-based simulation system for supporting planning and analysis of urban development, incorporating the interactions between land use, transportation, the economy, and the environment) and TRANUS[4] (software which simulates the location of activities in space, land use, the real estate market and the transportation system), that can run dynamic analysis of complex urban systems, CommunityViz is not operating on an integrated urban model. As a tool, it is similar to "what if" sketch planning. It is static in time. But at the same time, due to dynamic attributes of CommunityViz characteristics, it can be helpful in developing an open modeling framework. The basis of the mentioned attribute is formula that specifies the way the attribute is calculated. Any change in the analysis results in automatic updating of the value. Any kind of information or data like: the area function, mix of use or distance to the nearest infrastructure reevaluates every single land-use plan unit. The study area was described by 27 land-use predefined models (Fig. 5.2) which were given a name, symbols and detailed characteristics (set of attributes, many of them dynamic) representing building density, mix of use,

---

[3] http://www.urbansim.org/.

[4] http://www.tranus.com/.

Fig. 5.1 Conversion of local development polices

5 Holistic Assessment of Spatial Policies for Sustainable Management

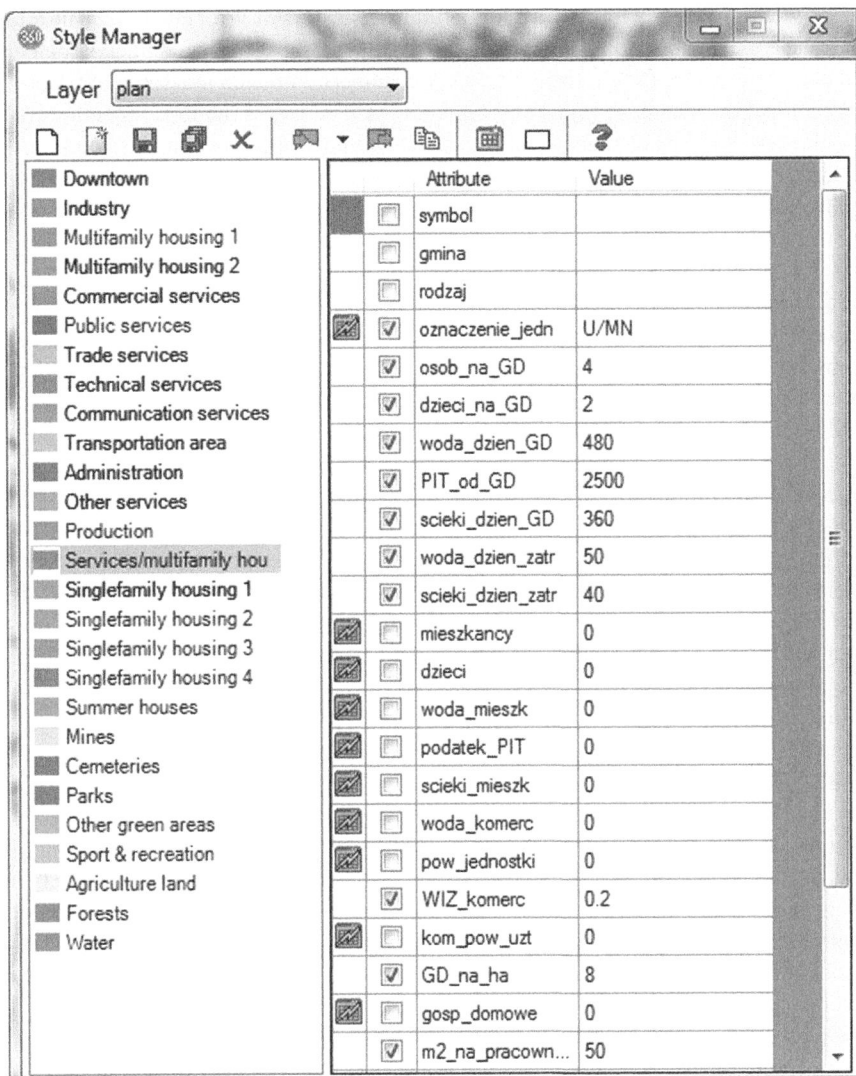

Fig. 5.2 Models used in research

resources utilization rates etc. While using Scenario Sketch tools of CommunityViz, by selecting each object, polygon is defined as a specific landuse. This operation change not only the color on the map but also all specified characteristics according to the landuse (attributes and values). New values which correspond with each object, change the parameters of calculations automatically. It is possible to use all landuse models provided with statistical characteristics predefined by the user.

CommunityViz is a framework for defining indicators of Wroclaw LUZ. The aim of indicators construction is to measure accumulated impact of spatial development plans and other modifiable scenarios. They ensure a cumulated measurement, based on each polygon landuse (attributes and values). The list of the indicators describing the demographic changes, build-out capacity under different zoning regulations, natural resources and energy consumption, waste production and increase of the indicative amount of cars as well as the number of their daily trips are presented in tabular form below (Table 5.1). Selected indicators were chosen by methodology based on four aspects. The first is accuracy which reflects if each indicator is built in a correct way and will be relevant. The second is directionality which shows if the changes of values are clearly understandable by the users. Reliability is the third feature which reflects if data used for those calculation are credible and data sources are easy to use. The last aspect is cost-effectiveness. Indicators which need additional data collection have a low level of cost-effectiveness in contrast to those which are using administration registers or documents as well as analyses of statistical offices. Indicators with the highest scores were taken into consideration in creating an indicator-based assessment of the model.

The purpose of defining an alternative, more conservative growth scenario with more concentrated development presented by higher density in already existing developed areas, (Fig. 5.3) was demonstrating the technical feasibility of CommunityViz. Using the second scenario as a comparison led to better evaluation of the potential of the current growth plan. Two variants are presented in Fig. 5.3. A more sustainable scenario on the right and current growth on the left. Designations of used symbols are presented in Fig. 5.2. The values of indicators are shown in Table 5.2.

The assessment of current policy revealed that the area where already 780,000 people are living is designed for over 2,600,000 citizens. Since the beginning of this century population of Wroclaw has been slightly decreasing (around 630,000 citizens in 2011), while suburban municipalities noted an increase at the level of approximately 25–30 % (around 150,000 citizens in 2011). Comparing spatial policies to demographical trends it seems that spatial development policies are highly overestimated. That is why it is crucial to support stakeholders to improve those policies and focus on those investments which can be useful for people. The model cannot replace stakeholders decisions but it can help them by assessing spatial policies in a quantified way. The point of creating revised policy in this research is not to find optimum policy of future development, but to show the way of changes which can lead to more sustainable solutions. According to contemporary changes caused by urban sprawl of Wroclaw, it was decided that revised policy should lead to reduction of transportation needs of citizens. That is why one of the biggest effort was concentrated on selecting those single-family housing areas which are not developed right now. Reclassification of those polygons into present landuse (mostly agriculture land, rarely forest) updated simultaneously values of all indicators. Because of the different assumptions defined in each landuse model (e.g. density specified by the number of people per hectare) reclassification of each

Table 5.1 List of calculated indicators

Indicator name	Description
Citizens	Total number of inhabitants.
Households	Total number of dwelling units.
Children	Children up to 19 years old.
Commercial area	Total commercial floor area.
Jobs	Total number of commercial jobs.
Income tax	Tax (PIT income tax) revenues for local commune budget.
Water consumption	Total water use associated with residential buildings in landuse plan layer.
Wastewater production	Waste water associated with residential buildings in landuse plan layer.
Energy consumption—households	Total annual energy used by residential buildings for all applications, including electricity and heating.
Energy consumption—commercial	Total annual energy used by commercial buildings in landuse plan layer for all applications, including electricity and heating.
Motorized trips	Total number of motorized trips taken each day, on average, by residential households (dwelling units).
CO emission	Total carbon monoxide emissions generated by vehicles associated with residential buildings in landuse plan layer.
CO2 emission	Total carbon dioxide emissions generated by vehicles associated with residential buildings in landuse plan layer.
Hydrocarbon emission	Total hydrocarbon emissions generated by vehicles associated with residential buildings in landuse plan layer.
NOx emission	Total emissions of oxides of nitrogen generated by vehicles associated with residential buildings in landuse plan layer.

polygon can cause different changes. The easiest way to reduce the number of citizens would be reclassification of multi-family housing area, because of the higher density. However, because of the defined goal to reduce transportation needs, the benchmark was to maintain more density housing area in the city (existing as well as the newly designed).

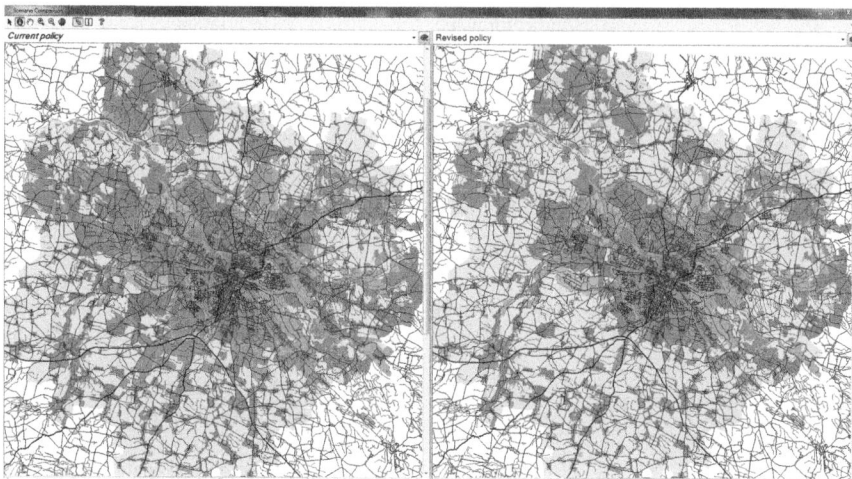

Fig. 5.3 Spatial comparison of two scenarios

Table 5.2 Indicator-based comparison of two scenarios

Indicator	Units	Current policy	Revised policy
Citizens	People	2,609,681	2,113,624
Households	Households	819,305	665,376
Children	People	1,073,127	864,943
Commercial area	Thousands of square meters	158,977	123,000
Jobs	People	1,894,185	1,474,444
Income tax	Million PLN	1951.64	1620.70
Water consumption	Cubic meters per day	269,856	212,517
Wastewater production	Cubic meters per day	212,866	167,627
Energy consumption—households	Million btu per year	75,539,933	61,347,659
Energy consumption—commercial	Million btu per year	155,720,831	120,480,035
Motorized trips	Trips per day	2,212,124	1,796,515
CO emission	Tons per year	12,111	9836
CO2 emission	Tons per year	238,594	193,768
Hydrocarbon emission	Tons per year	2665	2164
NOx emission	Tons per year	1817	1475

5.4 Results and Qualitative Analysis

As a result of reclassification made in revised policy scenario the number of population decreased by almost 500,000 citizens. The total housing area was reduced by around 50 % while the number of citizens decreased by almost 20 %. Such a change can be called leverage point as not such a big reduction of human development (in the number of new citizens) caused comparatively big profit for natural environment by maintaining open spaces. It is very important from the point of view of contemporary urban sprawl impact like landscape fragmentation or soil sealing. Changes in landuses also had an impact on the reduction of water or energy consumption, wastewater production and the number of motorized trips. Additionally shorter distances, caused by more dense development, decreased average length of a motorized trip. That is why emission caused by transportation reasons also became smaller.

The model which is presented in current article was shown to the local administration to highlight effects of current spatial policies. Mayors of three municipalities declared that they are interested in organizing workshops with use of CommunityViz for development departments in their offices. At this moment cooperation with local administration was based on presenting results of those calculations. So far the workshops for a group of stakeholders have been taken for NGOs and for participants of a scientific conference. Those experiences seem to be helpful in preparing similar workshops for local administration. Further steps assume preparation of different interactive workshops based on the problems and questions defined by participants as well as issues which administration has to face resulting from implementation of new strategies, policies and plans. This way of supporting decision-making process should lead to working out integrated and coherent spatial policy. The feedback received from workshop confirmed the usefulness of the model. First of all, analyses of current growth scenarios help to interpret the regional planning decisions. Another point shows that there is no balance between the infrastructure and landuse for development, as a result too much land is planned for development which is not realistic. Finally, Polish spatial planning system does not promote constrains to the problem of urban sprawl. The use of CommunityViz while presenting the model to the local administration as well as organizing interactive workshops highlighted advantages and disadvantages of that system. Participants filled in evaluation questionnaires which helped in assessing the features of that tool. Regarding simplicity of use, 40% of participants declared that CommunityViz is rather easy, the rest of them said that it is easy or very easy. 90% of them said that results are presented in a good way, so they are easy to understand. Only 10% of participants have a negative opinion about a speed of data processing. As far as the results of the questionnaires are concerned the biggest obstacle in using that tool is the price (70%), the next once are: lack of licenses for ArcGIS (55%) and lack of skills to work in GIS (30%). Most of the participants said that CommunityViz rather meets (75%) or meets (15%) user's expectations. Based on the evaluation questionnaires and workshops it can be concluded, that the use of that system should not be done by a person who does not have any experience in using GIS software. It seems very important to provide an expert for workshops to implement suggestions made by stakeholders into a model. Knowledge of this GIS expert might be very useful also during preparation of formulas of new indicators. Experiences from workshops show that participants can report the need of calculation of another indicator than those which were prepared in advance in the model. The big obstacle in Polish conditions is also lack of unitary parameters which are used in calculations. Knowledge about those values is essential because they are used as assumptions in formulas.

Finally, the most time-consuming obstruction is that spatial data (e.g. spatial policies, master plans, etc.) are not prepared or stored in any standard format. However, participants of workshops stressed that results of calculations were easy to interpret. Moreover, simultaneous change of indicator values for all scenarios showed the impact magnitude due to modifications. Furthermore, modification of assumption values or each landuse was intuitive. That is why participants could do it on their own which made them more involved in the workshop. Eventually, multi-scenario comparisons helped to make the problem less complex. Presentation of all the information side-by-side made it easier to compare spatial distribution of each landuse as well as numeric data. CommunityViz also provided the opportunity to compare all indicators on the charts, where a user did not have a chance to manipulate the scales. That helped to reduce communication risks caused by misinterpretation.

5.5 Conclusions

Implementation of the idea of sustainable development into spatial planning indicates a new approach in constructing future development visions and plans. One of the main issue is to understand the effects of human impact on natural environment. In parallel, the systems of such assessments should be possible on the over-local level which will make different spatial policies comparable. The model of spatial policies of Wroclaw LUZ, constructed in CommunityViz, proved that those documents are not suitable to the needs of the society and can cause adverse effects. Ability to modify the model allowed to revise policies by changing selected landuses and simultaneously assessed the impact of that development. As a result of workshops, where the main goal was to create compact region development, participants revised Wroclaw LUZ policies. Finally, many indicators have changed their values, which made the policies more realistic. The most notable change was decrease of the number of future population by almost 500,000 citizens. The aim of this workshop was to present the tool which can be useful in optimization of spatial policies and to show the way of possible changes to make those documents more realistic.

There are many techniques and tools that can be used to assess the impacts of spatial planning for geodesign. One of them is a GIS environment called CommunityViz, a reliable and efficient tool for forecasting, assessing parametric values (based on statistical assumptions) and monitoring spatial transformations. The participants of the workshop presented in their final feedback that such a model is helpful in solving many local problems, because CommunityViz is easy to use and clearly outlines the results of the planned activities. The participants pointed out that the scenarios with different alternatives to current spatial policies and impact of planning on the environmental capacity (the ability of absorbing new development by the ecosystem services) might be helpful before making a final decision on possible changes to the functional areas. The geodesign process with CommunityViz allows governments to estimate the potential of water or energy consumption, waste production, estimated load of the road network and many other features of the freely-defined demographic trends. Dynamic construction, analyses of impacts of various scenarios with indicators which are a combination of geoinformation and user-defined assumptions, simultaneous calculations and visualizations with modified assumptions—these are advantages of the method. The biggest obstacles in using CommunityViz are: the availability of proper data, the cost of the tool and the skills to operate in GIS. However, data in Europe are constantly improving by implementation of INSPIRE directive. Moreover, GIS is becoming very popular so more and more users have good skills to use it. As far as the financial obstacle is concerned, it might be worth thinking about equivalent of CommunityViz for open source GIS programs. Features of the model prepared in CommunityViz indicate that it can be useful in implementation of strategies and acts of law in the assessment of development policies.

References

Alexander, C., Ishikawa, S., & Silverstein, M. (1977). *A pattern language: Towns, buildings, construction (Center for environmental structure series)*. New York: Oxford University Press.

Arbury, J. (2005). *From urban sprawl to compact city—An analysis of urban growth management in Auckland*. Thesis, Auckland, University of Auckland.

Bhatta, B. (2010). *Analysis of urban growth and sprawl from remote sensing data*. Heidelberg: Springer.

Brown, G., Schebella, M. F., & Weber, D. (2014). Using participatory GIS to measure physical activity and urban park benefits. *Landscape and Urban Planning,121*(0), 34–44. doi:10.1016/j.landurbplan.2013.09.006.

Czochanski, J. (2009). Regional monitoring system as a tool of landscape research and spatial management. The Problems of Landscape Ecology, Vol. XXIII, 97–104.

Duany, A., Plater-Zyberk, Elizabeth., & Speck, Jeff. (2001). *Suburban nation: The rise of sprawl and the decline of the American Dream*. New York: North Point Press.

Fan, C., & Myint, S. (2014). A comparison of spatial autocorrelation indices and landscape metrics in measuring urban landscape frag3mentation. *Landscape and Urban Planning,121*(0), 117–128. doi:10.1016/j.landurbplan.2013.10.002.

Farina, A. (2000). *Landscape ecology in action*. Netherlands: Springer.

Goodchild, M. F. (2010). Towards geodesign: Repurposing cartography and GIS? *Cartographic Perspectives,66*, 7–22.

Gutry-Korycka, M. (2005). *Urban sprawl Warsaw agglomeration case study*. Warsaw: Warsaw University Press.

Hardt, E., dos Santos, R. F., & Pereira-Silva, E. F. L. (2014). Evaluating the ecological effects of social agent scenarios for a housing development in the Atlantic forest. *EcologicalIndicators,36*(0), 120–130. doi:10.1016/j.ecolind.2013.07.013.

Institute of Geography and Spatial Organization from Polish Academy of Sciences (2008). Raport o stanie i uwarunkowaniach prac planistycznych w gminach na koniec 2007 r. Warszawa: Institute of Geography and Spatial Organization from Polish Academy of Sciences.

Jenks M., & Burges R. (2004). *Compact cities: Sustainable urban forms for developing countries*. London: Taylor & Francis.

Kane, K., Tuccillo, J., York, A. M., Gentile, L., & Ouyang, Y. (2014). A spatio-temporal view of historical growth in Phoenix, Arizona, USA. *Landscape and urban planning,121*(0), 70–80. doi:10.1016/j.landurbplan.2013.08.011.

Kazak, J., Szewrański, S., & Decewicz, P. (2013a). Indicators-based assesment of landuse planning in Wroclaw Region with CommunityViz. In M. Schrenk, V. V. Popovich, P. Zeile, & P. Elisei (Eds.), *REAL CORP 2013 Proceedings/Tagungsband, 20–23 May 2013, Rome, Italy* (pp. 1247–1251). Schwechat: Eigenverlag des Vereins CORP.

Kazak J., Szewrański S., & Decewicz P. (2013b). Monitoring land use planning in Wroclaw region with CommunityViz. In F. Hoffmann, K. Charvat (Eds.) NNR Special Edition 2013, GI2013 X Border—GI/ GIS/ GDI—Forum. Proceedings, 29–30.04.2013, Dresden, Germany, 24–27.

Keranen K. & Kolvoord R. (2014). *Making spatial decisions using GIS and remote sensing: a workbook.* Esri Press.

KITCASP (2013). Key Indicators for Territorial Cohesion and Spatial Planning. Main report. ESPON.

Kozlowski, S. (red.) (2006). Zywiolowe rozprzestrzenianie sie miast. Narastajacy problem aglomeracji miejskich w Polsce, Studia nad zrównowazonym rozwojem, t. 2, Katedra Ochrony Srodowiska KUL, Komitet „Czlowiek i Srodowisko" przy Prezydium PAN, Bialystok-Lublin-Warszawa.

LaGro, J. A. (2011). *Site analysis: A contextual approach to sustainable land planning and site design.* New York: Wiley.

LeGates R. (2005). *Think globally, act regionally: GIS and data visualization for social science and public policy research.* Redlands: Esri.

Lerman, S. B., Nislow, K. H., Nowak, D. J., DeStefano, S., King, D. I., & Jones-Farrand, D. T. (2014). Using urban forest assessment tools to model bird habitat potential. *Landscape and urban planning, 122*(0), 29–40. doi:10.1016/j.landurbplan.2013.10.006.

Li, X., Tian, M., Wang, H., Wang, H., & Yu, J. (2014). Development of an ecological security evaluation method based on the ecological footprint and application to a typical steppe region in China. *Ecological Indicators, 39*(0), 153–159. doi:10.1016/j.ecolind.2013.12.014.

Lukowicz J., Kaczmarek I., & Iwaniak A. (2012). Semantic metadata in SDI for decision support systems in spatial planning, Global Geospatial Conference 2012 "Spatially Enabling Government, Industry and Citizens", Québec City, Canada.

Middel, A., Häb, K., Brazel, A. J., Martin, C. A., & Guhathakurta, S. (2014). Impact of urban form and design on mid-afternoon microclimate in Phoenix Local Climate Zones. *Landscape and Urban Planning, 122*(0), 16–28. doi:10.1016/j.landurbplan.2013.11.004.

Morales, P. K., Yunusa, I. A. M., Lugg, G., Li, Z., Gribben, P., & Eamus, D. (2013). Belowground eco-restoration of a suburban waste-storage landscape: Earthworm dynamics in grassland and in a succession of woody vegetation covers. *Landscape and Urban Planning, 120*(0), 16–24. doi:10.1016/j.landurbplan.2013.06.007.

Morrison C. (2007). Ecological Integrity in the Core Areas of Clayoquot Sound Biosphere Reserve and the Threat of Adjacent Land Use. http://www.clayoquotbiosphere.org/projects/2007/Ecological_Integrity.pdf. Accessed 3 Feb 2014.

Paradis, T., Treml, M., & Manone, M. (2013). Geodesign meets curriculum design: integrating geodesign approaches into undergraduate programs. *Journal of Urbanism: International Research on Placemaking and Urban Sustainability, 6*(3), 274–301. doi:10.1080/17549175.2013.788054.

Nassauer, J. I. (2012). Landscape as medium and method for synthesis in urban ecological design. *Landscape and Urban Planning, 106*(2012), 221–229. doi: 10.1016/j.landurbplan.2012.03.014.

Pichler-Milanovič, N., Gutry-Korycka, M., & Rink, D. (2008). Sprawl in the Post-Socialist city: The changing economic and institutional context of Central and Eastern European cities. In C. Couch, L. Leontidou, & G. Petschel-Held (Eds.), *Urban Sprawl in Europe* (pp. 102–135). Oxford: Blackwell. http://dx.doi.org/10.1002/9780470692066.ch4.

Rinne, J., Lyytimäki, J., & Kautto, P. (2013). From sustainability to well-being: Lessons learned from the use of sustainable development indicators at national and EU level. *Policy Use and Influence of Indicators, 35*(0), 35–42. doi:10.1016/j.ecolind.2012.09.023.

Saunders, D. (2010). *Arrival city: How the largest migration in history is reshaping our world.* New York: Knopf Doubleday.

Scoones, I. (1998). Sustainable Rural Livelihoods: A Framework for Analysis, Working Paper 72. Sussex: Institute for Development Studies.

Sebastien, L., & Bauler, T. (2013). Use and influence of composite indicators for sustainable development at the EU-level. *Policy Use and Influence of Indicators, 35*(0), 3–12. doi:10.1016/j.ecolind.2013.04.014.

Segan, D. B., Game, E. T., Watts, M. E., Stewart, R. R., & Possingham, H. P. (2011). An interoperable decision support tool for conservation planning. *Environmental Modelling & Software, 26*(12), 1434–1441.

Steinitz C. (2012). *A framework for geodesign: Changing geography by design*. Redlands: Esri.

Su, S., Ma, X., & Xiao, R. (2014). Agricultural landscape pattern changes in response to urbanization at ecoregional scale. *Ecological Indicators, 40*(0), 10–18. doi:10.1016/j.ecolind.2013.12.013.

Thiaw, I., & Munang, R. (2012). RIO + 20 outcomes recognize the value of biodiversity and ecosystems: Implications for global, regional and national policy. *Ecosystem Services, 1*(1), 121–122. doi:10.1016/j.ecoser.2012.07.013.

Thompson D. (2013). Suburban sprawl: exposing hidden costs, identifying innovations. http://thecostofsprawl.com/report/SP_SuburbanSprawl_Oct2013_opt.pdf. Accessed 3 Feb 2014.

Tian, Y., Jim, C. Y., & Wang, H. (2014). Assessing the landscape and ecological quality of urban green spaces in a compact city. *Landscape and Urban Planning, 121*(0), 97–108. doi:10.1016/j.landurbplan.2013.10.001.

Tsai, H.-T., & Pai, P. (2014). Why do newcomers participate in virtual communities? An integration of self-determination and relationship management theories. *Decision Support Systems, 57*(0), 178–187. doi:10.1016/j.dss.2013.09.001.

Turner, M. G., Gardner, R. H., & O'Neill, R. V. (2001). *Landscape ecology in theory and practice: Pattern and process*. Washington: U.S. Government Printing Office.

United Nations (2012). Report of the United Nations Conference on Sustainable Development. The Future We Want. Rio de Janeiro.

Valenzuela Montes, L. M., & Mataran Ruiz, A. (2008). Environmental indicators to evaluate spatial and water planning in the coast of Granada (Spain). *Land Use Policy, 25*(1), 95–105. doi:10.1016/j.landusepol.2007.03.002.

Visvaldis, V., Ainhoa, G., & Ralfs, P. (2013). Selecting Indicators for Sustainable Development of Small Towns: The Case of Valmiera Municipality. ICTE in regional development, December 2013, Valmiera, Latvia, 26(0), 21–32. doi:10.1016/j.procs.2013.12.004.

Vries, M. (2013). The challenge of good governance. *The Innovation Journal: The Public Sector Innovation Journal, 18*(1), Article 1.

Walker, D., & Daniels, T. (2011) *The planners guide to CommunityViz. The essential tool for generation of planning.* Chicago: APA Planners Press.

Wilson, E., & Piper, J. (2010). *Spatial planning and climate change*. New York: Taylor & Francis.

Chapter 6
Recent Applications of a Land-use Change Model in Support of Sustainable Urban Development

Eric Koomen and Bart C. Rijken

6.1 Introduction

Urban development is a complex dynamic process that is characterised by substantial spatiotemporal variation. Growth and decline coexist within neighbouring regions at short distances from each other. In the Netherlands, for example, population and employment are expected to grow in the so-called main-, brain- and greenports, whereas a decline is expected between these regions and, especially, in more peripheral parts of the country (PBL 2011). This makes steering urban expansion and intensification important policy issues in many regions, while preparing for decline and urban restructuring have become hot topics in others. These processes are driven by various interacting and sometimes even conflicting societal and economic forces that may impact regions differently. Globalisation, for example, is a dominant socioeconomic force that is associated with changes in the production and employment structure of countries and regions. It brings business opportunities and foreign investment to some regions. But at the same time, globalisation is partially responsible for rapid reductions in employment and outflows of high-skilled people in other regions. Both developments lead to increased differences between regions and give rise to the societal and political desire for regionalisation that emphasises 'own' identity and local interests and tries to limit globalisation (CPB et al. 2006).

Also on the governing side, we witness opposing forces: societal concerns such as safety, accessibility and economic development call for active and preferably centralised government control, but at the same time central government is increasingly

---

E. Koomen (✉) · B. C. Rijken
Faculty of Economics and Business Administration, Department of Spatial Economics, SPINlab,
VU University Amsterdam,
De Boelelaan 1105, 1081 HV Amsterdam, The Netherlands
e-mail: e.koomen@vu.nl

B. C. Rijken
PBL Netherlands Environmental Assessment Agency, 30314,
2500 GH The Hague, The Netherlands
e-mail: b.c.rijken@vu.nl

delegating its responsibilities to lower tiers of government (Kuijpers-Linde 2011). In fact, attention is shifting from government to governance and societal organisations (including businesses and non-governmental organisations) and individual citizens become more important in decision-making processes (Roodbol-Mekkes et al. 2012). This process of change is especially apparent in the renowned Dutch spatial planning system that is currently being stripped from its most prominent features, as top-down restrictive zoning polices and urban concentration polices are being abolished (I&M 2011; Kuiper and Evers 2011b). Also the underlying principles of distributive justice and solidarity between regions are being removed from spatial planning and the related allocation of funds. This is directly relevant for the management of regional population decline as it implies that classical, costly interventionist urban restructuring policies will not be feasible and calls for the development of other, more innovative strategies. The outcome of the ongoing and partially conflicting societal and economic processes is uncertain and can differ per region. Which processes dominate and prevail in a certain region and how governmental interventions of different governmental levels can help steer these developments is unclear.

This paper explores the extent to which an operational, much applied land-use model can support current planning issues. It starts by briefly introducing the land-use model that is used and then goes on to discuss three recent applications related to sustainable urban development. Based on these applications we discuss their potential role in a more iterative geodesign process and suggest possible improvements in the application of land-use models.

6.2 An Operational Land-use Modeling Framework

Models of land-use change can benefit the geodesign process as they help assess potential changes in land-use patterns that may result from general socio-economic or biophysical changes, proposed policies or spatial developments. They have become an established tool to help prepare and support spatial planning and can, for example, help formulate adequate spatial policies by simulating potential autonomous spatial developments or, perhaps more importantly, by showing the possible consequences of different policy alternatives. So they sketch the future context that decision makers have to take into account and they provide an indication of the spatial impacts the proposed policies or developments may have.

The Land Use Scanner model we apply here has its roots in economic theory. It simulates the competition between urban, natural and agricultural types of land use and thus offers an integrated view on spatial development. It was developed in 1997 and has been applied in many policy-related research projects in the Netherlands and abroad (e.g. De Moel et al. 2012; Hoymann 2010; Koomen et al. 2011; Te Linde et al. 2011). The model's basics and recent applications are described in a book that also contains numerous references to other publications (Koomen and Borsboom-van Beurden 2011).

Many applications visualise potential spatial patterns associated with specific scenario conditions or policy interventions. In that respect it is comparable to well-known rule based simulation models such as the original California Urban Futures (CUF) model and the What If? system (Landis 1994; Klosterman 1999). The model proved to be an especially valuable tool to inform policy makers about potential future developments in the context of strategic, national planning (Schotten et al. 2001; Borsboom-van Beurden et al. 2007). Other applications include ex-ante evaluations of specific policy alternatives, such as new locations for the Dutch national airport (Scholten et al. 1999) or strategies to limit flood risk (Van der Hoeven et al. 2009). More recently, the model was applied to help regional authorities formulate strategic visions for their territory (Jacobs et al. 2011; Koomen et al. 2011) and assess the potential for specific biofuel crops to help mitigate climate change (Kuhlman et al. 2013).

Demand and supply of land are balanced in the model using: (1) regional projections of land-use change (demand); (2) local definitions of suitability; and (3) an algorithm that allocates land (cells) to those land-use types that have the highest suitability, taking into account the regional land-use claim. Demands for land are specified for each land-use type and can be derived from, for example, sector-specific models of specialised institutes or policy-based ambitions. The local (cell-based) specification of suitable locations for the different land-use types typically incorporates many different spatial datasets referring to current land use, physical properties, operative policies and market forces generally expressed in distance relations to nearby land-use functions. Two different allocation algorithms are available that allocate land use either as fractions per cell (reflecting probabilities) or in a discrete way (filling each cell completely with one, most optimal type of land use).

6.3 Planning-Related Applications

To be able to understand the applicability of a land-use model in context of geodesign we briefly describe three recent applications that share the following characteristics. First, they deal with sustainability impacts of urban development like, for example, the loss of open space or the increased exposure of urban areas to flooding. Second they relate to contemporary planning issues, addressing questions as: what is the likely impact of decentralising the responsibility for spatial planning on future urbanisation patterns? To which extent is urban intensification possible to limit the ongoing expansion of urban areas? How will flood risk develop in the coming decades following socio-economic and climatic projections? Thirdly they reflect the three different types of applications generally found in planning related studies: what-if type of simulations, trend-based extrapolations and scenario studies (Koomen et al. 2008b).

6.3.1 Assessing the Impacts of a New National Spatial Policy

In 2011 the Dutch Ministry of Infrastructure and Environment formulated a new national spatial strategy that in many aspects differed substantially from its predecessors (I&M 2011). To simulate the possible future urbanisation patterns that might arise from the proposed policy change, Land Use Scanner was applied. These simulations were performed as part of the formal Strategic Environmental Assessment report and are briefly described here. For more information about the policy context and expected environmental impacts the reader is referred to the actual assessment report (Elings et al. 2011) and a dedicated publication (Koomen and Dekkers 2013).

The new national spatial strategy proposed three major changes in national spatial policy that are likely to affect land-use patterns:

1. abolishing national urban concentration and transformation policies;
2. limiting the extension of the National Ecological Network;
3. emphasising internationally unique cultural historic landscape values of, for example, UNESCO world heritage sites, abolishing buffer zones and decentralising national landscapes.

The outcome of these policy changes is by no means certain. Especially the impact of decentralising the responsibility for the National Landscapes is difficult to assess: provinces may decide to ease, continue or reinforce the current restrictive planning regime in these areas. This uncertainty is captured in the simulations by showing two potential, extreme outcomes: a reference situation in which current policies are fully maintained and a new policy alternative in which they are abolished. Neither outcome is necessarily more likely, but together they show the potential bandwidth of impacts. This scenario-based approach is common in strategic planning (Dammers 2000) and decision making (De Ruijter et al. 2011), but virtually absent in environmental assessment reports.

The table below summarises the main policy objectives and the associated spatial implications for the reference alternative and the new policy alternative grouped per policy domain (Table 6.1). These assumptions were translated in model input. First to obtain region-specific projections of the demand for new residences and business estates for each alternative in 2040. In a second step these regional demands were fed in Land Use Scanner together with alternative-specific, spatially explicit assumptions for local suitability related to, for example, the presence (or absence) of specific policy restrictions. PBL Netherlands Environmental Assessment Agency performed these model simulations as part of their ex-ante evaluation of the new policy report (Kuiper and Evers 2011a).

Figure 6.1 shows the simulated increase in urban area for both alternatives. It contains new urban areas according to current policy, new urban areas according to the new policy alternative and new urban areas according to both alternatives. The latter locations are likely to become urbanised irrespective of any changes in spatial policy. To represent the inherent uncertainty of the simulation outcomes, the outcomes are not directly shown at their initial, detailed 100 m resolution, but processed to highlight concentrations of similar values in within a 500 m environment.

Table 6.1 Overview of the main objectives and expected spatial implications of the reference alternative and the new policy alternative grouped per policy domain

Policy aspect	Reference alternative	New policy alternative
1 Urban development	Bundling zones and transformation zones maintained: 30% of new residences built within current urban areas	Bundling and transformation zones abolished: 20% of new residences within current urban area
	Location of new residences steered by supply	Residential preferences and accessibility (demand) dominate location of new residences
2 Nature development	National Ecological Network fully realised in 2018: 100,000 ha nature extra	Limited version of National Ecological Network: 20,000 ha. extra
3 Unique landscape values	Buffer zones and National Landscapes limit urbanisation	Buffer zones abolished, limited impact National Landscapes
	National Ecological Network and Natura2000 areas limit urbanisation	Only international obligations limit urbanisation (UNESCO, Natura2000)

The presented simulations offer indicative, almost caricatural images of potential changes that do not allow detailed impact assessments with environmental impact models. Yet, these simulations integrate the potential implications of domain-specific policies and help visualising the regional accumulation of the impacts associated with individual policy measures.

6.3.2 Comparing Trends and Ambitions in Urban Intensification

The second example of a land-use model application related to urban development focuses on the difference between policy ambitions and reality with respect to urban densification. Policy ambitions for the containment of urban development are ambitious in the Netherlands as is evident from national and local objectives to concentrate residential development within existing urban areas (VROM et al. 2004; Keers et al. 2011). These ambitions are formulated as target shares of the total net increase in housing stock that should be realised within designated urban area contours. A GIS-based analysis of local changes in housing stock between 2000 and 2008 shows, however, that especially in the already densely-populated western part of country, the realised urban intensification shares are below the specified ambition levels (Koomen et al. 2015).

Using these observations in combination with projections of the regional increase in housing stock we are able to define the land demand for new urban area in 2020. Using Land Use Scanner we then simulate urban development until 2020 according to two policy alternatives. In the first policy alternative, growth in urban area is based on policy ambitions, while in the second this is based on observed trends in the last decade. The urban area demand is calculated in two steps: first

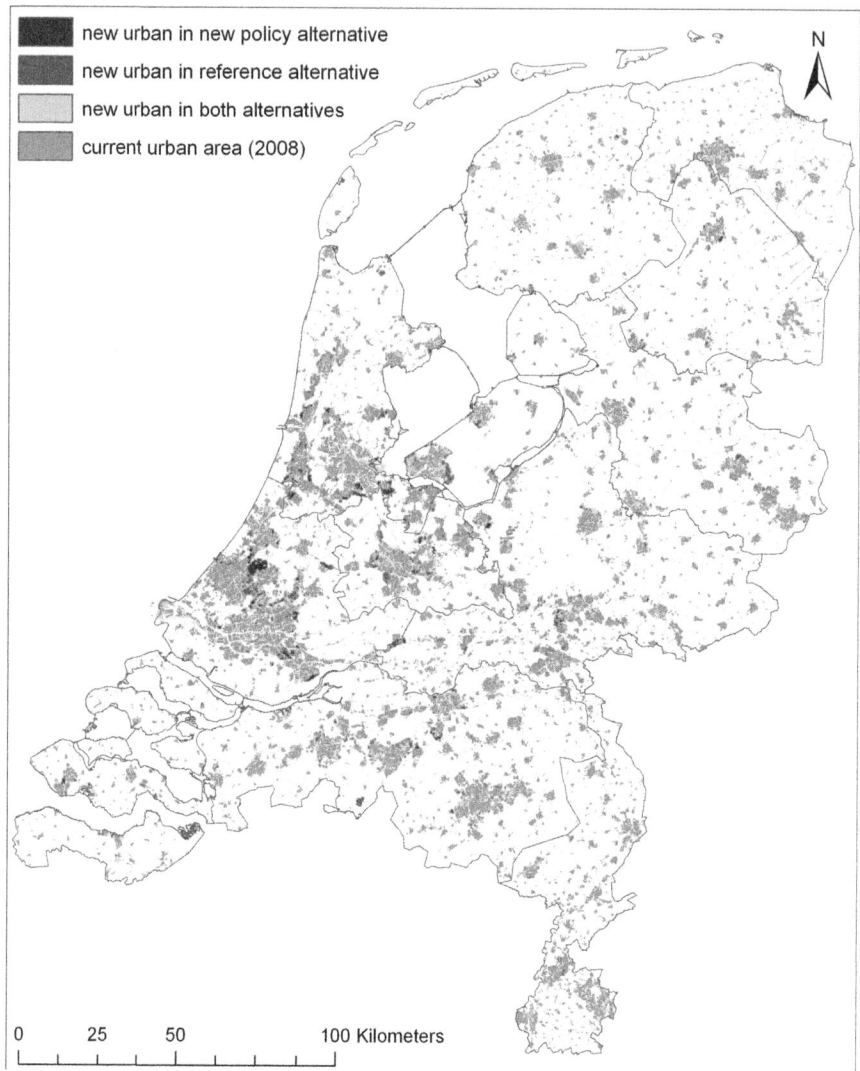

Fig. 6.1 Simulated changes in built-up area for the new national spatial strategy and current policy. (Adapted from Elings et al. 2011); *colour intensity* expresses expected degree of urbanization)

the intensification share (based on either the observed values or the policy ambitions) is subtracted from the regional housing demand to account for those houses that are expected to be realised within the urban area that exists in the base year of simulation. The remaining share of houses is expected to form new urban areas. In the second step the size of these areas is determined by dividing the corresponding number of houses by the residential densities observed in the urban extensions formed between 2000 and 2008. To account for uncertainty in the regional increases in number of houses we explore two alternative scenarios: Global Economy (GE) and Regional Communities (RC), provided by the TIGRIS XL model (Significance and Bureau Louter 2007; Zondag and Geurs 2011). These projections were made by PBL Netherlands Environmental Assessment Agency for the Delta study (see Rijken et al. 2013). The GE scenario assumes a higher population growth and stronger decrease in average number of people per household than the RC scenario and thus shows a higher net demand for new dwellings.

Growth is still expected in many parts of the country for the coming decades, but the projections differ substantially for individual regions (PBL 2011). Population decline is expected in several, mainly peripheral regions. This trend is already observed in some municipalities in border regions in the northeast, southeast and southwest. High growth is mainly expected for regions located within the densely-populated and already highly urbanised areas, forcing local policy makers to stimulate higher densities in urban development, preferably within the existing urban fabric. In the regions with population decline, ambitions regarding urban intensification are high for a completely other reason: more compact urban forms are expected to be better able to maintain current urban facilities levels (i.e. shops, restaurants, public services, etc.).

Figure 6.2 shows a snapshot of our simulations results for the centre of the country that faces substantial urban growth according to the GE scenario. These simulations are obvious simplifications of reality. For instance, we have only included a limited number of suitability maps and land-use types for these simulations, and no additional land-use demand for land-use types other than urban area and nature. The upper-left map presents the main land uses in the central part of the Netherlands in 2008. The urban footprint of the Randstad cities is displayed clearly.

The simulation result of the GE Trends alternative is displayed in the upper-right map. This is the most extreme scenario in terms of new urban extensions since it assumes a high population growth rate and a decreasing number of people per household with relatively high shares of new dwellings located outside of existing cities. The differences between the GE Trends simulation and the 2008 land use map are considerable. We see that urban settlements grow together and an almost-fully urbanized half-moon encompassing the main Randstad cities emerges. In parallel, a large land conversion process takes place in the Green Heart area, where relatively small cities increase their urban footprint by means of additional extensions.

The lower maps show the likely urban extension areas in the Ambitions alternatives (left) and the Trends alternatives (right). The expected increase in urban areas is shown with a colour intensity that reflects the probability of urban development. The maps clearly show that more urban extensification is expected in the GE scenario than in the RC scenario. The GE scenario shows a clear difference between the Ambitions and Trends alternatives. Current ambition levels will greatly help containing the large urban growth that is expected in the GE scenario. Under the more moderate conditions of the RC scenario this difference is less clearly visible.

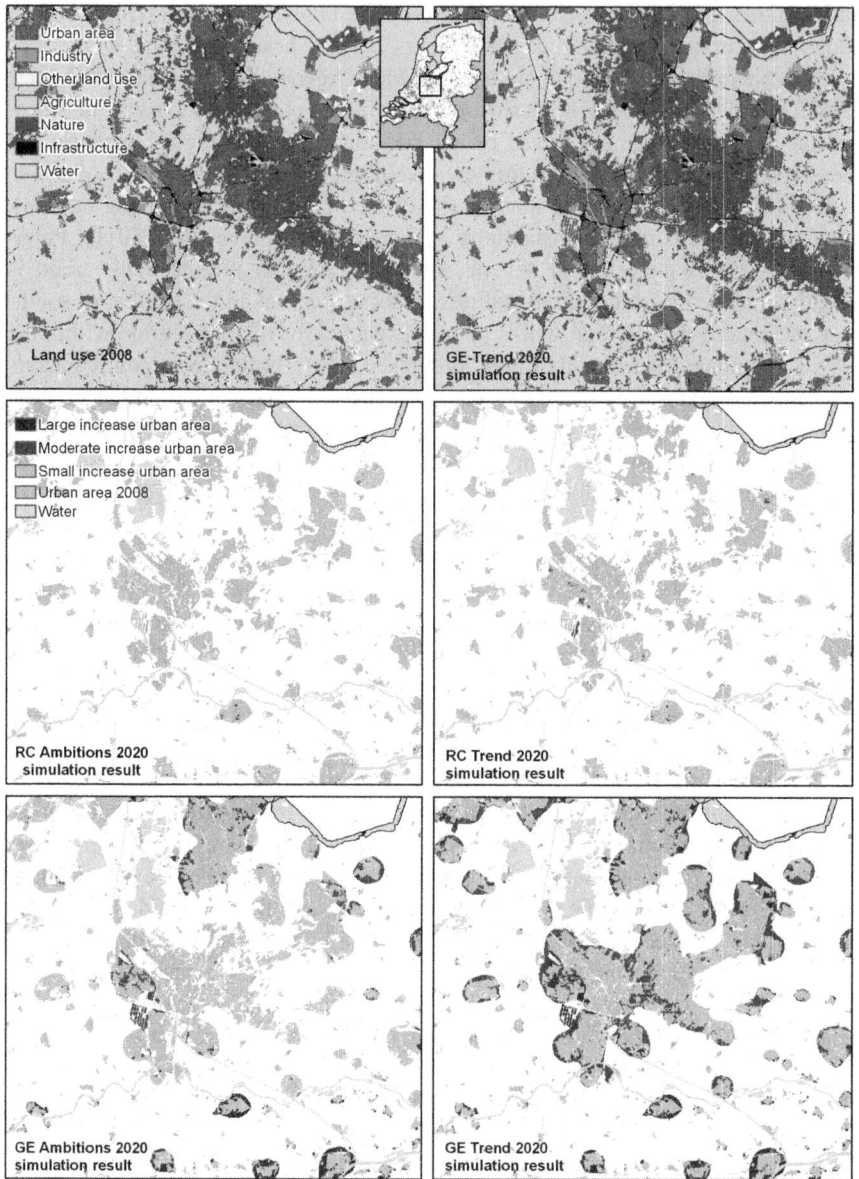

Fig. 6.2 Current (2008; *upper-left*) and simulated (2020; *upper-right*) land use according to the GE Trends alternative and simulated increase in urban area in 2020 for the RC & GE Ambitions and RC & GE Trend alternatives

6.3.3 Scenario-based Explorations of Climate Impacts

About a quarter of the Netherlands is situated below sea level. Current urban areas are concentrated in these areas, leaving the Dutch urban fabric very much exposed in the case of flooding. Most built-up areas are protected by extensive flood defence systems such as storm surge barriers or levies. Building and, especially, maintaining these systems is costly, so investments in these structures are at least partly based on the estimated costs of their failure. These costs are typically expressed in terms of human casualties, damage to physical property, etc. Such costs offer a description of flood risk and follow from the often distinguished combination of hazard, exposure and vulnerability (cf. Turner et al. 2003). Hazard, in this case refers to the occurrence of flooding, exposure depends on, amongst others, land use and vulnerability is usually described in dose-effect relations that link flood depth to damage (Klijn et al. 2007; Van der Hoeven et al. 2009).

Future urbanisation may raise these costs substantially. The question is where, when and to which extent flood risk, may thus increase. The Dutch national government is currently reappraising existing protection norms. The aim is to come up with a well-informed update in 2014. This reassessment takes place within the broader framework of the Delta Programme, a multi-departmental, multi-tier government initiative dealing with a wide range of water related issues.

The challenge is to prepare for investment decisions in flood protection (adaptation strategies) that are both efficient and robust, i.e., taking into account the likelihood of a wide range of probable extremes regarding future urbanization and climate change effects. Yet both climate change and urbanisation are highly complex processes. This is especially true for urbanisation. As indicated in the previous sections, urban dynamics are driven by the interplay of a wide range of both autonomous (demography, economy) and government-driven processes. The complexity rises with the required detail of the analysis. Yet, detail (up to the scale of 1 ha) is exactly what is required in this case.

To describe the uncertain context in which the envisaged flood protection measures must be evaluated the Delta Scenarios project was initiated. A consortium of research institutes cooperates in this multi-disciplinary subproject of the Delta Programme, to provide a set of scenarios for the Netherlands in 2050. This is done following the request of the Dutch Ministry of Infrastructure and Environment and the Ministry of Economic Affairs. The scenarios combine both socio-economic and climatic components. As for the socio-economic part of the scenarios, two main dimensions of uncertainty are distinguished: (1) high versus low economic and demographic growth, and; (2) liberal versus restrictive spatial policy. Steam and Busy are the high economic and ditto demographic growth scenarios corresponding to the Global Economy scenario discussed in the preceding section. Warm and Rest are the low growth scenarios corresponding to the Regional Communities scenario. On the policy dimension, Steam and Warm are the scenarios in which liberal spatial policy is assumed; restrictive policy is presumed to characterise the scenarios Busy and Rest. For a more extensive description of these scenarios see Bruggeman et al. (2013).

As indicated above, exploring future flood risk requires a high level of spatial detail. A similar level of detail is required for other themes of the Delta Programme such as fresh water supply for agriculture and nature. Therefore, Land Use Scanner is applied to consistently downscale the general story lines composed for the four

Delta Scenarios into local level indicators. This process is described by Rijken et al. (2013). The resulting land-use simulations provide the input for a wide range of (local) indicators, including fresh water demand, soil subsidence, soil sealing and flood risk. This section goes into the latter issue, illustrating the specific use of Land Use Scanner for exploring future flood risk. The focus is on the added risk emanating from residential extensions (green-field development). These developments are especially interesting as they can potentially be relocated by spatial planning measures at relatively low cost.

Figures 6.3a, b show a preliminary rendering of the two most extreme Delta Scenarios in terms of residential extensions between the year 2008 and 2050. The figures show the cells (1 ha) where, according to the scenarios, residential development and thus additional exposure is most likely to occur. To get insight in the increased flood risk emanating from these extensions local flood hazard is incorporated in the analysis (Fig. 6.3c). It is expressed here as the probability of flooding multiplied by the potential local damage of this event to the average single family home, with potential local damage being expressed as a factor, ranging from 0 to 1, depending on maximum inundation depth. Subsequently, the additional flood risk is calculated by combining this flood hazard with projections on regional average house prices for these average homes (that are expected to be the main type of construction) and projected residential densities. Figures 6.3d, e illustrate the results.

The simulations indicate that risk is concentrated in the few areas where residential extensions, high housing densities (not shown in the figure) and ditto flood hazard coincide. The figures also show that areas characterised by high risk are rather far and in between. Indeed, these small pockets of high risk are dispersed over a number of large flood protection areas (i.e. the areas within the blue lines in Fig. 6.3c). This is true for both the liberal, high growth scenario Steam as in the opposite Rest scenario—although total risk is evidently higher in Steam.

The dispersed patterns shown in the area selected in Fig. 6.3 can also be observed in other flood-prone parts of the Netherlands. From a flood risk management perspective, urban development at less risk-prone areas is preferred. Should that prove to be impossible, and should additional flood risk reduction by means of further fortification of the large-scale flood defence systems mentioned above be considered insufficient or inefficient, Dutch climate adaptation strategies can follow a localised approach tailored to the local idiosyncrasies constituting risk in a particular area. This approach would favour custom adaptation measures like, for instance, elevated or floating housing, spatial zoning or even evacuation plans. On account of the efficiency criterion, it would then be well advised to focus on areas in Figs. 6.3d, e showing the highest concentrations of risk. The robustness of the required investments would be served if priority would be given to those areas that show high risk in at least two scenarios. By way of example, these are marked with circles in the Figs. 6.3d, e.

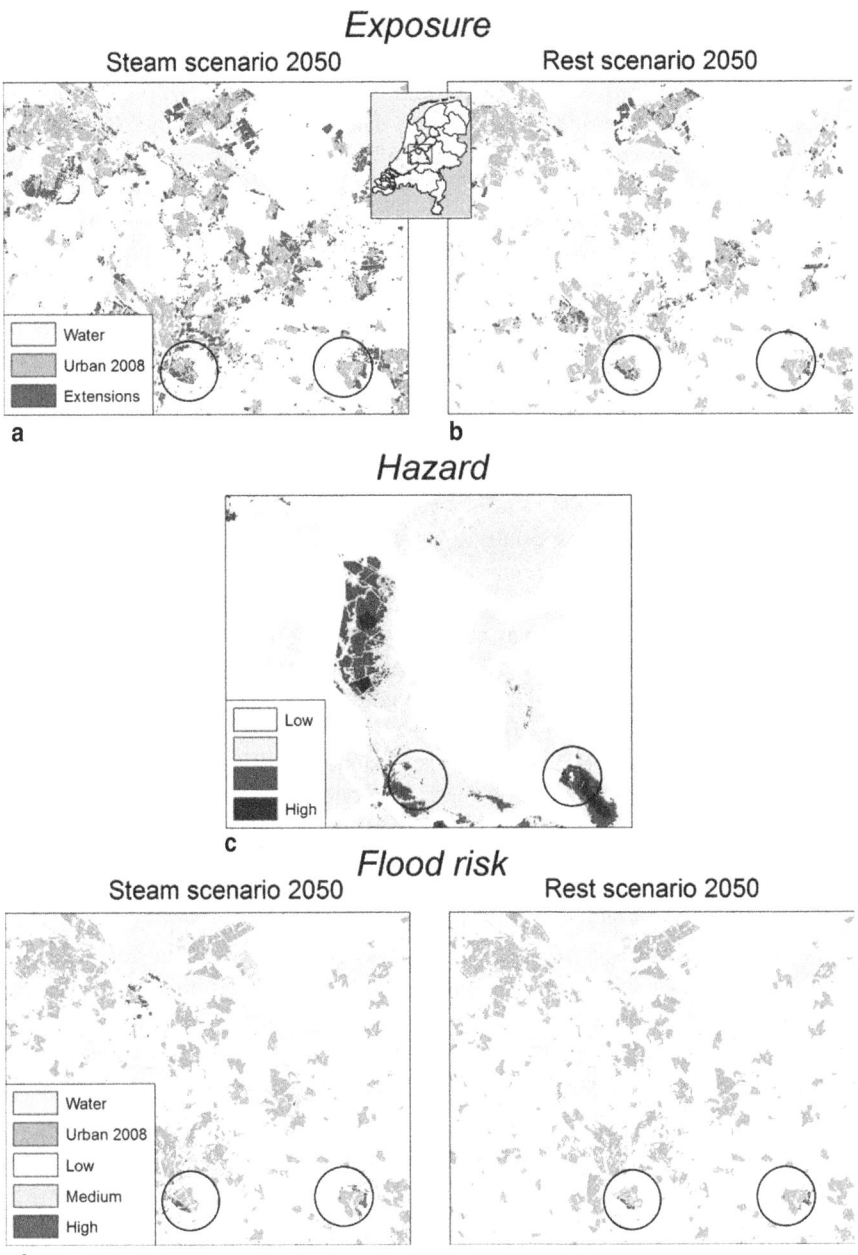

Fig. 6.3 Assessing flood risk in 2050 in two Delta Scenarios (**d** and **e**) based on a combination of exposure (simulated residential extensions; **a** and **b**) and flood hazard (**c**). (All scenario results adapted from: Bruggeman et al. 2013)

6.4 Conclusion and Discussion

Land-use simulation proves to be a useful tool in the evaluation of policy alternatives and the exploration of future scenarios. The first example shows that a land-use model can successfully be used to explore likely outcomes of policy changes in a Strategic Environmental Assessment report dedicated to the newly proposed national spatial strategy for the Netherlands. This what-if type of simulation typically relies strongly on expert judgement as it has to describe the impact of policies that have not yet been implemented and whose effects cannot be observed. These expert opinions can, to some extent, be tested in GIS-based analyses of the effectiveness of similar policies as described in other studies (Koomen et al. 2008a). To incorporate the notion of uncertainty in our ex-ante policy evaluation we developed scenario-based simulations of potential impacts of policy changes.

The second example shows that current ambition levels are needed to prevent extensive loss of open space. Should urban development (in terms of intensification share and extensions' densities) follow past trends, large-scale urban extensions are likely to occur. This is especially true when future socio-economic conditions resemble the Global Economy scenario.

Finally, the third case study shows that the Land Use Scanner can be a helpful in the field of climate adaptation as well. By simulating future residential development at the highly detailed level of 1 ha grid cells, it demonstrates that adaptation to flood risk emanating from future residential extension can be limited to highly localised strategies like spatial zoning. Cost can be minimised by focusing local adaptation strategies on the areas where risk is concentrated. Prioritising the places where these concentrations occur in more than one scenario enhances the robustness of adaptation.

The presented, recent Dutch land-use model applications focussed on simulating future urban expansion patterns. This is for many regions a very real scenario and Land Use Scanner is a powerful instrument to explore the impacts of new urban development in terms of, for example, loss of open space and flood risk. In the near future other spatial policy challenges may become prominent. These may relate to steering urban intensification in regions of urban growth and urban restructuring in regions of population decline (Kuijpers-Linde 2011). Intensification has the potential to preserve open space, keep down transportation costs, maximise urban economic productivity etc. Effective restructuring, in turn, could minimise the damaging effect shrink may have on the economic health of urban areas.

To allow the simulation of processes such as intensification and restructuring we are currently incorporating a residential land-use density layer in Land Use Scanner. Ultimately, this requires a distinction between actors (residents), objects (residences) and land use (residential land) and the addition of information on each of these three layers in the model. Initial attempts to incorporate residential density (as described in the second example in Sect. 6.3) have treated density change as exogenous model input, based on expert judgement and the extrapolation of past trends. This implies that, while all kinds of assumptions can be made regarding future dynamics on these issues, and numerous local indicators can be derived from the resulting output, the underlying phenomena remain outside the model, preventing them to be simulated in an integrated way that is most probably require. Making these phenomena endogenous model parameters by incorporating the underlying mechanisms into the Land Use Scanner model would at once improve the explanatory power of the model regarding these highly relevant policy issues, and enhance the detail of the relevant model output (i.e., local density). This would greatly enhance the potential for subsequent impacts assessment such as local flood risk assessments discussed in the third example in Sect. 6.3.

Obviously, the presented land-use change model can easily be expanded to explore other scenarios or strategies, and accommodate additional spatial restrictions or other types of development (e.g. agriculture, industry/commerce, recreation). Such simulations have been performed as part of other studies focusing on, for example, the environmental impacts of a new regional strategic vision (Koomen et al. 2011) or the potential for new biofuels in the country (Kuhlman et al. 2013). Land-use modelling thus offers a useful contribution to geodesign processes as they help describe how the study area operates, and how it is likely to change in the future through autonomous developments or implemented policies. In the design process of spatially explicit policies and developments they can act as the process and change models distinguished in the geodesign framework (Steinitz 2012). In addition simulated land-use patterns can be used as input in impact models as was demonstrated in the explorations of climate impacts in this chapter.

The presented examples show the application of a land-use model in classic sequential planning processes that do not show the iterative character advocated in recent geodesign definitions (Flaxman 2010; Ervin 2011). This is, however, inherent to current practice in most planning processes and not characteristic for the applied model. In regional applications of the model focussed on developing strategic spatial visions for regional authorities we experimented with a more iterative work process (Koomen et al. 2011). In a series of workshops, land-use simulations were first created to show expected socio-economic developments in the region, discuss the need for policy intervention and provoke the definition of spatially-explicit ambitions and related spatial policies for various planning domains (e.g. water management, health and safety). Based on these initial reactions we simulated the potential spatial impact of the proposed policies and used these outcomes in joint discussions on a more integrated set of policy objectives. In a final simulation run the combined set of selected spatial policies was then fed into Land Use Scanner to obtain the required input for the formal environmental impact assessment procedure. Our land-use modelling framework thus proved flexible and fast enough to provide input to an iterative planning process.

Currently, researchers indeed observe that decision making related to spatial development is moving away from a straightforward, sequential process (Roodbol-Mekkes et al. 2012). Public debates, storytelling, the generation of new ideas (visioning) and spatial investments are becoming incremental and iterative processes. More and more, the debate about regional development takes place in non-related discourses. Moreover, the consistency between medium and long term relevant policies is becoming weaker or turns out to be absent. Furthermore, many actors are involved that move away from a rational-logical approach to decision making. Traditional tools for evaluation and visualisation may not be fully suitable in this new context, mainly because they lack the flexibility to change spatial scale, time horizon or area selection. They usually are slow in responding to new policy strategies and their output is often difficult to understand or even irrelevant for the various non-expert stakeholders that are involved in today's spatial development processes. An important element that is absent in most spatially-explicit decision support tools is financial information. Especially since the recent financial crises, this is a major shortcoming because spatial planning discussions are currently being dominated by financial debates about investment.

A possible way forward is to combine existing land-use models, evaluations frameworks (such as cost benefit analysis or dedicated impact assessment framework) with new visually attractive user interfaces (such as touch tables) to overcome part of the shortcomings of existing evaluation tools (Dias et al., 2013). In doing so it may become possible to bring together the values and arguments of different actors in decision-making processes in the built environment that differ in spatial scale, time horizon and focus concerning people, planet and profit aspects.

Acknowledgements This research is funded by the research programme Urban Regions in the Delta (URD), part of the VerDuS-programme (Verbinding Duurzame Steden) of the Netherlands Organisation for Scientific Research (NWO). We would like to thank PBL Netherlands Environmental Assessment Agency for providing the spatial data and initial land-use simulations that were incorporated in the analyses discussed in this chapter. The work described in Sect. 6.3.1 was performed as part of a project one of the authors carried out for the Ministry of Infrastructure and Environment in cooperation with Royal Haskoning and Alterra Wageningen University and Research Centre. The former colleagues and partners in that project are thanked for the pleasant and effective cooperation and fruitful discussions that provided the results discussed here. In addition we thank Dani Broitman for running the land use simulations discussed in Sect. 6.3.2 and Jasper Dekkers for providing valuable input to construct the Trends alternatives in that section.

References

Borsboom-van Beurden, J. A. M., Bakema, A., & Tijbosch, H. (2007). A land-use modelling system for environmental impact assessment; Recent applications of the LUMOS toolbox. Chapter 16. In E. Koomen, J. Stillwell, A. Bakema, & H. J. Scholten (Eds.), *Modelling land-use change; progress and applications* (pp. 281–296). Dordrecht: Springer.

Bruggeman, W., Dammers, E., Van den Born, G. J., Rijken, B. C., Van Bemmel, B., Nabielek, K., Beersma, J., Van den Hurk, B., Polman, N., Lindenhof, V., Folmer, C., Huizinga, F., Hommes, S., & Te Linde, A. H. (2013). Deltascenario's *voor 2050 en 2100; Nadere uitwerking 2012–2013*. Delft: Deltares/PBL/LEI/CPB.

CPB, MNP, & RPB. (2006). *Welvaart en Leefomgeving. Een scenariostudie voor Nederland in2040*. Centraal Planbureau. Den Haag: Milieu- en Natuurplanbureau en Ruimtelijk Planbureau.

Dammers, E. (2000). *Leren van de toekomst. Over de rol van scenario's bij strategische beleidsvorming*. Delft: Uitgeverij Eburon.

De Moel, H., Koks, E. E., Dekkers, J. E. C., Lassche, M. R., & Bouwer, L. M. (2012). *Methods for future land-use projections for Flanders*. Amsterdam: Institute for Environmental Studies, VU University Amsterdam.

De Ruijter, P., Stolk, S., & Alkema, H. (2011). *Klaar om te wenden; handboek voor de strateeg*. Schiedam: Scriptum.

Dias, E. S., Linde, M., Rafiee, A., Koomen, E., & Scholten, H. J. (2013). Beauty and brains: Integrating easy spatial design and advanced urban sustainability models. Chapter 27. In S. Geertman, F. Toppen, & J. C. H Stillwell (Eds.), *Planning support systems for sustainable urban development* (pp. 469–484). Berlin: Springer.

Elings, C., Zijlstra, R., Koomen, E., & De Groot, S. (2011). *Milieueffectrapport ontwerp structuurvisie infrastructuur en ruimte in opdracht van ministerie van infrastructuur en milieu*. Amsterdam/Nijmegen: Geodan/Royal Haskoning.

Ervin, S. (2011). A system for geodesign. In E. Buhmann, S. Ervin, C. D. Tomlin, & M. Pietsch (Eds.), *Teaching landscape architecture—Prelimenary proceedings* (pp. 145–154). Bernburg: Anhalt University of Applied Sciences.

Flaxman, M. (2010). Fundamentals of geodesign. In E. Buhmann., M. Pietsch, & E. Kretzler (Eds.), *Digital landscape architecture 2010, digital landscape architecture2010* (pp. 28–41). Berlin: Wichmann, VDE.

Hoymann, J. (2010). Spatial allocation of future residential land use in the Elbe River Basin. *Environment and Planning B: Planning and Design,37*(5), 911–928.

I&M. (2011). *Ontwerp structuurvisie infrastructuur en ruimte; Nederland concurrerend, bereikbaar, leefbaar en veilig*. Den Haag: Ministerie van Infrastructuur en Milieu.

Jacobs, C. G. W., Koomen, E., Bouwman, A. A., & Van der Burg, A. (2011). Lessons learned from land-use simulation in regional planning applications. Chapter 8. In E. Koomen & J. Borsboom-van Beurden (Eds.), *Land-use modelling in planning practice* (pp. 131–149). Heidelberg: Springer.

Keers, G., Smeulders, E., & Teerlink, T. (2011). *Onderzoek: Toekomst voor de stedelijke woningbouw?* P15360. RIGO Research en Advies/Bouwfonds Ontwikkeling.

Klosterman, R. E. (1999). The what if? Collaborative planning support system. *Environment and Planning B: Planning and Design, 26,* 393–408.

Klijn, F., Baan, P., Bruijn, K., & Kwadijk, J. (2007). *Overstromingsrisico's in Nederland in een veranderend klimaat verwachtingen, schattingen en berekeningen voor het project nederland later*. Bilthoven: Milieu- en Natuurplanbureau (MNP).

Koomen, E., & Borsboom-van Beurden, J. (Eds.) (2011). *Land-use modeling in planning practice. GeojournalLibrary* (vol. 101). Heidelberg: Springer.

Koomen, E., & Dekkers, J. E. C. (2013). The impact of land-use policy on urban fringe dynamics; Dutch evidence and prospects. In D. Malkinson, D. Czamanski, & I. Benenson (Eds.), *Modelling of land-use and ecological dynamics* (pp. 9–35). Berlin: Springer.

Koomen, E., Dekkers, J., & Van Dijk, T. (2008a). Open-space preservation in the Netherlands: Planning, practice and prospects. *Land Use Policy,25*(3), 361–377.

Koomen, E., Rietveld, P., & De Nijs, T. (2008b). Modelling land-use change for spatial planning support; Editorial. *Annals of regional science,42*(1), 1–10.

Koomen, E., Koekoek, A., & Dijk, E. (2011). Simulating land-use change in a regional planning context. *Applied Spatial Analysis and Policy, 4*(4), 223–247.

Koomen, E., Dekkers, J. E. C., & Broitman, D. (2015). Analysing and simulating urban density; exploring the difference between policy ambitions and actual trends in the Netherlands. In J.-C. Thill (Ed.), *Spatial analysis and location modeling in urban and regional systems*. Londan: Springer.

Kuhlman, T., Diogo, V., & Koomen, E. (2013). Exploring the potential of reed as a bioenergy crop in the Netherlands. *Biomass and Bioenergy,55,* 41–52.

Kuijpers-Linde, M. (2011). A policy perspective of the development of Dutch land-use models. Chapter 10. In E. Koomen & J. Borsboom-van Beurden (Eds.), *Land-use modelling in planning practice* (pp. 177–189). Heidelberg: Springer.

Kuiper, R., & Evers, D. (2011a). *Ex-ante evaluatie Structuurvisie Infrastructuur en Ruimte*. Den Haag: Planbureau voor de Leefomgeving.

Kuiper, R., & Evers, D. (2011b). Structuurvisie Infrastructuur en Ruimte perkt nationale belangen in. *ROM Magazine, 29*(6), 10–13.

Landis, J. D. (1994). The California urban futures model: A new generation of metropolitan simulation models. *Environment and Planning B: Planning and Design, 21,* 399–420.
PBL. (2011). *Nederland in 2040: een land van regio's.* Den Haag: Planbureau voor de Leefomgeving.
Rijken, B., Bouwman, A., Van Hinsberg, A., Van Bemmel, B., Van den Born, G. J., Polman, N., Lindenhof, V., & Rijk, P. (2013). *Regionalisering en kwantificering verhaallijnen Deltascenario's 2012. Technisch achtergrondrapport.* Den Haag: Planbureau voor de Leefomgeving & LEI Wageningen UR.
Roodbol-Mekkes, P. H., Van der Valk, A. J. J., & Korthals Altes, W. K. (2012). The Netherlands spatial planning doctrine in disarray in the 21st century. *Environment and Planning A, 44*(2), 377–395.
Scholten, H. J., Van de Velde, R. J., Rietveld, P., & Hilferink, M. (1999). Spatial information infrastructure for scenario planning: The development of a land use planner for Holland. In J. Stillwell, S. Geertman, & S. Openshaw (Eds.), *Geographical information and planning* (pp. 112–134). Berlin: Springer.
Schotten, C. G. J., Goetgeluk, R. W., Hilferink, M., Rietveld, P., & Scholten, H. J. (2001). Residential construction, land use and the environment. Simulations for The Netherlands using a GIS-based land use model. *Environmental Modelling and Asessment, 6*(2), 133–143.
Significance and Bureau Louter. (2007). *Toepassen TIGRIS XL binnen de studie 'Nederland Later'.* Bilthoven: PBL Netherlands Environmental Assessment Agency.
Steinitz, C. (2012). *A framework for geosign; Changing geography by design.* Redlands: ESRI Press.
Te Linde, A. H., Bubeck, P., Dekkers, J. E. C., De Moel, H., & Aerts, J. C. J. H. (2011). Future flood risk estimates along the river Rhine. *Natural Hazards and Earth System Sciences, 11*(2), 459–473.
Turner, B. L., Kasperson, R. E., Matson, P. A., McCarthy, J. J., Corell, R. W., Christensen, L., Eckley, N., Kasperson, J. X., Luers, A., Martello, M. L., Polsky, C., Pulsipher, A., & Schiller, A. (2003). A framework for vulnerability analysis in sustainability science. *Proceedings of the National Academy of Sciences U S A, 100*(14), 8074–8079.
Van der Hoeven, E., Aerts, J., Van der Klis, H., & Koomen, E. (2009). An integrated discussion support system for new Dutch flood risk management strategies. Chapter 8. In S. Geertman & J. C. H. Stillwell, (Eds.), *Planning support systems: Best practices and new methods* (pp. 159–174). Berlin: Springer.
VROM, LNV, V&W, & EZ. (2004). *Nota Ruimte. Ruimte voor ontwikkeling. Ministeries van Volkshuisvesting, Ruimtelijke Ordening en Milieubeheer, Landbouw, Natuur en Voedselkwaliteit, Verkeer en Waterstaat en Economische zaken.* Den Haag: SDU uitgeverij.
Zondag, B., & Geurs, K. (2011). Coupling a detailed land-use model and a land-use and transport interaction model. Chapter 5. In E. Koomen & J. Borsboom-van Beurden (Eds.), *Land-use modeling in planning practice* (pp. 79–95). Dordrecht: Springer.

Chapter 7
Using Geodesign to Develop a Spatial Adaptation Strategy for Friesland

Ron Janssen, Tessa Eikelboom, Jos Verhoeven and Karlijn Brouns

7.1 Introduction

Collaborative workshops are common in land use planning. Stakeholders using maps to design plans is not something that started in recent years. Initially, workshops were supported using large hard copies of maps in combination with sheets of tracing paper to sketch attributes of the proposed plan or plan area (Burrough and McDonnell 1998). In following years, with the arrival of Geographical Information Systems (GIS), the transparent tracing map sheets were replaced by map layers presented within a GIS on a computer screen (Longley et al. 2005). This has now proceeded further towards interactive map interfaces with direct interaction between participants and information. Along with these technical developments, the involvement of stakeholders in spatial planning has changed over the years. In the early years, the emphasis was on informing the public. In later years this shifted to participation where active involvement of stakeholders was required (Sieber 2006). At present, the focus is on collaboration, with stakeholders actively working together to reach the best plan.

Geodesign tools can be used to support collaborative processes. Typical tools combine different methods, such as simulation models, spatial multi-criteria analysis, visualization, and optimization. User-friendly interfaces allow multiple users to provide input and generate real-time output to support negotiated spatial decisions (Geertman and Stillwell 2009; Pelzer et al. 2013; Pettit 2011).

R. Janssen (✉) · T. Eikelboom
Institute for Environmental Studies, VU University Amsterdam, Amsterdam, The Netherlands
e-mail: ron.janssen@vu.nl

T. Eikelboom
e-mail: tessa.eikelboom@vu.nl

J. Verhoeven · K. Brouns
Ecology and Biodiversity, Utrecht University, Utrecht, The Netherlands
e-mail: j.t.a.verhoeven@bio.uu.nl,

K. Brouns
e-mail: k.brouns@uu.nl

The Province and Water board of Friesland have decided to develop a long-term adaptation strategy for the peat meadow area of the province. Primary activities in this region are highly productive dairy farming, nature conservation, recreation and housing. The region is currently mainly used for commercial dairy farming but is also important for its high natural, cultural and historical values. Important problems in the region are soil subsidence causing damage to buildings and infrastructure, deterioration of landscape values, inefficient water management, poor water quality, and the changing perspectives for dairy farming (Janssen et al. 2013). A planning process with all stakeholders has been initiated to develop an adaptation strategy. The Province and Water board have described three strategies in general terms: 1. *Business as usual*; 2. *Parallel tracks* and 3. *New horizons*. The *Business as usual* strategy included technical measures to reduce soil decline. These technical measures could only be applied if they did not create limitations for agriculture. *Parallel tracks* was based on zoning and separation of the different types of land use. Buffer zones were created to separate the different types of land use. *New horizons* involved a major transformation of agriculture. New types of crops were introduced that could adapt to the wet conditions needed to prevent soil decline. None of these strategies included a spatial allocation of the proposed measures.

As part of this policy process stakeholder workshops were conducted that had the following objectives:

- Exchange of information
- Validate information
- Design three spatial adaptation strategies.

The workshops are exploratory and do not involve a choice for one of the strategies. The workshops were held in three regions, Hommerts, Groote Veenpolder and Buitenveld. These regions were assumed to be representative for the peat meadow area in the province.

An interactive mapping device (the Touch Table) was used as a common interface (Fig. 7.1). The use of the Touch Table made it possible for participants to have direct access to tools and information and provided a common platform for discussion. The geodesign application developed for these workshops included an evaluation tool and a design tool. Both tools were dynamic and provided immediate feedback on the effects of any change made by the participants. These tools were used in the workshops to design spatial strategies by allocating water management measures, such as changes in water levels and drainage systems, and land use types specific to each strategy.

The workshops followed the steps of the geodesign framework as defined by Steinitz (2012). The framework for geodesign is compatible with the adaptation framework of Willows and Connell (2003) that assumes that the development of an adaptation strategy is a circular process containing eight stages and was found useful for linking tools to stakeholder tasks (Eikelboom and Janssen 2013).

This chapter describes the use of geodesign tools to support the development of a spatial adaption strategy for peat meadow areas in Friesland. The evaluation and design tools are described in the next two sections. The third section describes the use of these tools in the workshops. The chapter concludes with a

7 Using Geodesign to Develop a Spatial Adaptation Strategy for Friesland   105

Fig. 7.1 Workshop participants around the Touch Table

discussion of the effectiveness of the tools in exchanging information and generating plans for each strategy. A full report of the workshops is available in Janssen et al. (2013).

7.2 The Approach

The objective of the workshops was for the stakeholders to develop three spatial strategies at the local scale based on an implementation of the three general strategies for the whole region (*Business as usual, Parallel tracks, New horizons*). The workshops were exploratory in nature and no strategy was selected as the preferred choice.

The workshops followed the steps of the geodesign framework as defined by Steinitz (2012). Steinitz distinguishes three iterations through the framework phrasing the questions as "Why?" questions in the first iteration, as "How?" questions in the second iteration and as "What?, where? and when?" questions in the third and final iterations. The workshops go through these iterations in one afternoon (Fig. 7.2). To be able to do this so quickly, the workshops relied heavily on stakeholder input. No extensive field work was done prior to the workshops. It was accepted that the information about the region was not complete and maybe even incorrect but that stakeholders will correct this information.

Participants of the workshops were invited by the Province and Water board of Friesland and were based on prior involvement in the planning process. Some participants had a direct stake in the planning process as they owned a farm or house in the region, others represented interest groups such as farmers or nature conservationists.

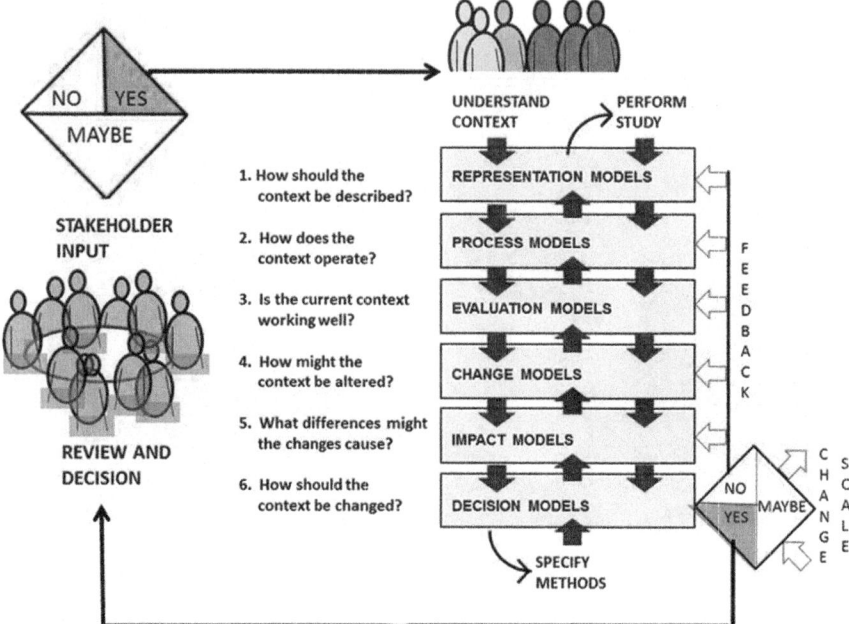

Fig. 7.2 The stakeholders, the geodesign team, and the framework for geodesign. (Source: Steinitz 2012)

The workshop started with an introduction to the region and communication of information needed for the assignments (representation). The complicated relations between land use, ground water level and soil subsidence were communicated by an exercise where participants experimented by changing land use and water levels and observe the result on soil decline. Only very simple models were used to model the relations between water management, soil decline, nature and agriculture. This made it possible to explore and even adapt all model relations during the workshop. About half the available time was used to communicate all information. The second half was used for three assignments where participants were asked to design three strategies for the region.

The first assignment around the Touch Table was an introductory assignment to get a feel for the Touch Table and to learn about the mechanics behind soil subsidence. The participants were asked to change land use and water levels, to observe the impacts on soil subsidence and to inspect relevant map layers. This was followed by three assignments linked to the three strategies. The participants were asked to design a land use and water management plan for the following three policy strategies:

- *Business as usual:* low impact technical measures only, no changes in land use;
- *Parallel tracks:* create buffers to separate conflicting functions such as agriculture, housing and nature;
- *New Horizons:* introduction of new crops, large changes in land use and water management.

The evaluation, change, impact and decision model were integrated in the design tool. The design tool allows to make changes and to get immediate feedback on the resulting changes on policy objectives such as agricultural production, prevention of soil decline and quality of nature. The physical impacts were calculated but not presented to the participants as they would not be immediately understandable. To provide more easy to use feedback to stakeholders, impact and evaluation were combined. The underlying model for evaluation was based on multicriteria analysis (Arciniegas et al. 2011). This model can be used on various spatial scales and can be adapted to specific decision conditions (Eikelboom and Janssen in press).

The value of the objectives had a linear relation with the water level in the ditches. This relation was different for each type of land use. Each objective value responded differently to increasing water levels. A high water level results in high objective values for soil subsidence and nature, but in low values for agriculture. Similarly, extensive grasslands have a higher value for nature due to land management compared to intensive grassland and with similar water level conditions nature areas have the highest values for nature. A high value for soil subsidence means low subsidence rates, high values for nature means high quality and low values for agriculture means low productivity. During the process the stakeholders zoomed in to different scales and locations within the area. They kept iterating between evaluation, change, impact and decision until there was consensus about the plan. In doing this they have answered the why, how and what/where questions. The when question was not represented on the map but appeared in the notes of the meeting. No field research was conducted in preparation of the workshops as the approach is based on input from the stakeholders. This made it possible to move quickly but made the approach very dependent on the available knowledge of the stakeholders present during the workshop.

7.3 Geodesign Tools

The workshops were supported by an evaluation and a design tool. The tools were developed using Community Viz 4.3 which is an extension to ArcGIS 10. Community Viz is a software package that allows for spatial programming of dynamic attributes and indicators (http://placeways.com/communityviz/ last accessed on October 10, 2013). The Samsung SUR40 with Microsoft Surface 2.0 provided the interface between the participants and the tools.

A large amount of spatial information is available for the regions. Current and historical land use, and land ownership were obtained from the Provincial authorities. Water levels and an elevation map were obtained from the Water board. Spatial information about the soil was supplied by Alterra University and Research Centre, Wageningen. Most of the information is collected in the years 2012 and 2013.

The information is first translated to value maps based on expert judgement. Figure 7.3a shows a value map for agriculture (see also Arciniegas et al. 2011). The value in Fig. 7.3a represents the production conditions for agriculture and depend on both land use and water level. Values above 0.80 are considered acceptable and are therefore coloured green. Values between 0.70 and 0.80 are considered problematic (orange) and values below 0.70 unacceptable (red). In Fig. 7.3b the value maps for agriculture, soil and nature are combined. The value of the objectives is

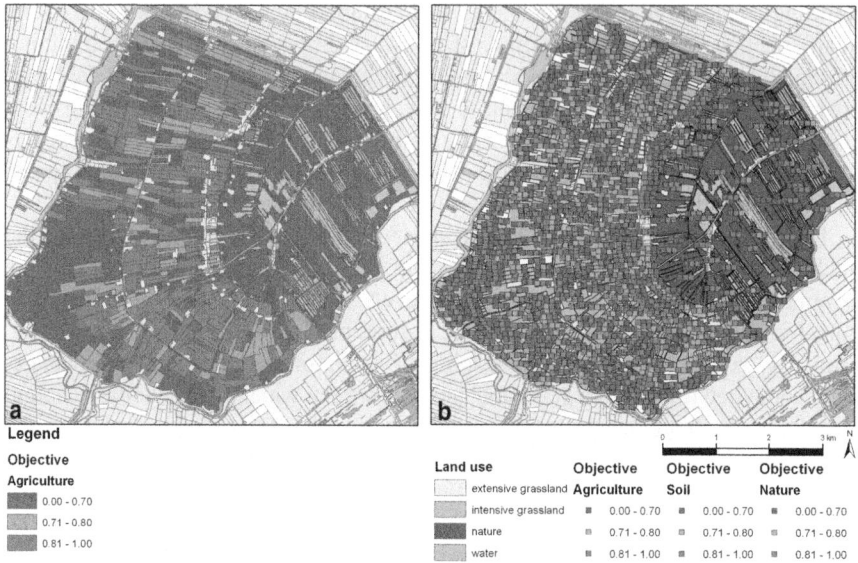

Fig. 7.3 Value map agriculture (**a**) and value map agriculture, soil and nature (**b**) in the Groote Veenpolder (current situation)

presented as a traffic light where red is low, white is average, and green represents a high value. The traffic light shows for each parcel the value for agriculture (left) soil (middle) and nature (right). For example a traffic light with three red boxes means a low value for all three objectives.

The traffic lights makes it possible to project the main indicators on top of other maps. During the workshops, the objective values were mainly shown in combination with land use and water level, because the objective values change when water level or land use is changed. This tool enabled participants to monitor the performance for each objective under different circumstances. The evaluation tool is used in combination with the design tool. Figure 7.4 shows the design tool for one of regions. The map on the left shows land use and the map on the right water levels. The tool provides next to the map a list of potential land uses, a list of potential water levels and a list of water management measures such as drainage and storage. Participants can apply these measures to one or more parcels to improve one or more objectives. Changes in values are shown immediately as changes in the colours of the traffic lights in each parcel. Participants can also change the land use of each parcel. Changes in land use and water management are shown on the map. Any map participants consider relevant can be used as a background for the traffic lights. A division is made between intensive and extensive grassland. Intensive grasslands are characterized by a high density of cattle, use of heavy machinery and use of pesticides and fertilizers. Extensive grassland have a lower production, do not need heavy machinery and no or little fertilizer and pesticides. As a result extensive grassland are more suitable for higher water levels.

7 Using Geodesign to Develop a Spatial Adaptation Strategy for Friesland

Fig. 7.4 The use of the design tool for the region 'Hommerts': combined with land use (**a**) and water levels (**b**)

7.4 Workshop Results

Workshops were organized in three regions, Hommerts, Groote Veenpolder and Buitenveld. These regions were assumed to be representative for southeast Friesland. The workshops started with an introduction to the region and communication of information needed for the assignments. An effective way to introduce the region is a comparison of the current and historic topographical maps (Fig. 7.5). Swiping the two maps showed how the current map has evolved from the past, especially in peat meadow areas were many of the current issues can be traced back to the past.

The introduction was also used to validate the information presented. As the preparation did not involve extensive field visits not all information was up to date. Figure 7.6 (a) shows the information as available from the topographical map and

Fig. 7.5 De Groote Veenpolder 1860 (**a**) and 2012 (**b**)

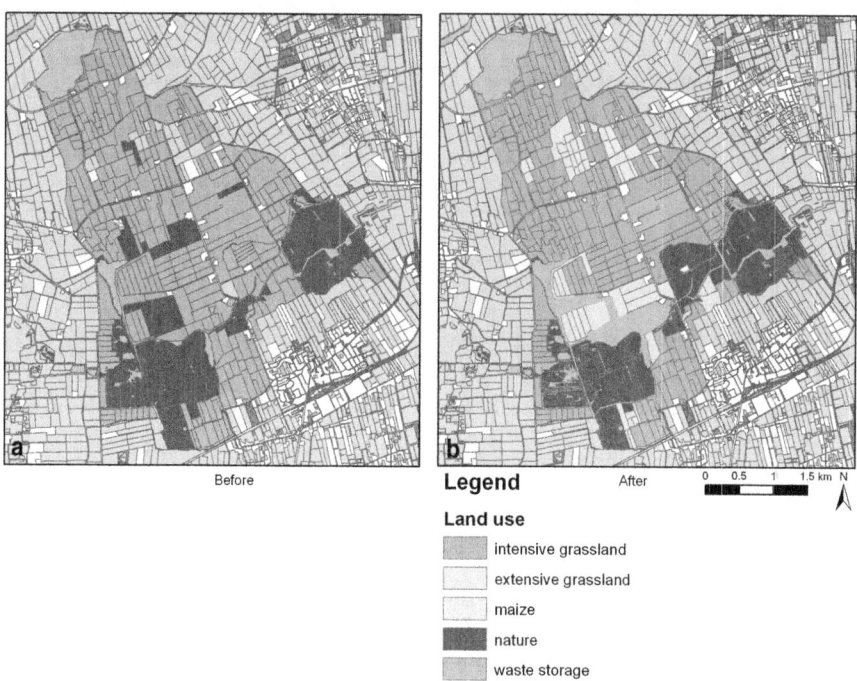

Fig. 7.6 Land use map Buitenveld before (**a**) and after the workshop (**b**)

aerial pictures. Figure 7.6 (b) shows the same map after corrections from the participants. It appeared that, recently, open water had been added, more nature had been created and some parcels had been changed to extensive grassland. Also the former waste dump was not included in the original map.

7 Using Geodesign to Develop a Spatial Adaptation Strategy for Friesland 111

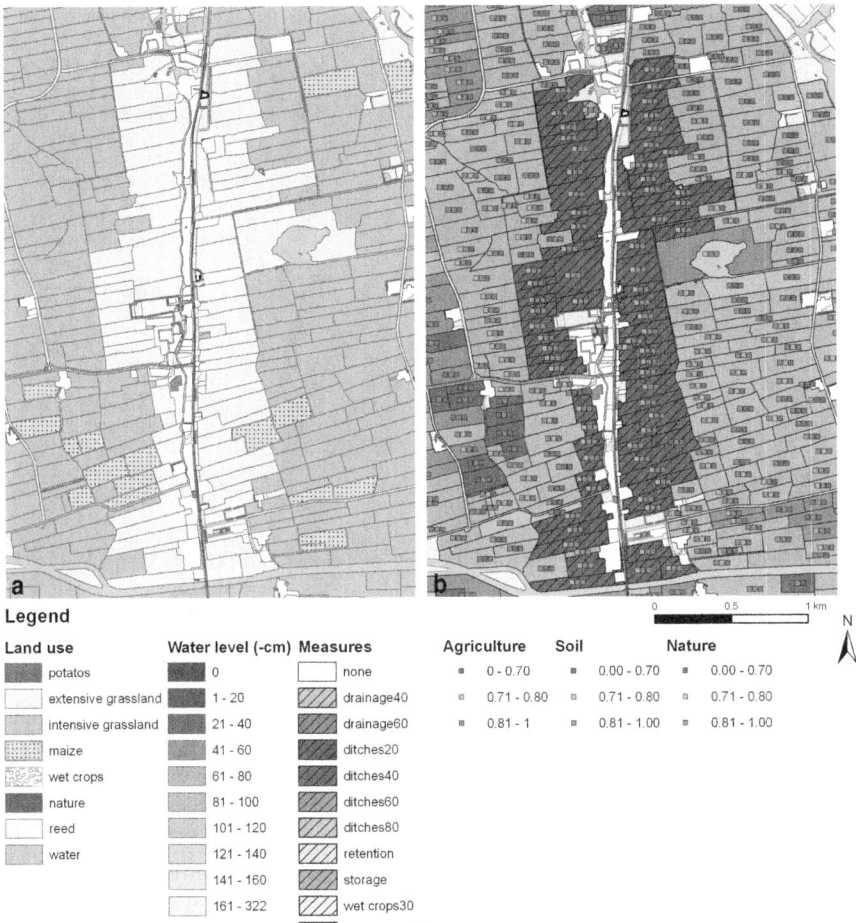

Fig. 7.7 Creating a buffer in strategy *Parallel tracks*: (**a**) Change in land use, (**b**) Change in water level and performance

Next, the participants were asked to design a land use and water management plan for the following three policy strategies:

- *Business as usual:* low impact technical measures only, no changes in land use;
- *Parallel tracks:* create buffers to separate conflicting functions such as agriculture, housing and nature;
- *New Horizons:* introduction of new crops, large changes in land use and water management.

This meant that for each region three plans were developed. Figure 7.7a shows the result for *Parallel tracks* for the region Hommerts. Participants identified damage to houses resulting from soil subsidence as a problem. First, participants zoomed in on a long strip of houses on both sides of the central road. As can be seen on the new

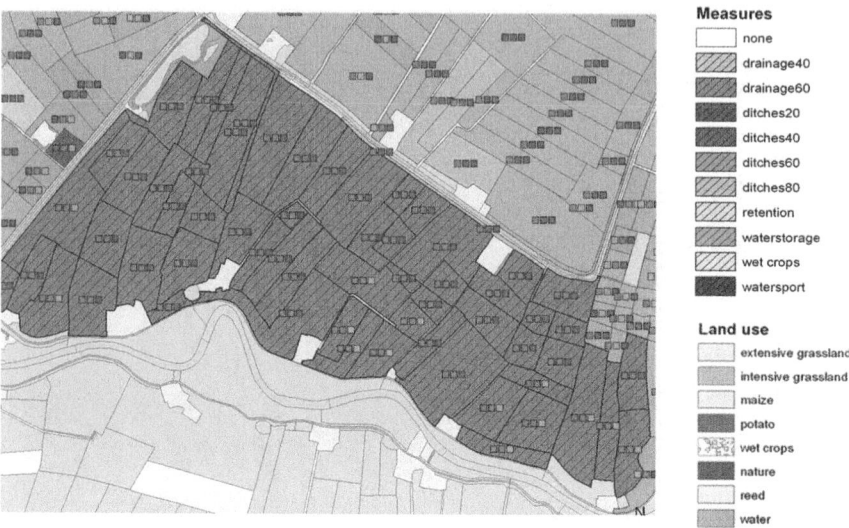

Fig. 7.8 The effect of underwater drains for the south east corner of the Groote Veenpolder

land use map (left) they created a zone of extensive agriculture along these houses. Figure 7.7b shows that in this zone the value of agriculture decreased to average but the scores for soil subsidence and nature increased, which solves the problem for the houses.

The next map shows the result for strategy *Business as usual* for the Groote Veenpolder. Before applying any measures, as a first step, participants zoomed in to a specific area. To do this they used the map library to identify parts of the region with specific characteristics. As one of the available measures could only be applied in areas without upward seepage participants used the hydrology map to identify the relevant sub region (Fig. 7.8). As is shown in the figure participants applied drainage at -40 cm to large part of this area. The map showed that for this part of the region drainage at 40 cm was beneficial both for agriculture (left box) and prevention of soil subsidence (middle box).

The design tool allows for detailed, small scale, changes but can also support a total redesign of the region. This is what is called for by the *New horizons* strategy. This strategy assumes substantial changes from the current situation. A good example was the *New Horizon* plan designed for the Groote Veenpolder (Fig. 7.9).

A major problem in this polder is the net loss of water from the nature reserve located east of the lower situated agricultural area in the west. Participants decided that only a radical measure could solve this problem. As can be seen in the maps in this strategy (Fig. 7.9), a large area to the west of the nature reserve will be flooded to be used for water sports or aquaculture. In the agricultural area to the west of this buffer, water levels were increased and the land use was changed from intensive to extensive grassland. Wet crops and common reed were introduced along the border between water and agriculture. This radical change led to a total stop of soil

7 Using Geodesign to Develop a Spatial Adaptation Strategy for Friesland

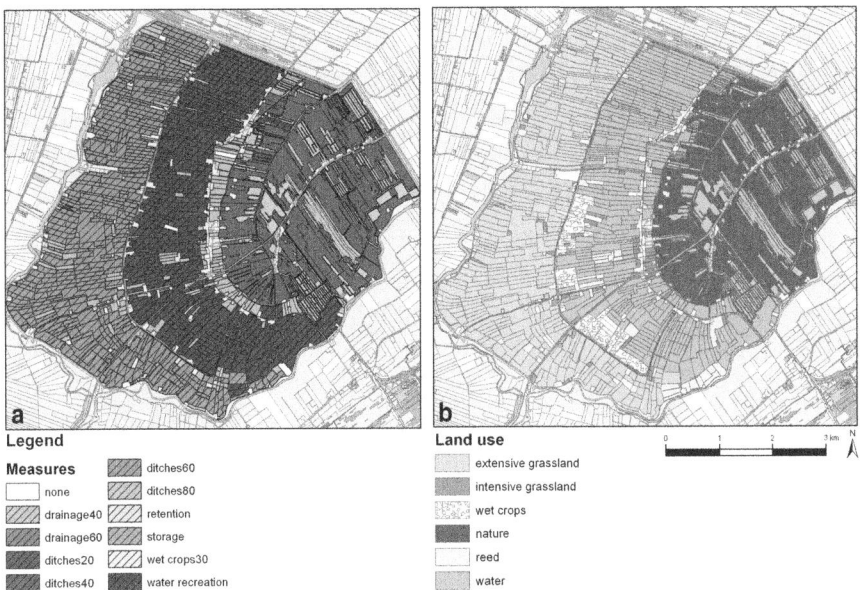

Fig. 7.9 Strategy *New Horizons* for the Groote Veenpolder: (**a**) measures, (**b**) land use change

subsidence and created high values for nature, while intensive agriculture was no longer possible.

Surveys were conducted before and after the workshops. Results from these surveys suggested that participants find the traffic light presentation easy to understand and are able to use feedback provided by the tool to perform tasks such as changing land use or changing water management. Results also indicated that no more than three indicators should be presented and that the calculation of the scores should be simple (Eikelboom and Janssen in press). The surveys also showed that eight persons considered information from the Touch Table most important, while also eight persons valued information from participants and seven from the experts present during the workshop. This shows that workshop did well in facilitating exchange of information (Fig. 7.10).

An important objective of the workshops was to bring the strategies to life. It is remarkable that only a small proportion of the participants think that Business *as usual* is the most realistic. This proportion is even smaller after the workshop (Fig. 7.11b).

Feedback from the participants indicated that they found the exchange of information very important. The opportunities to switch between maps to allowed learning by doing were considered useful. Participants reported an increased understanding of these relations and a raised awareness of the perspectives of other participants. These results are consistent with studies that stressed that presenting spatial information through interactive map interfaces during planning workshops facilitated exchange of views on spatial decision problems (e.g., Andrienko et al. 2007; Bacic et al. 2006).

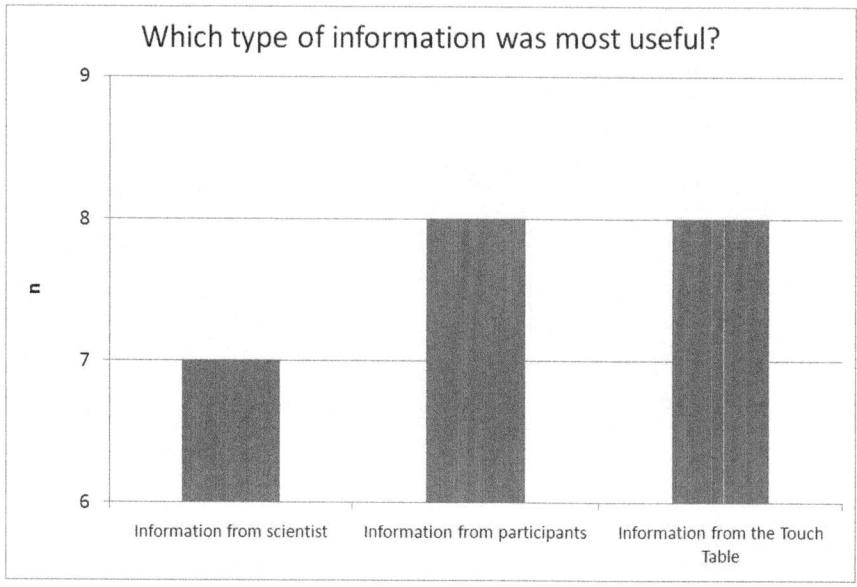

Fig. 7.10 Participants evaluation of the usefulness of different types of information (n=25)

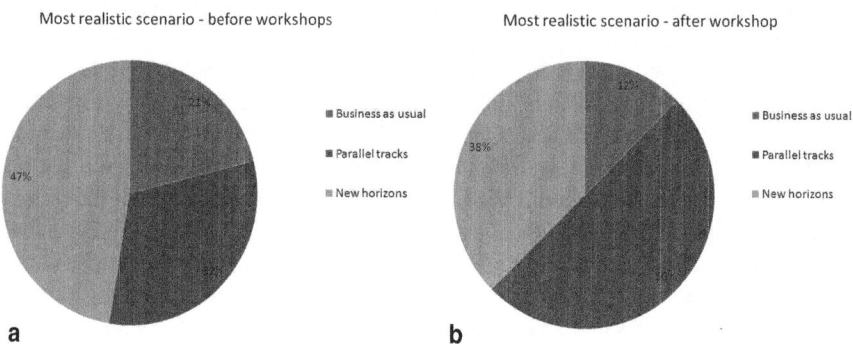

Fig. 7.11 Most realistic strategy according to the participants before (*left*) and after (*right*) the workshop (n=25)

7.5 Conclusions

The decision support tools provided in the design workshops combined maps, a simple simulation model, an evaluation tool, a design tool, basic GIS functions and a shared interactive map interface. Maps were used to obtain feedback from participants. The interactive map interface made it possible to learn about the complex relations in the model by trial and error.

Organizing policy workshops is a learning process that requires substantial effort in terms of preparation, logistics and technical challenges. Preparation is especially crucial as there should be no technical problems and all relevant information must be readily available. There is little tolerance from the participants for technical or methodological errors.

A first lesson is that in the Netherlands it is not so easy to convince policy-makers to include workshops in their planning process. There is fear for the unknown and fear that bringing the problem outside the usual setting would change the level of control on the process. In this process the workshops were actively supported by the Province and Water board. Results of this phase are currently being used as the starting point of the next phase of the process. This phase will involve selection of the most appropriate strategy for each region.

A second lesson is that selection of the participants is essential. Participants of the workshops were invited by the Province and Water board of Friesland. Although this increased commitment of the participants it was not always easy to attract participants to the workshops. It proved important to ensure that there was some benefit for all the participants invited and that the workshop was a positive experience. The main advantage of the presented approach was to get different types of people in the same conversation to optimize opportunities for knowledge exchange between different expertise.

The research results suggest that the approach depends highly on a cooperative attitude by participants. This worked well in a Dutch context as it suits the Dutch consensus-oriented way of decision-making. Other examples of this approach can be found in Arciniegas and Janssen (2012) and Alexander et al. (2012). However, it is uncertain if the same approach would also work in contexts of sharp conflict or with a more power-based style of decision making. For this study the use of geodesign as part of interactive workshops proved to be a useful instrument for facilitating group work around spatial information. Using interactive maps as tools for communication and interaction proved to be an effective method to support planning meetings with stakeholders with different backgrounds.

References

Andrienko, G., Andrienko, N., Jankowski, P., Keim, D., Kraak, M. J., Maceachren, A. M., & Wrobel, S. (2007). Geovisual analytics for spatial decision support: Setting the research agenda. *International Journal of Geographical Information Science, 21*(8), 839–857.

Alexander, K. A., Janssen, R., Arciniegas, G. A., O'Higgens, T. G., Eikelboom, T., & Wilding, T.A. (2012). Interactive Marine Spatial Planning: Sitting tidal energy arrays around the Mull of Kintyre. *PLoS ONE, 7*(1), 1–8.

Arciniegas, G. A., & Janssen, R. (2012). Spatial decision support for collaborative land use planning workshops. *Landscape and Urban Planning, 107*(3), 332–342.

Arciniegas, G. A., Janssen, R., & Omtzigt, A. Q. A. (2011). Map-based multicriteria analysis to support interactive land use allocation. *International Journal for Geographical Information Science, 25*(12), 1931–1947.

Bacic, I. L. Z., Rossiter, D. G., & Bregt, A. K. (2006). Using spatial information to improve collective understanding of shared environmental problems at watershed level. *Landscape and Urban Planning, 77*(1–2), 54–66.

Burrough, P., & McDonnell, R. (1998). *Principles of geographic information systems.* Oxford: Oxford University Press.

Eikelboom, T., & Janssen, R. (2013). Interactive spatial tools for the design of regional adaptation strategies. *Journal of Environmental Management, 127,* S6–S14.

Eikelboom, T., & Janssen, R. (in press). *Comparison of geodesign tools to communicate stakeholder values.* Group Decision and Negotiation.

Geertman, S. C. M., & Stillwell, J. (2009). *Planning support systems: New methods and best practice.* New York: Springer.

Janssen, R., Eikelboom, T., Brouns, K., Jansen, P., Kwakernaak, C., & Verhoeven, J. T. A. (2013). Verslag workshops Friese Veenweidevisie, Kennis voor Klimaat, Utrecht (in Dutch).

Longley, P. A., Goodchild, M. F., Maguire, D. J., & Rhind, D. W. (2005). Geographic information systems and science. New York: Wiley.

Pelzer, P., Arciniegas, G., Geertman, S., & Kroes, J. (2013). Using MapTable to learn about sustainable urban development. In S. Geertman, F. Toppen, & J. Stillwell (Eds.), *Planning support systems for sustainable urban development* (pp. 167–186). Berlin: Springer.

Pettit, C. J., Raymond, C. M., Bryan, B. A., & Lewis, H. (2011). Identifying strengths and weaknesses of landscape visualization for effective communication of future alternatives. *Landscape and Urban Planning, 100*(3), 231–241.

Sieber, R. (2006). Public participation geographic information systems: A literature review and framework. *Annals of the Association of American Geographers, 96*(3), 491–507.

Steinitz, C. (2012). *A framework for geodesign.* Redlands: Esri press.

Willows, R. I., & Connell, R. K. (2003). Climate adaptation, risk, uncertainty and decision-making, UK-CIP technical report.

Chapter 8
Geodesign to Support Multi-level Safety Policy for Flood Management

Sanneke van Asselen, Henk J. Scholten and Luc Koshiek

8.1 Introduction

Low-lying deltas are increasingly vulnerable to flooding. This vulnerability is the result of different factors such as global sea-level rise, variability in river discharge and rainfall intensity, storm surges, soil subsidence due to sediment compaction, sediment trapping in upstream reservoirs and floodplain engineering (e.g., land reclamation). Moreover, many of the world's largest deltas are densely populated and heavily farmed: close to half a billion people live on or near deltas (Syvitski et al. 2009). The Rhine-Meuse delta in The Netherlands is an example of such a densely populated low-lying delta. In this area floodplains are embanked, land is reclaimed (polders) and river discharge and sediment transport are influenced and controlled by dams and reservoirs. Capital and inhabitants in the floodplains need to be protected against flooding. In The Rhine-Meuse delta this has so far been mainly done by taking preventive measures: testing, standardizing and reinforcing dikes.

In recent years, the use of a multi-level safety approach for flood management is increasingly discussed in The Netherlands (Wit et al. 2010). In a multi-level approach, flood management aims to reduce flood risk by both preventive and effect-restrictive measures (Kolen et al. 2012; Leskens et al. 2013). The objective of preventive measures is to reduce the flood probability (level 1). Effect-restrictive measures aim to reduce effects of floods, and comprise measurements related to both spatial planning (level 2) and crisis management (level 3). Examples of spatial planning measurements are elevating roads and (new) residential

S. van Asselen (✉) · H. J. Scholten
Geodan, President Kennedylaan 1, 1079 MB Amsterdam, The Netherlands
e-mail: sanneke.van.asselen@geodan.nl

H. J. Scholten
e-mail: henk.scholten@geodan.nl

L. Koshiek
Hoogheemraadschap Hollands Noorderkwartier, Bevelandseweg 1,
1703 AZ Heerhugowaard, The Netherlands
e-mail: l.kohsiek@hhnk.nl

areas, and building secondary dikes in polders (compartmenting). In the case of managing a disaster, technology can help to provide, collect and store essential information, and share this information among people and organizations involved. For example, in case of a flood it is very efficient to quickly visualize the extent and travel speed of a flood, inundation depths, population density, and the location of for example vulnerable objects and evacuation possibilities. A net-centric approach enables to share actual information online that is accessible for different organizations involved in a disaster (e.g., Bharosa et al. 2009; Santen et al. 2009). They should see the same up-to-date map with the boundaries of a disaster, important locations such as the location of a dike breach, locations of ground staff, evacuees or possible casualties. This knowledge is described as situational awareness. It is important that the information, both spatial and textual, is up-to-date and can be relied upon. The information should be synchronized among all organizations as soon as a change occurs. This shared actual view of a disaster is called the Common Operational Picture (COP). A COP facilitates collaborative planning of operations and assists all staff levels to achieve situational awareness (Neuvel et al. 2012).

Because of the complexity of a densely populated delta such as the Rhine-Meuse delta, where many different processes and stakeholders are involved, designing a flood management plan is a big challenge. Points of discussion in this context usually include the type of investments (preventive and/or effect-restrictive measurements), cost-efficiency and the validity of predictions of natural processes and flood risks (Wit et al. 2010). The geodesign framework may be used to efficiently design and implement a secure flood protection plan, that is acceptable for all stakeholders. In short, geodesign is the process of change (geography) by design (Steinitz 2012). The framework, and how it can be used for flood management, is explained in more detail in the next sections.

8.2 Geodesign

8.2.1 *The Geodesign Framework*

Geodesign can be described as the development and application of design-related processes intended to change a (geographical) study area (Steinitz 2012). The framework is based on six questions:

1. How should the study area be described?
2. How does the study area function?
3. Is the current study area working well?
4. How might the study area be altered?
5. What difference might the changes cause?
6. How should the study area be changed?

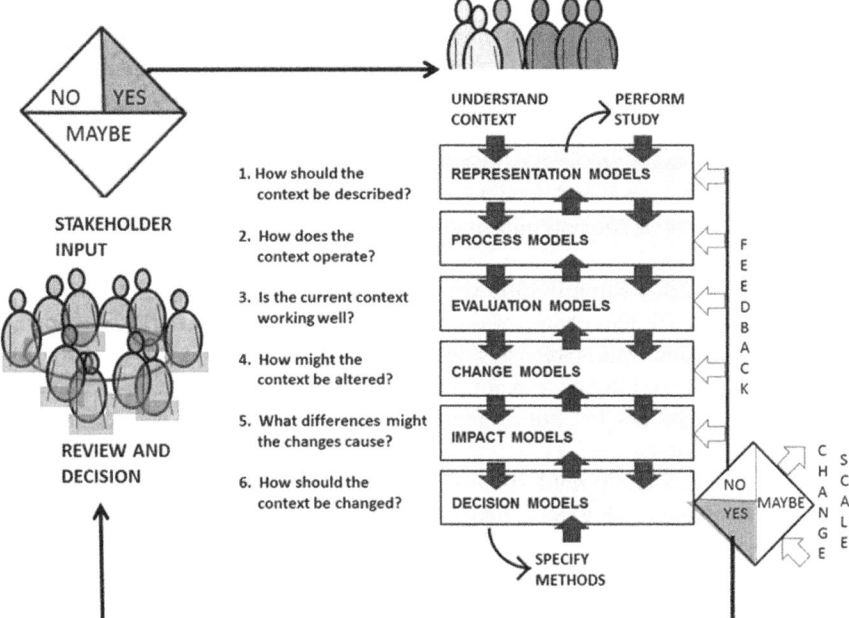

Fig. 8.1 The geodesign framework. (Steinitz 2012)

Solving these questions is necessary for many design-problems in the world. In many cases these problems are complicated, because processes act on different spatial scales, many actors are involved that have conflicting views and needs, and problems are not well understood. To ensure the best solution, people from different disciplines should collaborate, like people from the place, scientists, design professionals and technologists. The geodesign framework effectively facilitates this collaboration. The different steps (questions) are solved iteratively, allowing to learn from previous steps and from feedback from stakeholders (Fig. 8.1).

By visualizing different types of data, processes, scenarios and impacts, the geodesign framework gives one integrative picture of reality. Impacts of changes in the landscape can immediately be explored and visualized.

8.2.2 From Past to Future—The Noord-Holland Case–Study

The emergence and use of the geodesign framework is closely related to increasing human impact on the landscape and the increasing availability of data and tools. These factors, together with a changing flood perception through time, makes the geodesign framework increasingly valuable for developing flood management plans. This is illustrated by three palaeogeographical maps representing three past situations in the province of Noord-Holland, The Netherlands.

Figure 8.2A shows the situation around the year 800 AD. At that time, human impact was still minor and only some simple situational maps were available (Fig. 8.3). People described the landscape in a simplistic way, for example distinguishing 'wet areas' and 'dry areas'. The lower floodplains and peatland areas were often wet. The dunes and coastal barriers along the coast were drier areas. This situation may be described as *nature designed*. If people wished to deal with flooding their main options or were to retreat or defend, for example by building small dikes or artificial hills. They did not yet do any impact analyses, although soil subsidence started to become a problem.

In the period of 800 until 1500 the landscape has changed by human impact (Figs. 8.2B and 8.3). This change was mainly caused by drainage of the land, which lowered the ground water table causing soil subsidence by compaction and oxidation of peat. This resulted in a major loss of land. At the other side people started to take defending measures. For example, the West-Friesland dike has been built in the thirteenth century.

In the period of 1500 until 2000 the change of the landscape was dominated by huge reclamations (Fig. 8.2C and 8.4). This was mainly done to fulfill the increasing demand for agricultural production, and was facilitated by new techniques (e.g., wind mills). Along with the increase in population the wish for protection grew stronger, and people continued reclaiming land by building (stronger) dikes. Although the problem of compaction was recognized as a major problem, people still did not carried out impact analyses regarding this topic. Decisions were mainly focused on building dikes for land reclamation. Reclamation projects were initiated both from the private and public sector. For example, the Beemster area was reclaimed by rich traders from Amsterdam. Furthermore, more (accurate) data was available, such as more detailed topographical maps and water heights (Fig. 8.4).

At the end of the twentieth century the geodesign framework has changed again, in accordance with increasing knowledge and changing perceptions (Fig. 8.5). Much detailed data is available, on a range of themes such as climate, water, soil and demography. Also, more knowledge of and insight into landscape processes allows us to describe different aspects of the landscape more accurately, e.g. by models. Besides flood protection more wishes have arisen such as building with nature (Vriend and Koningsveld 2012). In addition, perceptions have changed as flood protection is more and more based on risk acceptance (IPDD 2013). Besides building stronger dikes, scenario's include increasing the pumping capacity, water retention, sand supply and nature development. Impact analyses allow to assess the consequences of such flood protection measures on for example society, economy and nature. Decisions focus on either (1) increasing protection levels (preventive measures like strengthening dikes), (2) a 'safe' spatial planning, or (3) improving emergency response. This is a multi-level approach which is further explained in Sect. 8.3.

The change in flood management systems and perception in The Netherlands is driven by a number of historical events (IPDD 2013). In the period 1860–1910 river regulation was focused on ensuring sufficient flow and discharge (e.g. by dredging), to prevent upstream floods and enable ships to pass through. These measures were hence mostly driven by economic reasons.

8 Geodesign to Support Multi-level Safety Policy for Flood Management 121

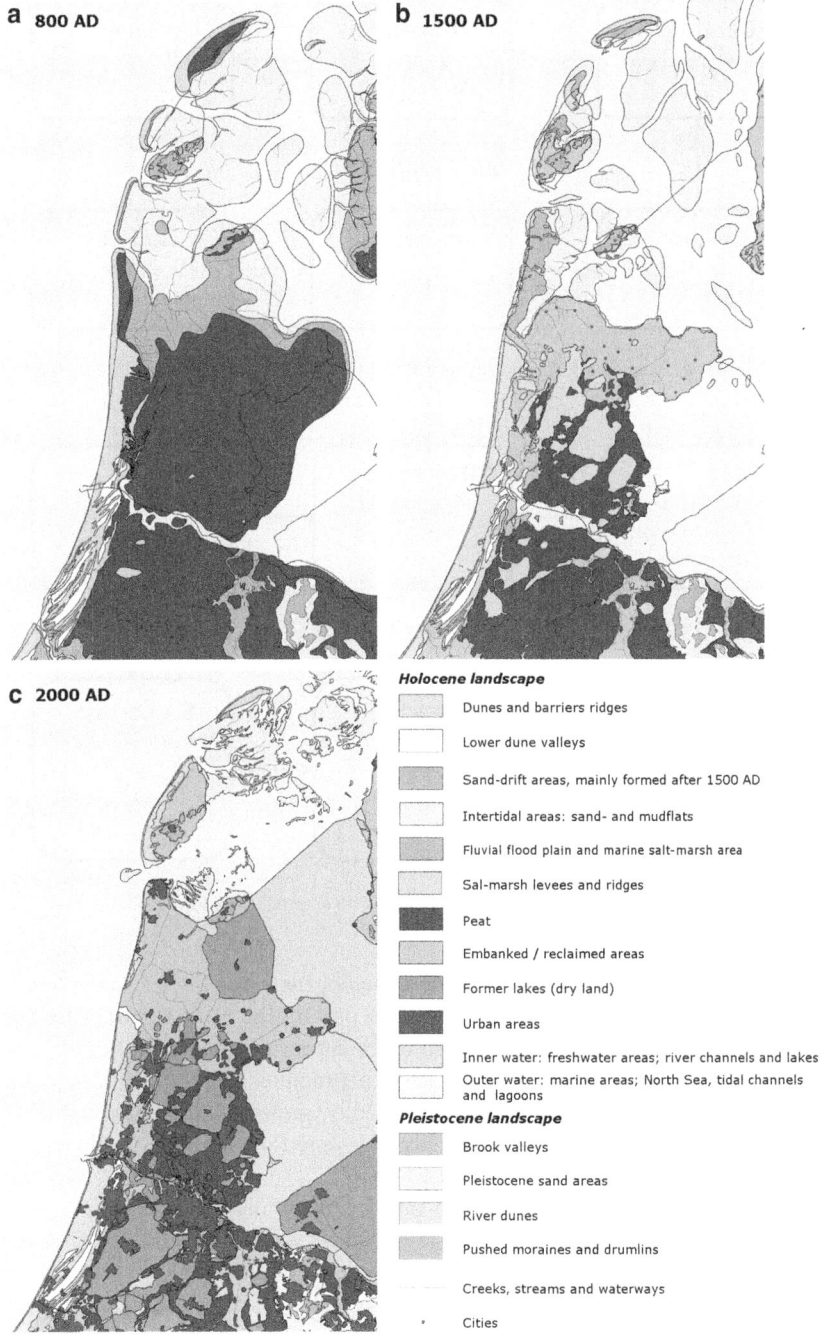

Fig. 8.2 Palaeogeographical maps of the year (**A**) 800 AD, (**B**) 1500 AD and (**C**) current situation. (From Deltares)

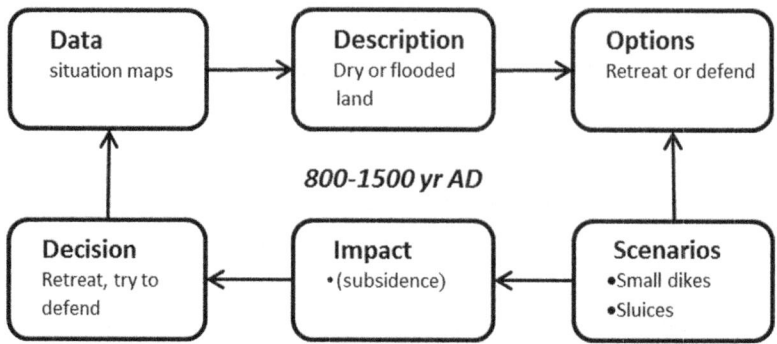

Fig. 8.3 Geodesign framework for the period 800–1500 yr AD in Noord-Holland

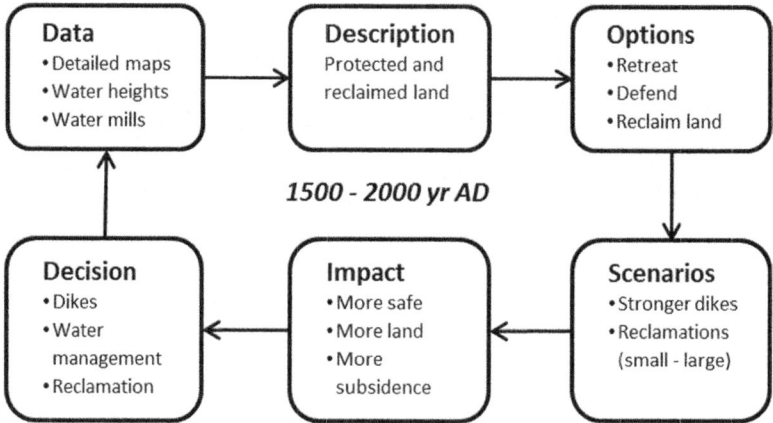

Fig. 8.4 Geodesign framework for the period 1500–2000 year AD in Noord-Holland

An important event *that influenced flood management in The Netherlands* was a flood catastrophe in 1953 in the southwestern part of The Netherlands. This flood was a consequence of badly maintained dikes during the preceding decades. The flood triggered an extensive flood protection program, the *Deltawerken*, which closed the open connection to the sea and thereby protected the hinterland (IPDD 2013). The main objectives of the *Deltawerken* were flood protection and regional economic growth.

After flood events in 1993 and 1995 the maximum allowed discharge capacity, a standard used to 'design' rivers and river dikes, was increased from 12,000 to 16,000 m^3/s (IPDD 2013). Furthermore, more attention was given to nature values and environmental quality. This change in perception resulted in the *Ruimte voor de Rivier* program ('Room for the River'), which focused on water safety along with spatial and nature development. The main measure taken was widening river beds, thereby increasing the discharge capacity.

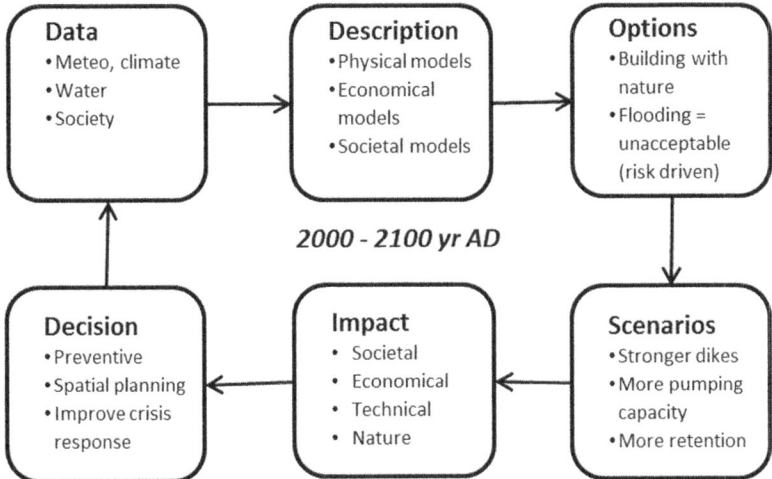

Fig. 8.5 Geodesign framework for the period 2000–2010 yr AD in Noord-Holland

In 2009 the *Deltaprogramma*[1] (Delta program) started. This program deals with long term climate uncertainty regarding the water system and water safety in the twenty-first century. Major decisions regarding flood safety standards and fresh water supply will be taken in the summer of 2014.

8.3 Geodesign and Multi-level Safety for Flood Management

In this chapter it is illustrated how the geodesign framework can be used to design a multi-level safety policy for flood management (prevention, spatial planning, crisis management). The different steps of the geodesign framework are described in the next paragraphs.

8.3.1 Data Inventory

The first step involves collecting data that describe the water system, topography and the built environment. For example, the location, type and area of buildings can be visualized based on the BAG[2], which is a Dutch governmental key register and is open data. Another key register is the Top10NL[3], which provides

---

[1] http://www.rijksoverheid.nl/onderwerpen/deltaprogramma.

[2] (In Dutch) Basisregistratie Adressen en Gebouwen: https://www.kadaster.nl/bag.

[3] (In Dutch) https://www.kadaster.nl/web/artikel/productartikel/TOP10NL.htm.

information regarding infrastructure (roads, railway, waterways) and terrain. Data on dikes and other water structures is mostly provided by water boards. Lithological data, which can be used to for example assess the strength of dikes, is provided as open data by the *DINOloket* of TNO[4]. Another important dataset is the Digital Elevation Model of The Netherlands (AHN2[5]; Fig. 8.6). This can be used to calculate, using flood models, which parts and/or buildings in an area will be flooded and which parts stay dry. More generally, it shows low-lying areas that are more vulnerable for flooding. In the province of Noord-Holland, more than 70 % of the area is lying below sea level and therefore vulnerable for flooding (Fig. 8.6). Regarding crisis management datasets such as the location of vulnerable buildings and demography are important. These can be obtained by the IPO Risk Map[6] and the CBS Statistics Netherlands[7] respectively.

The static information is used in all levels of the multi-level safety approach. Storing data in a Spatial Data Infrastructure (SDI) is vital for creating interoperable systems and enabling exchange of geographical data and information (Williamson 2004). This is a critical aspect of risk and emergency management activities (Neuvel et al. 2012).

8.3.2 Processes

In this step processes that are part of or influence the water system are defined. The following models and data can be used to describe these processes:

- Groundwater and surface water models.
- Flooding and calamity models.
- Information on precipitation (time-series).
- Geological composition (for example, vital information for assessing the strength of dikes)
- Location and functioning of dikes, water pump stations, etc.
- Climate models.

The collected data and model output can be used to create different views of the water system. For example, by running a flood model multiple times for dike breaches at different locations and combing the resulting inundation maps with digital elevation maps, it can be specified what are high risk areas (those that are in all or most cases affected by the flood). Such areas deserve extra attention during a crisis.

The collected data, and information obtained from model runs and other analyses, can be uploaded in a net-centric crisis management system such as Eagle[8] CMS

---

[4] www.dinoloket.nl.

[5] (In Dutch) www.ahn.nl.

[6] (In Dutch) www.risikokaart.nl.

[7] www.cbs.nl.

[8] http://www.geodan.nl/producten/eagle-cms/#t_introductie_tab.

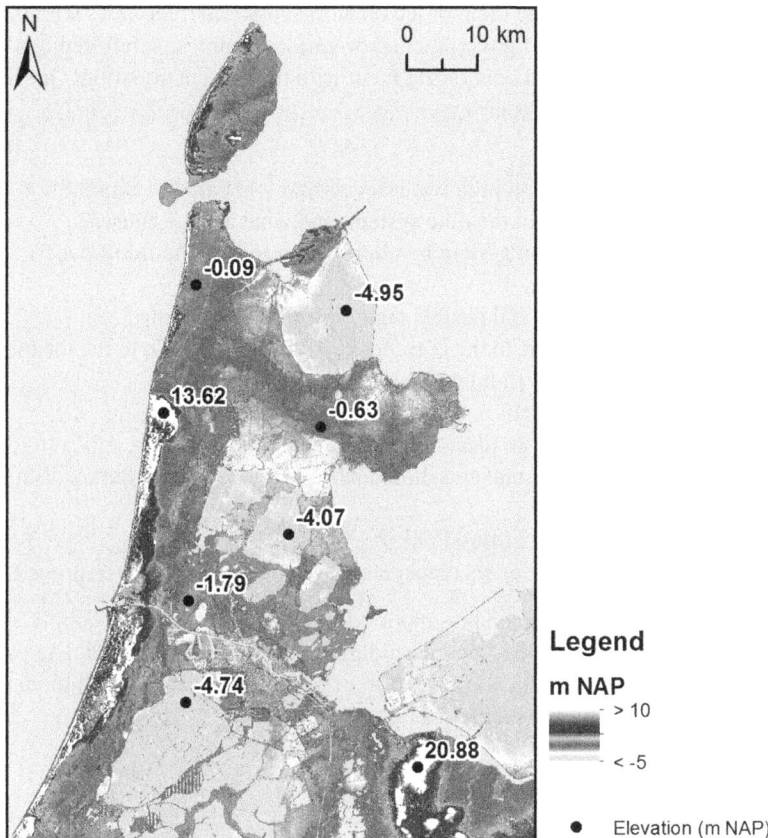

Fig. 8.6 Digital Elevation Model of the province Noord-Holland, clearly showing the location of low-lying polders

(Scholten et al. 2008). This system serves as a platform to work effectively in a net-centric way, to share data and to realize a Common Operational Picture. A huge advantage of Eagle CMS is its use of interactive maps showing actual geographical information that is part of the Common Operational Picture. Spatial data displayed on a map with the relevant context and background information is a huge asset in understanding the current situation, allowing all users to make faster, better decisions and react more swiftly to changes in the disaster situation.

8.3.3 Capability and Sustainability

The third step in the geodesign framework concerns the question whether or not the current water system functions in accordance to our desires regarding economy, sustainability, education and nature, and what are the options to alter the system. Undesired processes related to all three levels need to be identified and acted upon.

This can be done by assessing them based on workshops and interviews with different stakeholders, like (local) governments, private companies, safety regions (fire rescue, hospital, police), and comparing them with the current situation. Questions to be asked are for example (per level):

- Level 1: Prevention
  - Are there problems with high water levels and what are the causes?
  - Are there weak spots in the dike system, and what are the causes?
  - How can the protection system be improved to reduce the flood risk?
- Level 2: Spatial planning
  - What are future plans and desires regarding spatial planning?
  - How does this compare to the current situation, for example to the location of areas that are prone to flooding?
- Level 3: Crisis management
  - What is the outline of an ideal crisis management plan?
  - Which items are important and should be visualized (evacuation roads, hospitals, etc.)?
  - Are inhabitants aware of flood risks?
  - How could the current crisis management plan (if it exists) be improved?

Outcomes of such workshop can be supported by data analyses. For example, weak spots in dikes may be identified based on digital elevation models and data on the composition of dikes. Like mentioned before, flood models can be used to identify areas that are especially vulnerable for floods.

8.3.4 Scenarios

In this step different scenarios regarding all three levels are developed and visualized. Examples of measures, which can be visualized in a geographical information system, are:

- Preventive measures:
  - Strengthening/improving a dike.
  - Building new dikes.
  - Increasing pumping capacity.
- Spatial planning measures:
  - Elevating objects (roads, buildings).
  - Compartmenting polders (building extra small dikes in a polder).
  - Sustainable (new) residential/industrial areas.
- Crisis management:
  - Elevated evacuation roads.
  - Program for vertical evacuation.
  - Developing a net-centric crisis management information system (e.g. Eagle CMS).

The scenarios can be visualized on different platforms (e.g. desktop, mobile device, touch table), which allows collaboration and discussion between different stakeholders. Also, stakeholders can propose additional scenarios.

8.3.5 Impact Analysis

Impacts of the different scenarios are analyzed in step 5 of the geodesign framework. Preventive measures aim to reduce the probability of a dike failure, by counteracting different failure mechanisms like overflow, piping, heave and instability of dike parts[9,10,11]. In The Netherlands, the change in probability of a dike failure, as a result of preventive measures, is estimated based on manuals of testing the safety of dikes ('Leidraad toetsen op veiligheid regionale waterkeringen' (STOWA 2007), and 'Voorschrift Toetsen op Veiligheid' (MVW 2007). These manuals are used for the six-yearly assessments of dikes in The Netherlands, based on various methods and models to calculate hydraulic parameters.

Effects of spatial planning measures may be estimated and visualized using a flooding model and combing the result with geographical data describing the topography (e.g., Digital Elevation Models, topographical maps). A 3D flood model has recently been developed by the 3Di consortium[12]. This model is designed to rapidly calculate and visualize the progression of a flood in 3D. In the model, multiple geographical data layers are combined and translated into hydraulic and hydrologic parameters, which are used for model calculations. It is possible to model interactively. This means it is possible to pause model calculations, which allows a stepwise analysis (and possibly adaption) of model results. An example of a model result visualization is presented in Fig. 8.7. Such visualizations are extremely important for a good communication during a flood.

Model result visualizations show the impact of a flood on spatial planning. Also, inundation depths can be calculated and used to assess the inundation depth of buildings. This information, in combination with data on the type of building and demography, may be used for estimating flood damage. Regarding crisis management, such information may be used for assessing possibilities of for example vertical evacuation, and whether or not it is possible to use roads. Vertical evacuation can be a very effective evacuation strategy in case preventive evacuation is not possible (Kolen and Helsloot 2012).

Integrating the 3Di flood model with an incident and crisis management system like Eagle CMS allows to directly visualize flood propagation and inundation depths during a flood and to act upon unexpected situations, for example if a second dike breach occurs. In such a case evacuation routes might have to be changed quickly.

---

[9] http://www.enwinfo.nl/asp/content.asp?niveau=2&DocumentID=4 .

[10] www.helpdeskwater.nl/publish/pages/4942/trob-h06.pdf.

[11] http://www.dijkverbetering.waterschaprivierenland.nl/dijkverbetering/waarom.

[12] www.3di.nu.

Fig. 8.7 Visualization of a 3Di model run. (From: www.3di.nu)

The main entry of Eagle CMS is a map, showing the current geographical information. The system contains functionality to edit both spatial data, such as incident locations or the extent of a flood, and textual data on a variety of relevant subjects. There are also functions for analysis of data, messaging, and issuing orders to ground staff.

8.3.6 Decision

Based on the results of previous steps decisions can be made. One decision could be to go for a higher level of prevention. Another decision could be to situate vital infrastructure more above sea level, or built an extra compartmentation dike. Another possibility is to improve the calamity response using a crisis management system like Eagle CMS. Noord-Holland will probably decide for a combination of these possibilities.

8.4 Multi-level Safety Measure Examples

8.4.1 Compartmenting Polders

A well-known example of an effect-restrictive spatial planning measure for flood protection is compartmenting polders by building an extra dike. In Fig. 8.8 a digital elevation model and inundation depths simulated by the 3Di model are shown of

Fig. 8.8 A digital elevation model and inundation depth, simulated by 3Di, of an area in Noord-Holland

a part of Noord-Holland. Low-lying areas (dark green) and higher elevated former river channels (yellow) are clearly visible. A simulation of a flood initiated by a dike breach (red arrow) shows that the area north of the secondary dike, running from Hoorn to Enkhuizen, is not affected by the flood even though this is a low-lying area. This illustrates the protective functionality of secondary dikes within polders.

8.4.2 Disaster (Flood) Management

The Waterboard of Noord-Holland is implementing a new version of Eagle (Neuvel et al. 2012). This system facilitates sharing of actual geographical and textual information during a flood, which is vital for taking operational decisions. Thereby, integrating 3Di and Eagle facilitates to quickly visualize and act upon unexpected situations like a second dike breach. Also, 3Di can create different views (maps) of water system situations, which can directly be shared among the different parties by using Eagle CMS. It is likely that such a system will improve crisis management. Still, there are a number of challenges in managing a disaster, because the nature of a disaster management system is substantially different from a day-to-day information system. Eagle CMS that is currently being developed intends to meet all of these challenges, which are at the same time challenges of each geodesign project:

Situational awareness: During a crisis it is very important that all organizations involved are aware of the situation on the ground in the disaster area (boundaries of the disaster area, location of specific incidents, etc). Nearly all the data collected or needed in a disaster management information system have an important spatial component. It is imperative that all staff have access to the same, actual geographic information.

Real-time location awareness: When trying to manage a disaster, it is especially important to know at any time the real-time locations of staff, citizens, victims, volunteers or response teams. But also it is crucial to know the status of the most important assets (the dikes themselves, but also the pumping systems, the electricity hubs, etc). Tracking and tracing of assets, people and equipment is essential. The ability to count potential casualties in affected areas is also extremely helpful.

Sharing data among different organizations: When a disaster occurs, various organizations will respond instantly to manage and contain the impact of the disaster in situ. Other organizations provide information or knowledge, such as governmental or meteorological institutes, or utility companies. Those different organizations should work together and this requires management of communication and exchange of information. All data (spatial, textual, imagery) should be shared between all organizations involved, who have a need to know of such information.

Large data flows: Often there is a lot of data that needs to be made available. The data can be static like roadmaps, as well as dynamic like meteorological data and operational data. This information has to be aggregated and/or filtered depending on the type and scale of the disaster, and the amount and type of organizations involved.

Different networks: Organizations have often different physical networks. There are usually several LAN, WAN or mobile networks for emergency response teams, each with their own management and restrictions. Information should be shared across the boundaries of these networks. The organizations will already have networking equipment for their daily work and are generally reluctant to invest in completely new equipment for emergency response only.

Different levels: Organizations can be involved in different levels: strategic, tactical and operational levels, and each have specific requirements for information types (reports, maps, images, videos, etc.) and scope (generalized, detailed, subject).

Unreliable network connections: In the case of a disaster, the network connections can be unreliable. The network can be overloaded because of large data flows and heavy use. The physical network can be severed. A disaster management information system must be able to deal with these situations by ensuring alternative access to data, ensuring a good backup strategy.

8.5 Conclusions

In a low-lying country as The Netherlands, a sound flood management strategy is vital. Flood management perceptions have changed over time in this country. Also, the trends described in this article show a tremendous increase of information through time. In this perspective, decision makers have a more easy job than in the early days. However, when you look from the perspective of possible number of causalities or economic damage by flooding the risks are increased significantly through time. So, although the wealth of information, decisions have more risks and therefore are more complicated.

The multi-level approach is for this moment the Dutch approach to cope with the expected climate change. In this approach prevention will be most important: most capital is still invested in preventive measures (~90%) compared to investments in spatial planning and crisis management (both ~5%). Investing more in spatial planning and crisis management potentially reduces flood damage and/or the risk of being affected by a flood for individuals. Examples are compartmenting polders (level 2) and developing net-centric incident and crisis management systems like Eagle CMS, integrated with the 3Di model.

Changing a (water system into a) multi-level water safety system is facilitated by the geodesign framework, which (1) supports collaboration between different stakeholders, (2) helps to effectively collect and store data and information, (3) define ambitions, potentials and scenarios, and (4) performs impacts analyses. Stakeholders are involved in all geodesign steps and can continuously give feedback. In this way, finally a well-considered decision regarding the water system can be made that is widely supported by the stakeholders. However, having better data, better models, improved approaches for decision-making does not mean that it has become easier to make decisions regarding our highly complex water system.

References

Bharosa, N, van Zanten, B., Janssen, M., & Groenleer, N. (2009). Transforming crisis management: Field studies on the efforts to migrate from system-centric to network-centric operations. *Lecture Notes in Computer Science, 5693*, 65–75.

IPDD. (2013). *Integrated planning and design in the Delta. Nieuwe Perspectieven voor Een Verstedelijkte Delta, Naar Een Methode van Planvorming en Ontwerp*. Delft: Urban Regions in the Delta program.

Kolen, B., Zethof, M., & Maaskant, B. (2012). Toepassing basisvisie afwegingskader meerlaagse veiligheid; een methode om mee te werken in de praktijk. *STOWA Rapport, 66*, 2012–2023.

Kolen, B., & Helsloot, I. (2012). Time needed to evacuate the Netherlands in the event of large-scale flooding: Strategies and consequences. *Disasters, 36*, 700–722.

Leskens, A., Boomgaard, M., & van Zuijlen, C. (2013). *Meerlaagse veiligheid: hoe maken we dat concreet?* Den Haag: Vakblad H_2O, Vakartikelen.

MVW. (2007). *Voorschrift Toetsen op Veiligheid Primaire Waterkeringen*. Den Haag: Ministerie van Verkeer en Waterstaat.

Neuvel, J. M., Scholten, H. J., & van den Brink, A. (2012). From spatial data to synchronised actions: The network-centric organisation of spatial decision support for risk and emergency management. *Applied Spatial Analysis and Policy, 5*, 51–72.

Santen van, W., Jonker, C., & Wijngaards, N. (2009). Crisis decision making through a shared integrative negotiation mental model. *International Journal of Emergency Management, 6*, 342–355.

Scholten, H. J., Fruijtier, S., Dilo, A., & van Borkulo, E. (2008). Spatial data infrastructures for emergency response in the Netherlands. In: S. Nayak & S. Zlatanova (Eds.), *Remote sensing and GIS technologies for monitoring en prediction of disasters*. Berlin: Springer.

Steinitz, C. (2012). *A framework for geodesign. Changing geography by design*. Redlands: Esri Press.

STOWA. (2007). *Leidraad toetsen op veiligheid regionale waterkeringen*. Utrecht: STOWA.

Syvitski, J. P. M., Kettner, A. J., Overeem, I., Hutton, E. W. H., Hannon, M. T., Brakenridge, G. R., Day, J., Vörösmarty, C., Saito, Y., Giosan, L., & Nicholls, R. J. (2009). Sinking deltas due to human activities. *Nature Geoscience, 2,* 681–686. doi:10.1038/ngeo629.

Vriend de, H., & Koningsveld van, M. (2012). *Building with nature. Thinking, acting and interacting differently*. Dordrecht: EcoShape.

Williamson, I. P., Rajabifard, A., & Feeney, M. E. F (Eds.). (2004). *Developing spatial data infrastructures: From concept to reality*. Boca Raton: CRC Press.

Wit de, S., Jongejan, R., van der Most, H. (2010). Niet bij preventie alleen. Simulatie en analyse van een debat tussen voor- en tegenstanders van meerlaagsveiligheid. In: H. van der Most, S. de Wit, B. Broekhand, & W. Roos (Eds.). *Kijk op waterveiligheid*. Delft: Uitgeverij Eburon.

Chapter 9
The Multi-Layer Safety Approach and Geodesign: Exploring Exposure and Vulnerability to Flooding

Mark Zandvoort and Maarten J. van der Vlist

9.1 Introduction

A storm surge in the North Sea on the night of the 1st of February 1953 caused flooding of large coastal areas in the Netherlands, Belgium and the United Kingdom (d'Angremond 2003). This was the onset of the development of a risk based approach for the Dutch flood risk management system (d'Angremond and Kooman 1986). This risk based approach builds upon the formula: risk = probability *consequences (exposure*vulnerability) and is currently also evolving in flood risk management policy in amongst others the United Kingdom (Sayers et al. 2002), Belgium (Kellens et al. 2013), the USA, and Japan (IWR 2011). In this risk based approach geoinformation and analysis can be of vital importance to inform the design task in an area. The design task is to change the landscape towards a more sustainable physical layout of the area with respect to flood risk. Geographical characteristics of an area determine to a large extent the components of flood risk, thus a geodesign approach, seen as "including ways of solving spatial design problems in any way and with any technology" (Steinitz 2012, p. 8), can help assess and determine the effectiveness of flood risk reducing measures.

Past decades, scientists and policy-makers increasingly applied Geographic Information Systems (GIS), made possible by the wide-spread availability of computing power and increasingly larger and more sophisticated hydraulic models. However, in applying GIS for flood risk management mostly the probability side of the equation was emphasized, as is also visible in the advice of the first Delta Committee in the Netherlands (Deltacommissie 1960) and the Dutch guideline for

---

M. Zandvoort (✉)
Wageningen UR Landscape Architecture Group, P.O. Box 47,
6700 AA Wageningen, The Netherlands
e-mail: Mark.zandvoort@wur.nl

M. J. van der Vlist
Wageningen UR Landscape Architecture Group, Rijkswaterstaat,
Wageningen, The Netherlands
e-mail: Maarten.vander.vlist@rws.nl

dike design (TAW 1985). Long since, prevention is the main approach to deal with high discharges and sea levels and it informs the current flood risk safety standards for designing Dutch dikes (Ten Brinke et al. 2008). The incorporation of the consequences of a flood (exposure and vulnerability) in flood risk management is still in its infancy, and surrounded with uncertainties (De Moel and Aerts 2010). The aspects 'exposure' and 'vulnerability' need to be clear, and flood risk managers need insight in the necessary information to integrate these aspects in risk reducing measures. In the Netherlands this integration is stimulated by a discussion concerning the so-called multi-layer safety (MLS) approach which was described in the National Water Plan in 2008 (Min I&M 2008). From the perspective of flood risk, this approach consist of three layers:

1st layer: prevention
2nd layer: sustainable spatial planning, and;
3rd layer: disaster management (id. p. 71).

The aim of layer one is to lower the probability of a flood by means of preventive measures like dikes and dredging river branches. In the second and third layer flood risk managers primarily deal with exposure and vulnerability. In the Dutch context, perception of flood risk is shifting and these two aspects are going to be included in the safety standards for the dikes by differentiating their design (height, width, strength) (Min I&M 2013). As a consequence, geoinformation concerning assets, whereabouts of humans and livestock in the hinterland, and damage on capital by possible floods is necessary. Information which is also highly relevant for measures to reduce exposure and vulnerability. These measures demand more detailed geoinformation about the whole area potentially affected by a flood. The effects of a flood can occur throughout the area, thus, measures are not limited to the dimensions of preventive measures near the river but effective throughout the area.

In order to provide geoinformation for flood risk managers and policy-makers, the multi-layer safety approach is translated into a geographic information tool. This tool can be used to support exploratory studies and analysis of areas, and can be used to consider the relation between probability, vulnerability and exposure reducing measures. Although the tool is still in development (its design component is not yet applicable and it is only based on aggregated information) its use already shows the value of geoinformation and geodesign for flood risk management from a multi-layer perspective. We explore how this tool can be used, based on the current state-of-the-art of geodesign thinking, and attempt to relate the multi-layer safety approach to geo-information and geodesign in order to improve current practice of flood risk management.

The remainder of this chapter focuses on the role geodesign can have in reducing the components exposure and vulnerability of humans and assets to floods. Probability is beyond the scope of this chapter, because the aim to provide insight in the role of geodesign in the aspects vulnerability and exposure is challenging enough. We illustrate this with a Dutch riverine case, 'Rivierenland', wherein we focus on evacuation strategies for healthcare institutions in the case of two different flood scenarios. We look especially at exposure related aspects in assessing and designing evacuation strategies for healthcare institutions, and indicate possible measures for the regional, local and building-specific level of healthcare institutions. We conclude with suggestions for advancing the role of geoinformation and geodesign in reducing exposure and vulnerability for flooding in layers two and three of the multi-layer safety approach.

9.2 Exposure and Vulnerability to Flooding

There are three relevant characteristics of exposure in the context of flood risk. First the duration of exposure to a flood. This can be defined in hours, days or weeks in a year or over a lifetime. The second characteristic is the amount of water humans and assets are exposed to. This can be equaled to the inundation depth, expressed in height. Third, the velocity of the water flow is relevant. The water can be stagnant or flow with tens of kilometers an hour. The velocity also determines the arrival time of a flood and the moment a flood reaches a specific depth.

In geodesign, duration, height and speed are the main characteristics that have to be taken into account when considering a design-based reduction of exposure. For calculating exposure, flood scenarios and the physical lay-out of the exposed area are important. These determine the exposure itself (for example the depth and velocity of a flood), but also the opportunities to decrease exposure for individuals (for example the elevation and location of roads to get away from the flood). In a geodesign approach a tradeoff is possible between scenario altering and exposure reducing measures.

Reducing exposure depends on the time, movability of humans and assets in front of, or through the flood, and availability of resources and transport routes. Regarding real estate, duration, height and speed of the flood[1] are important regarding stability and strength of a construction (Nadal et al. 2010). Measures to reduce exposure on the long term, thus altering the scenario, include compartmentalization, local flood defenses, heightening or strengthening real estate, measures to slow the flood down and reduce height at strategic locations to create buffers (retention). For *ad hoc* exposure reducing measures, time of arrival related to depth needs to be determined. This indicates whether flood risk managers are able to evacuate humans and assets, to reduce exposure by sand-bagging or can take other instant exposure reducing measures. It also indicates the possibility to elevate assets or humans to first or second floors.

The notion of exposure is sometimes equaled with, or being an aspect of, vulnerability (Adger 2006; De Bruin and Klijn 2009; Fuchs et al. 2011), however, we take the definition of Renn (2005) and distinguish exposure and vulnerability accordingly. Renn (2005) defines vulnerability from the perspective of risk and hazard as: "the various degrees of the target to experience harm or damage as a result of the

---

[1] Note that the salinity of water, related to the source of a flood (sweet, brackish or salt), is also important for deterioration of real estate and crops.

exposure" (p. 15). We use this definition of vulnerability[2], in accordance with the use of others regarding flood risk (De Moel et al. 2011; Kazmierczak and Cavan 2011; Sitzenfrei et al. 2011).

Applying the concept of vulnerability to humans, it can be related to individuals (social vulnerability) and the functioning of society which depends on the vulnerability of critical structures (societal vulnerability). Social vulnerability characteristics include the autonomy of individuals and amount and cohesion of societal networks, determining possible help to evacuate or seek protection (Birkmann 2006). For example elderly, children and sick people are more vulnerable since they are more fragile and more reliant on others for help. Thus, they are more likely to be wounded or to die if exposed to a flood. On a societal level vulnerability is determined by the critical structures for the functioning of society. Such critical structures include roads, underground infrastructures for utilities like water, gas and telecom, and critical locations such as homes for elderly, hospitals, power plants and sewage treatment plants.

Societal vulnerability is related to the vulnerability of assets, which includes the ability of a structure to withstand a flood, and the vulnerability and value of (the content of) the structure. Assets can be damage-prone (compared to fragility of humans) or, in a systems perspective, highly valuable for the normal functioning of society or for its functioning during a flood (related to societal vulnerability) (Van de Ven et al. 2011). Vulnerability is the opposite of robustness (Mens et al. 2011) and design for robustness can be regarded as the same as design for a reduction of vulnerability.

9.3 The Design Problem

From the perspective of flood risk, the design problem can be related to the components of exposure and vulnerability. The objective in the geodesign process is to find opportunities to reduce exposure and vulnerability for flood risk management, which can be divided into the domains of spatial planning and emergency planning. By taking exposure and vulnerability into account, better evacuation strategies and physical plans for flood risk management can be created. The aspects which need to be assessed from the exposure side are time of arrival, duration, inundation depth, and velocity of a flood. For vulnerability of individuals, society and assets, flood risk managers need to assess the opportunity for self-reliance regarding evacuation and resources, characteristics of local structures (buildings), characteristics of the physical environment, and the available opportunities to take vulnerability reducing measures on different spatial scales.

---

[2] See for an extensive discussion on the concept vulnerability and its relation to exposure in the risk approach Birkmann (2006).

Table 9.1 Time estimates to evacuate. (Adapted from Barendregt et al. 2005 & Rijkswaterstaat and HKV 2008)

Evacuation process	Minimum time required to initiate evacuation (h)
Decision making	4
Warning population	3
Response time	3
Hospital evacuation	24
Total	34

The geodesign approach is used to visualize, simulate, model and measure flood related exposure and vulnerability in order to explore possible contributions of this approach for decision-making and designing related to flood risk management. We assessed the case of Rivierenland regarding the relevant three components of exposure: time, depth and velocity and related arrival time of a flood, and specified this for three healthcare institutions. We indicate how, for each of the three locations, the exposure related information can be assessed for evacuation strategies in two different flood scenarios. This informs the design of future land use strategies but also current planned evacuation routes.

Two types of information were most important in relation to the assessment. First, the necessary time to decide to evacuate, and actually evacuate healthcare institutions in a region facing a flood. Second the depth until it is still possible to evacuate over land. Estimates for necessary time to decide and inform the public to evacuate an area are provided by Barendregt et al. (2005), and actual evacuation times for the riverine area are modelled by Rijkwaterstaat and HKV (2008). The latter include also the necessary capacity of material to evacuate healthcare institutions and other non-self-reliant people out of an area, and indicate that at least 12 h, but more safely 24 h are needed in order to evacuate all patients to safe areas. Together these numbers are indicated in Table 9.1, leading to at least 34 h for an evacuation after a flood started. Estimates for depth are provided by the Dutch Council for the Living environment and Infrastructure (RLI 2011) (Table 9.2). These are rough estimates as indication for this geodesign exercise.

9.4 Case Study Rivierland, Dike Ring 43

Dike ring area 43, the Betuwe, in Rivierenland, the Netherlands was studied, referred to as 'Rivierenland' in the remainder of the text. The case study area was chosen because of the flooding data available for this area and the characteristics of the area. This allowed a look in to different flood scenarios, providing insight in the main characteristics of exposure to a flood. For our results we partly use the analysis of a student who wrote a master thesis under our supervision (Van der Biezen 2013).

Table 9.2 Remaining traffic possibilities during a flood. (Adapted from RLI 2011)

Water level (m)	Remaining traffic possibilities
0-0,2	Cars are able to move at walking pace
0,2-0,5	People are reachable on foot
0,5-0,8	Military vehicles (and emergency services) are still able to drive
0,8-2	First floor of an institution remains safe
2-5	Second floor of an institution remains safe
5<	Only higher levels remain safe

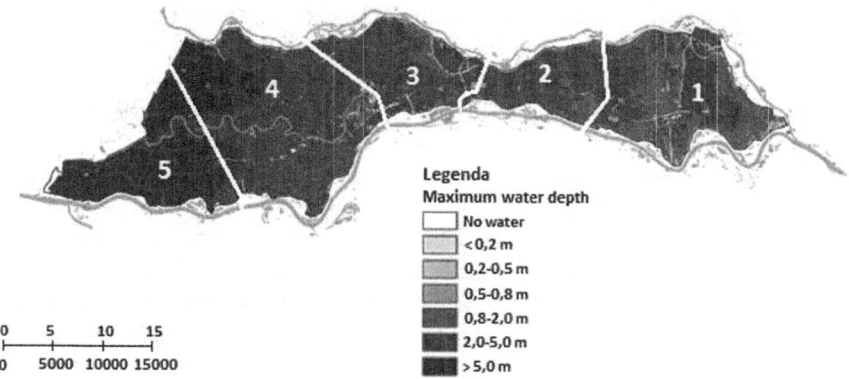

Fig. 9.1 The area with an indication of the downward slope and the five subareas characterized by difference in height (m) and population density. (Province of Gelderland and Min V&W 2010, p. 12)

9.4.1 Rivierenland

The area Rivierenland is part of the eastern riverine area of the Netherlands, and is surrounded by the main tributaries of the Rhine delta (Fig. 9.1). We looked at dike ring 43, which borders the lower Rhine (the east side of this river branch) and the Lek (west side of the river branch) on the northern side, while the south side borders the largest Rhine branch, the Waal. The canal of Pannerdensch is located on the eastern side of the area. On the west side, dike ring 43 is closed by a dike called the Diefdijk, which connects the dikes along the northern and southern river branches. In this area the small river Linge flows from east to west, and the Amsterdam Rhine Canal runs on a north-south axis. This canal is built for transport purposes and a VIb class canal according to the European ECMT classification (ECMT 1992). The canal intersects the area as do some roads and railways. All these intersections affect flood scenarios in the area by hindering free flow in the east-west direction.

Before flood defence systems were built, the area regularly flooded from the river branches. This led to clay deposits and higher banks with sand deposits. Due

Table 9.3 Comparison of key characteristics of each of the five subareas in dike ring 43, Rivierenland. Five point scale indicating the relative importance from a risk perspective. (Province of Gelderland and Min V&W 2010, p. 18)

Area	Population density	Depth	Velocity of flood	Risk of damage	Risk of casualties
1	++	--	++	++	++
2	-	-	+	-	0
3	-	0	0	-	0
4	+	+	+	+	+
5	--	++	-	0	-

to the very fertile soil provided by these deposits, the land between the residential areas is used for agriculture, including high revenue crops. The area can be characterized as open and flat. Its height differs slightly, with a total difference of a little over 10 m.

In a study of the Province and the Ministry of Traffic and Water Management (Min V&W), the area was divided into five subareas (Fig. 9.1), mainly based on the height of the area and population density (Province of Gelderland and Min V&W 2010). Several main urban centres and villages are located in the area with a higher population density in the east compared to the west and an estimated population of 220.000–250.000 inhabitants (Provincie Gelderland and Min V&W 2010, p. 12). The different characteristics of each sub area are indicated in Table 9.3.

9.4.2 Simulating Two Flood Scenarios

A key aspect in determining the flood risk for this area is the slope of the area and the location of a dike breach. Due to this slope, water will always flow to the west side of the area. The location of a dike breach determines the velocity of the flooding. The velocity depends on the characteristics of the area behind the breach, but, more importantly, on the velocity of the water in the river branch (Provincie Gelderland and Min V&W 2010). The Waal on the south side of the area flows roughly three times as quick as the Lower Rhine and Lek on the northern side of the area.

To assess the area's exposure to flooding, we simulated two scenarios using the platform Flood Lizard (Nelen-Schuurmans 2013). Flood Lizard is a GIS based, open source platform with a user friendly (browser based) interface for running flood simulations. With the two scenarios we aimed to demonstrate the benefit of a geodesign process in assessing and managing flood risk related decisions such as moment, type and time until evacuation and to indicate the role geodesign can have in reducing vulnerability and exposure to flooding. In this study, three basic flood parameters (time, depth and velocity) for dike ring 43 are assessed. The relevant information for the decision if, and when to evacuate three healthcare institutions in the area is distilled from the scenarios and translated into principles for evacuation strategies.

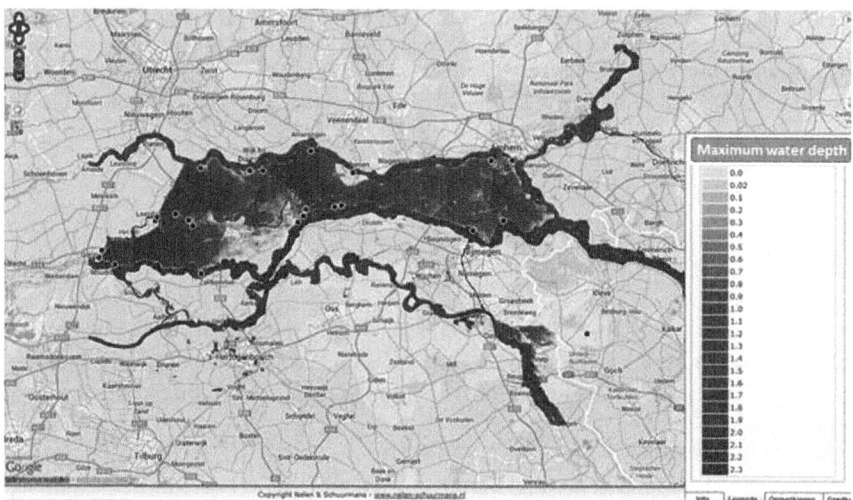

Fig. 9.2 Maximum depth (m) of the Angeren flood scenario (Flood Lizard, Van der Biezen 2013)

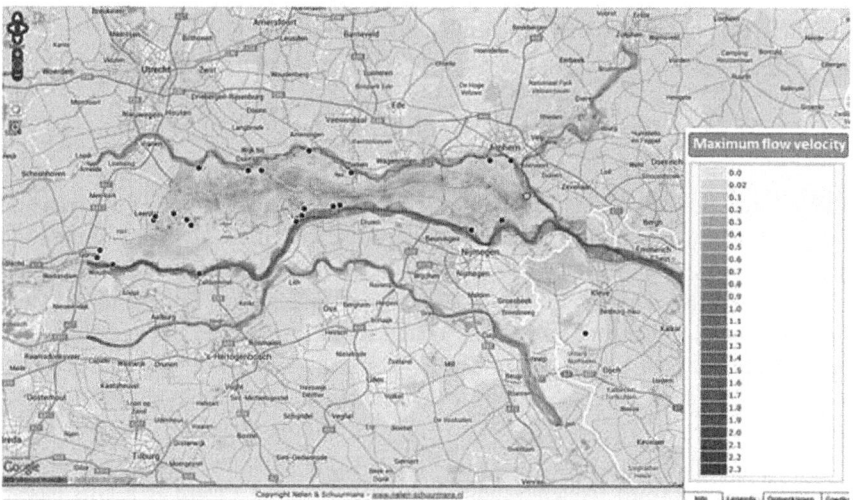

Fig. 9.3 Maximum velocity (m/s) of the Angeren flood scenario (Flood Lizard, Van der Biezen 2013)

The first scenario starts with a breakthrough of the dike near Angeren in the far eastern part of the area. Figs. 9.2 and 9.3 indicate the maximum inundation depth and the velocity of this specific scenario in the whole dike ring area. Due to the small slope of the area, it takes several days before the flood arrives in the west. The peak of the flood arrives in the most western location on the 11th day after flooding has started (Van der Biezen 2013). The flood will diminish because the water

9 The Multi-Layer Safety Approach and Geodesign

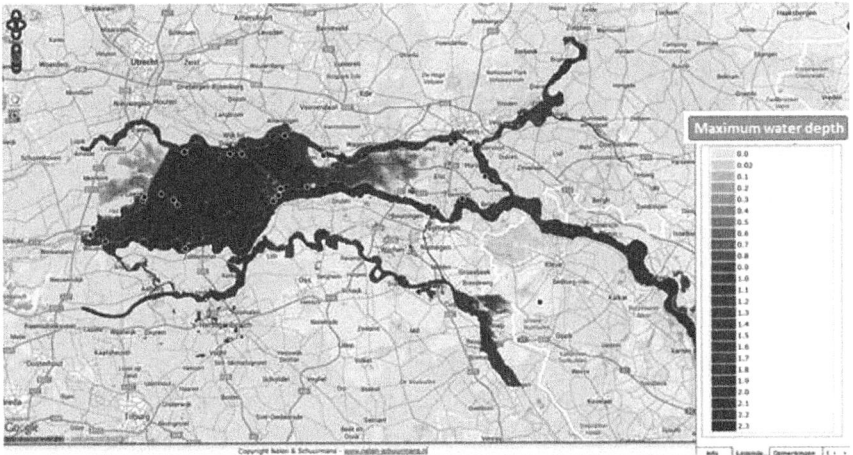

Fig. 9.4 Maximum depth (m) of the IJzendoorn flood scenario (Flood Lizard, Van der Biezen 2013)

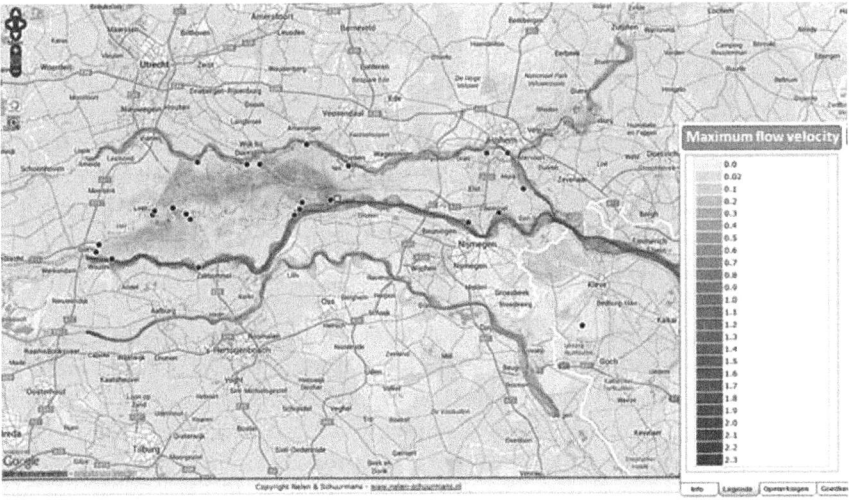

Fig. 9.5 Maximum velocity (m/s) of the IJzendoorn flood scenario (Flood Lizard, Van der Biezen 2013)

infiltrates, and pumps for normal water management, which are located throughout the area, drain the water back into the river system.

The second scenario starts with a breach near IJzendoorn, halfway along the southern side of the area (Figs. 9.4 and 9.5). A breach in the dike along the Waal, the main tributary, will result in a higher velocity. This is visible when comparing the velocity of flooding after a breach at Angeren (Fig. 9.3) with the velocity of the scenario IJzendoorn (Fig 9.3). This will result in higher damage in the vicinity of the dike breach (Kreibich et al. 2009). The flooded area, however, is smaller since the water only floods the western part of the area (Fig 9.4). Compared to the scenario Angeren, the flood from IJzendoorn did not show a wave pattern throughout the area. Instead, several areas are flooded within 5 days up to 2.5 m and remain flooded for several days (Van der Biezen 2013). Since timing and inundation depth are essential for the possibilities for evacuation, the possibilities differ per scenario. A stagnant inundation for several days also has consequences for the damage of assets, since these remain in the water for several days, increasing the impact compared to a short wave which travels on after a day.

9.4.3 Results at Three Healthcare Institutions

From the described scenarios more specific details were distilled for three healthcare institutions. These details include height, time before arrival and duration of the flood. The healthcare institutions are located in Kesteren, Tiel and Culemborg, with Kesteren in the middle of the area located near the Nederrijn-Lek side, Culemborg to the west and located near the Nederrijn-lek, and Tiel adjacent to the Waal. Figure 9.6 shows that it is possible to evacuate the healthcare institution in each situation before inundation becomes too severe for traffic (at 34 h). The figure show a great differentiation between the scenarios, especially with respect to the moment the water starts to inundate the healthcare institutions. The scenario Angeren is less severe regarding both the moment of inundation and depth. In this scenario inundation of each institution starts later, because it is further away from the assessed healthcare institutions. Note that these results only show the depth and arrival time at the location. We did not take into account the effects of depth of flooding on the infrastructure necessary to evacuate patients to safe areas.

Regarding evacuation, emergency managers can choose for horizontal evacuation or vertical evacuation (higher floors or locations and possibility to wait until the flood is over) (Kolen 2013). In all cases horizontal evacuation is possible, regarding depth and arrival time of the flood at each location. This is also the best option regarding depth, except in the scenario Angeren for the location Tiel. In this case (Fig. 9.6c) the water level stays almost all the time below 0.2 m and, since we did not plot activities to reduce the water level or reduce exposure by sandbagging, it seems better to retreat to higher levels in the building if the healthcare institution allows this.

An important aspect in evacuating healthcare institutions is the availability of ambulances to transfer patients to safe facilities. In order to use capacity available in an area efficiently geoanalysis can provide valuable information. By plotting each of the scenarios on the three locations it is possible to see the sequence of flooding (Fig. 9.7 and 9.8). Capacity-planning for the transfer of the hospital can be optimized by evacuating in order of arrival of the flood. For a flood from Angeren (Fig 9.8) the evacuation capacity should first go to Kesteren, then to Culemborg

Fig. 9.6 Maximum inundation depth (m) in Kesteren (**a**), Culemborg (**b**), Tiel (**c**) after a dike breach near Angeren to the east and a flooding from the Waal at IJzendoorn in the middle of the area plotted for 12 days. Traffic possibilities (Table 9.2) are also plotted

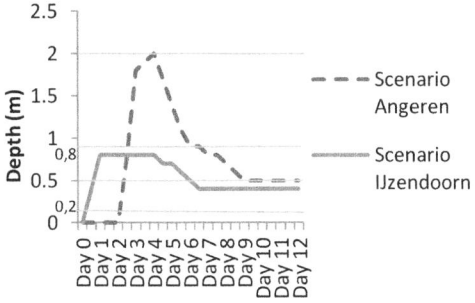

a

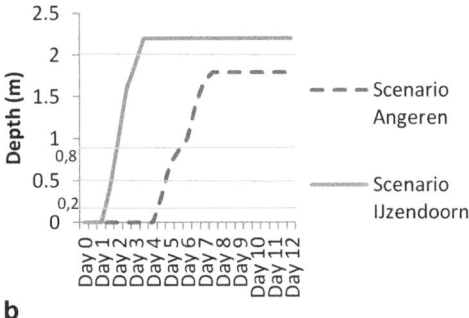

b

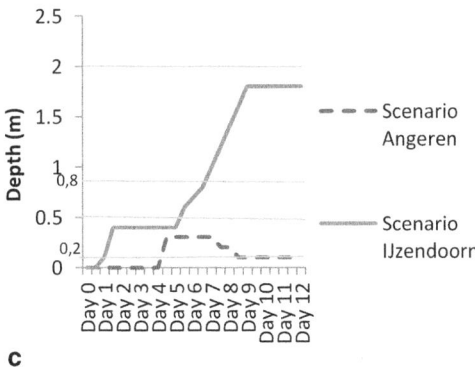

c

and finally to Tiel (if horizontal evacuation is chosen). In the case of scenario IJzendoorn the order is harder to determine since Kesteren is affected earlier but less extreme compared to Culemborg, where the flood inundates the institution later but more severe (Fig. 9.7). In each case Tiel can be evacuated last.

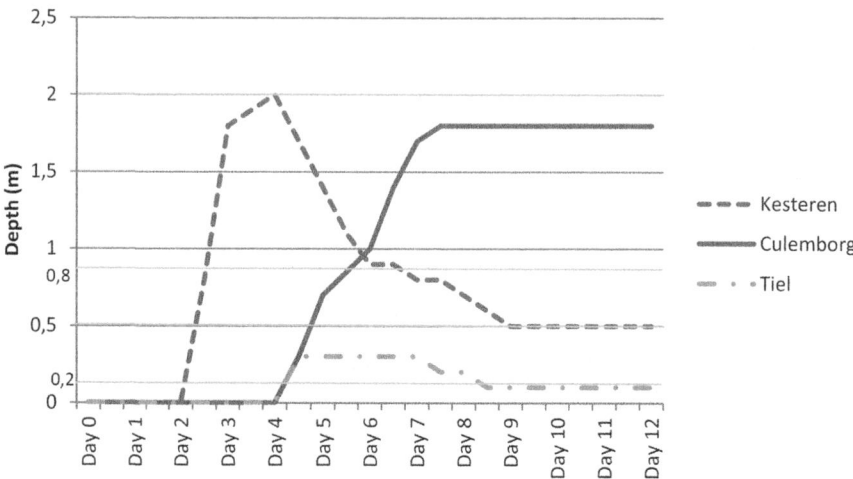

Fig. 9.7 Time sequence of the arrival of a flood (days) and inundation depth (m) from Angeren, relevant for steering evacuation capacity through the region. Traffic possibilities (Table 9.2) are also plotted

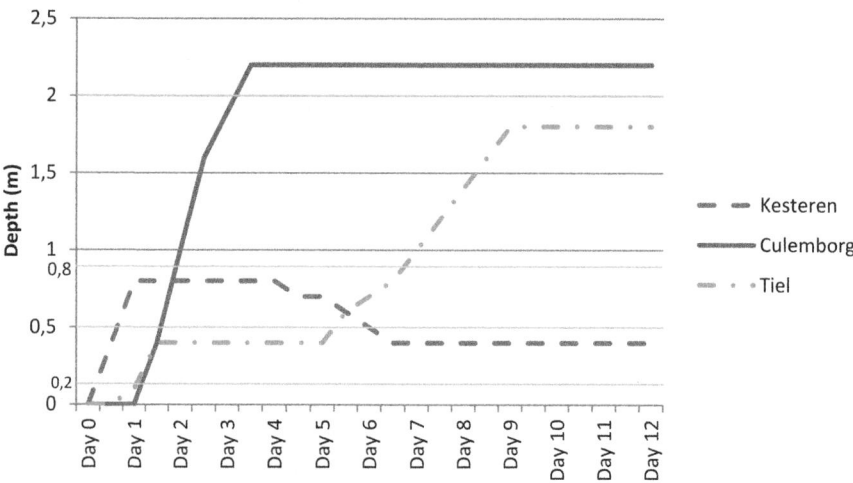

Fig. 9.8 Time sequence of the arrival of a flood (days) and inundation depth (m) from IJzendoorn, relevant for steering evacuation capacity through the region. Traffic possibilities (Table 9.2) are also plotted

9.5 Discussion

In this research, only components of exposure were taken into account. The deficit of not including vulnerability comes pressingly to the fore by comparing Kesteren and Culemborg. The first institution only has a ground floor, thus, the patients can't retreat to a first or second floor, while Culemborg is a large hospital were this is an opportunity. The facility in Kesteren is more vulnerable to flooding compared to the facility in Culemborg. This shows that exposure and vulnerability should go together in deciding effectively about evacuation strategies and other flood risk reducing measures.

Advancement of geoinformation is necessary, and the summarizing figures contain only the best possible estimates at this stage of research. Also, the relevant sector of decision-making differs for each figure. While Fig 9.6 is more important for spatial planners for allocating a new hospital or managers of individual hospitals to assess their susceptibility for a flood (and opportunities to decrease vulnerability), Figs. 9.7 and 9.8 are useful especially for disaster risk managers in case of a flood. Thus, simulations of flood can contribute to preliminary assessment, shaping risk-perception, and advanced design decisions on different spatial scales. It can also inform quick assessments in order to steer capacity for evacuation in case of a disaster. Both uses contribute to a further reduction of risk in case of flooding.

Our analysis is still limited by data concerning the strength of the buildings, the details of the build-up area with the related effects on flooding scenarios and evacuation strategies. It is necessary to have more detailed data about the heights of the objects in the affected regions (the height of roads, location of tunnels, height and vulnerability of buildings and so on) to have full insight in the relation between flooding and possibilities to evacuate from a specific location. Currently it is not yet possible to have a full size assessment of possible evacuation strategies, although our analysis can already inform the creation regarding timing of evacuation, and sequence of evacuation compared to other health care institutions. Also, more detailed information should be included about the velocity of the flooding and the effects on traffic, buildings and other objects in the environment. This can increase our understanding of flooding and use geodesign to further inform flood risk managers in preparing and coping with a flood. Lastly, more detailed flood scenarios should provide hour-to-hour information (or even more detailed), since each hour counts in case of a flood and detailed capacity planning schemes for evacuation (as in Table 9.1) are available to assess hour-based information.

We showed that the physical layout of an area is a key to differentiate the extent of exposure and vulnerability where it concerns the healthcare institution. Coupling our analysis to the second and third layer of the multilayer safety approach (sustainable spatial planning and disaster management), reduction of exposure and vulnerability can take place in each of these layers (Table 9.4) and opportunities for further research can be ordered accordingly[3]. From Table 9.4, the limitations of this study

---

[3] Since probability and prevention were not discussed we made it a 2×2 matrix instead of the possible 3×3 matrix

Table 9.4 Matrix with topics for further research to increase the capacity of health care institutions to cope with flood risk, based on multilayer safety and the components exposure and vulnerability. Aspects in bold were dealt with in this chapter

	Exposure	Vulnerability
Sustainable spatial planning	Effects of spatial layout on flood scenarios	Strength of buildings, amount of floors
	Choice of location of health care institutions	Availability and robustness of utilities like water, electricity, sewage
	Design of evacuation routes (height, capacity)	Vulnerability of transport lines, provision of water, food, electricity, medicine
	Physical configuration of health care institutions	
Disaster management	**Exact timing before arrival**	Amount of people to be evacuated with ambulances, other necessary capacity
	Velocity and depth of floods at specific locations	Necessity of evacuation of an institution (height of building, provision of resources, medical employees, etc.)
	Relation with other institutions (moment of exposure, capacity, etc.)	Relation with other institutions (order of vulnerability, capacity, etc.)
	Alternatives for evacuation routes	

also become visual. In the case we only focussed on exposure to flooding, related to disaster management (capacity planning and evacuation strategies). Although there are links with vulnerability and spatial planning (i.e. the capacity of ambulances is related to amount of vulnerable people and information about depth and velocity is needed for sustainable spatial planning), geodesign could be advanced to incorporate all aspects related to the components in the risk approach and the layers.

9.6 Conclusion

Flood risk defined by probability, exposure and vulnerability, is a highly complex phenomenon with which flood risk managers are confronted. Geodesign for a multilayer safety approach to flood risk mitigation and evacuation planning helps to concretize ideas for managing such complex phenomena. However, it is also a challenging task because it shows the complexity of the decision at hand. Regarding evacuation the availability of roads depends on the total traffic out of an area, and a lower part in the road can hinder an evacuation route or even make it inaccessible. Also vehicle availability determines the possibilities for evacuation and the necessary time after deciding to evacuate. Deciding to stay demands facilities, personnel and the availability of utilities like drinking water, electricity and food for the duration of the flood and recovery. In deciding about such complex situations, where lives are at stake, a geodesign approach can be beneficial to design and assess the best evacuation strategies for floods.

Geodesign for flood risk management requires detailed information for creating effective evacuation strategies and capacity management in case of a disaster, but also for design of evacuation routes and physical structures like healthcare institutions (answering questions concerning the benefit of creating possibilities for vertical evacuation, the best location to locate new facilities, etc.). Despite its limitations, just a few technical parameters concerning depth and time of arrival of a flood can provide insight for evacuation strategies, as was shown for the case Rivierenland. The relation of geodesign with multilayer safety in this case, provides new research directions and helps to structure information with the different layers of multi-layer safety and the risk approach to flooding. Since the layers are not similar to the components of the risk approach these two concepts can be used complementary to each other. By using geodesign to assess exposure and vulnerability to flood risk in a multi-layer safety approach, geodesign can provide additional opportunities for flood risk managers to reduce flood risk.

References

Adger, W.N. (2006). Vulnerability. *Global Environmental Change, 16,* 268–281.

Barendregt, A., van Noortwijk, J.M., van der Doef, M. & Holterman, S.R. (2005). *Determining the time available for evacuation of a dike-ring area by expert judgement.* Proceedings of ISSH Stochastic Hydraulics, 23–24 May 2005.

Birkmann, J. (Ed.). (2006). *Measuring vulnerability to natural hazards: Towards disaster resilient societies.* Tokyo: UN University Press.

d'Angremond, K. (2003). From disaster to delta project: The storm flood of 1953. *Terra et Aqua, 90,* 3–10.

d'Angremond, K., & Kooman, D. (1986). Eastern scheldt storm surge barrier: Management aspects of a multibillion dollar project. *International Journal of Project Management, 4*(3), 149–157.

De Bruin, K.M., & Klijn, F. (2009). Risky places in the Netherlands: a first approximation for floods. *Journal of Risk Management, 2,* 58–67.

Deltacommissie. (1960). *Rapport Deltacommissie: Eindverslag en Interim adviezen.* Report of the Delta Committee: final report and interim reports, The Hague.

De Moel, H., & Aerts, J.C.J.H. (2010). Effect of uncertainty in land use, damage models and inundation depth on flood damage estimates. *Natural Hazards, 58,* 407–425.

De Moel, H., Aerts, J.C.J.H., & Koomen, E. (2011). Development of flood exposure in the Netherlands during the 20th and 21st century. *Global Environmental Change, 21,* 620–627.

ECMT. (1992). *Resolution No. 92/2 on new classification of inland waterways.* Athens: Resolution of the European Conference of Ministers of Transport.

Fuchs, S., Kuhlicke, C., & Meyer, V. (2011). Editorial for the special issue: vulnerability to natural hazards-the challenge of integration. *Natural Hazards, 58,* 609–619.

IWR. (2011). *Flood risk management approaches: As being practices in Japan, Netherlands, United Kingdom, and United States.* Institution for Water Resources (IWR), United States Army Corps of Engineers: IWR Report No.: 2011-R-08.

Kazmierczak, A., & Cavan, G. (2011). Surface water flooding risk to urban communities—analysis of vulnerability, hazard and exposure. *Landscape and Urban Planning, 103,* 185–197.

Kellens, W., Vanneuville, W., Verfaillie, E., Meire, E., Deckers, P., & De Maeyer, P. (2013). Flood risk management in flander: Past developments and future challenges. *Water Resource Management, 27,* 3585–3606.

Kolen, B. (2013). *Certainty of uncertainty in evacuation for threat driven response.* Nijmegen: Dissertation, Radboud University.

Kreibich, H., Piroth, K., Seifert, I., Maiwald, H., Kunert, U., & Schwarz, J., et al. (2009). Is flow velocity a significant parameter in flood damage modelling? *Natural Hazards and Earth System Sciences, 9,* 1679–1692.

Mens, M.J.P., Klijn, F., De Bruijn, K.M., & Van Beek, E. (2011). The meaning of system robustness for flood risk management. *Environmental Science and Policy, 14,* 1121–1131.

MinI & M. (2008). *National water plan 2009–2015(Dutch National Water plan).* The Hague: Ministry of Infrastructure and the Environment.

MinI & M. (2013). *Delta programme2014.* The Hague: Ministry of Infrastructure and the Environment.

Nadal, N.C., Zapata, R.E., Pagán, I., López, R., & Agudelo, J. (2010). Building damage due to riverine and coastal floods. *Journal of Water Resources Planning and Management, 136*(3), 327–336.

Nelen-Schuurmans. (2013). *Flood Lizard, stable version 2.0.* Utrecht.

Provincie Gelderland, & Min V & W. (2010). *Verkenning waterveiligheid Betuwe, Tieler-en Culemborgerwaarden (dijkring 43).* (Exploration water safety dijkring 43). Arnhem: DHV B.V.

Renn, O. (2005). White paper on risk governance: Towards an integrative framework. In O. Renn & K. Walker (Eds.), *Global risk governance: Concept and practice using the IRGC framework* (pp. 3–73). New York: Springer-Verlag.

RLI. (2011). *Tijd voor waterveiligheid–strategie voor overstromingsbeheersing (Time for water safety–strategy for flood risk management).* The Hague: Report of the Dutch Council for the Living environment and Infrastructure.

Rijkwaterstaat, & HKV. (2008). *Capaciteitenplanning–Ergst denkbare overstromingsscenario's. (Capacityplanning–Worst case flood scenarios).* Netherlands: Lelystad.

Sayers, P.B., Hall, J.W., Meadowcroft, I.C. (2002). Towards risk-based flood hazard management in the UK. *Proceedings of ICE, Civil Engineering, 150,* 36–42.

Sitzenfrei, R., Maïr, M., Möderl, M., & Rauch, W. (2011). Cascade vulnerability for risk analysis of water infrastructure. *Water Science and Technology, 64*(9), 1885–1891.

Steinitz, C. (2012). *A framework for geodesign: Changing geography by Design.* Redlands: ESRI.

TAW. (1985). *Leidraad voor het ontwerpen van rivierdijken: Deel 1: Bovenrivierengebied.(Guideline for the design of riverine dikes. Part 1: Upper catchment area).* The Netherlands: Report Technische Adviescommissie Water, Rijkswaterstaat.

Ten Brinke, W.B.M., Saeijs, G.E.M., Helsloot, I., & Van Alphen, J. (2008). Safety chain approach in flood risk management. *Proceedings of ICE, Municipal Engineering, 161,* 93–102.

Van der Biezen, V. (2013). *What are vulnerable features?* Wageningen: Masterthesis Wageningen UR, Land Use Planning group.

Van de Ven, F.H.M., Gersonius, B., de Graaf, R., Luijendijk, E., & Zevenbergen, C. (2011). Creating water robust urban environments in the Netherlands: Linking spatial planning, design and asset management using a three-step approach. *Journal of Flood Risk Management, 4,* 273–280.

Chapter 10
Interactive Spatial Decision Support for Agroforestry Management

André Freitas, Eduardo Dias, Vasco Diogo and Willie Smits

10.1 Introduction

Forests are important ecosystems that are able to support productive functions (e.g. supply of wood products and non-timber forest products) and protective functions such as climate regulation, air pollution filtering, regulation of water resources, conservation of biodiversity and protection from wind erosion, coastal erosion and avalanches (FAO 2005). In the last decade, around 13 million ha of forest have been ruined or converted to other uses each year, compared to 16 million ha per year in the 1990s (FAO 2010). Despite this decrease, deforestation rates are still alarmingly high. Therefore, there is a need to globally improve the management of forest resources, and particularly to take into account additional forest values (such as biodiversity and social functions) towards long-term sustainable management (Varma et al. 2000).

Paletto et al. (2013) define Sustainable Forest Management (SFM) as a dynamic concept with the main purpose of maintaining and enhancing the economic, social and environmental value of forests, for the benefit of present and future generations. Agroforestry is regarded as a promising approach for sustainable forest management (Schoeneberger and Ruark 2003). Agroforestry systems are practiced in

E. Dias (✉) · A. Freitas
Geodan, President Kennedylaan 1, 1079 MB Amsterdam, The Netherlands
e-mail: eduardo.dias@geodan.nl

A. Freitas
Faculdade de Ciências e Tecnologia, Universidade Nova de Lisboa, Qta da Torre, 2829–516 Caparica, Portugal

E. Dias · V. Diogo
Faculty of Economics and Business Administration, Department of Spatial Economics, SPINlab, VU University Amsterdam, De Boelelaan 1087, 1081 HV, Amsterdam, The Netherlands

W. Smits
Institute of Technology Minaesa (ITM Tomohon), Jl. Stadion Selatan Walian, Tomohon, Sulawesi Utara 95439, Indonesia

tropical and temperate regions and include traditional and modern land-use systems in which trees are managed together with crops for multiple benefits. These systems allow communities to produce food, contributing to food and nutritional security, and to achieve productive and resilient cropping environments. Moreover, they can provide a range of forest products, including fuel-wood and non-timber products, increase biodiversity, protect water resources and reduce soil erosion. On a large scale, agroforestry systems can also prevent the occurrence of extreme weather events, such as floods and drought (FAO 2013).

However, agroforestry projects present very complex and interdependent economic, technical, political and social challenges, with its sustainability ultimately depending on the extent to which a well-coordinated land management strategy is designed and implemented (Sampson 1998). Decision support systems (DSS) have been an important tool in forest management since the early 1980s (Reynolds 2005). Segura et al. (2014) suggest that the future development of DSS for forest management should place stronger emphasis on economic models integrating the value of environmental services and collaborative decision making of multiple decision makers and stakeholders. In addition, decision support and management should be augmented with spatially explicit analysis as the costs and opportunities for different solutions have intrinsic geographic variability. A dynamic approach to land-use planning is needed to evaluate the long-term effects of present management decisions (Mönkkönen et al. 2014; Varma et al. 2000).

Spatially explicit systems for forest management have been developed in the past, but mainly based on individual tree growth (Phillips et al. 2003) or forest succession (Gustafson et al. 2000; He and Mladenoff 1999), not taking into account economic parameters. Van der Hilst et al. (2010) studied the potential, spatial distribution and economic performance of regional biomass chains, using attainable yields for biophysical suitability. Kosonen et al. (1997) studied the financial, economic and environmental profitability of reforestation of Imperata grasslands in Indonesia, a monoculture in a similar area to the case study of this paper. Furthermore, soil erosion and biodiversity indices were also developed for different vegetation covers, where the slope and the richness of bird and tree species were the principal components considered.

From a financial perspective, Hinssen and Rukmantara (1996) built a cost Comparison Model for budgeting of reforestation projects. While Chertov et al. (2005) used geo-visualization of forest simulation modelling on a case study of carbon sequestration and biodiversity. Wang et al. (2010) presents an integrated assessment framework and a spatial decision support system as a tool to support forestry development with consideration of carbon sequestration.

Vierikko et al. (2008) studied the interrelationships between ecological, social and economic sustainability at the regional scale, analysing their trade-offs. Segura et al. (2014) compared different decision support systems (DSS) for forest management and concluded that the majority of DSS do not include environmental and social values, focusing mainly on market economic values.

Stakeholders are generally uninformed of the benefits of agroforestry and the factors that determine the adoption of agroforestry practices (FAO 2013). The lack of awareness of the consequences and benefits of agroforestry projects may lead to unsustainable forest management. Therefore the tool presented can provide the needed awareness of the outcomes for different innovative agroforestry solutions including the complicated small scale permaculture approaches.

Despite the recent progress, a DSS for agroforestry management that is able to combine spatially explicit information with non-spatial factors and perform integrative assessments of the economic, social and environmental aspects has not been developed so far. In this work, we propose an interactive system that makes use of advanced visualization and analysis of spatially and temporally related data that attempts to inform and support decision making in the field of agroforestry management. It is intended to be useful for planners, stakeholders and managers in order for them to understand consequences of their spatial plans. The system was developed as an implementation of the emerging concept of geodesign (Steinitz 2012; Ervin 2011). In geodesign, proposed landscape changes are directly evaluated against impact models previously defined, so the design ideas and solutions can be iteratively and collaboratively created by different stakeholders and domain experts (Dias et al. 2013). The plans are continuously evaluated against multiple objectives and continuously evolved to develop fitter (less impact) and more robust (more benefits) solutions.

The goal of this chapter is to present a spatially explicit DSS fully integrating the economic, environmental and social dimensions of agroforestry systems streamlined by the geodesign framework. The system demonstrated in this chapter provides:

- An approach to identifying the most beneficial locations for agroforestry projects based on the biophysical properties and evaluate its economic, social and environmental impact;
- A simulation environment that enables evaluation via a simple dashboard and with the opportunity to perform straight forward sensitivity analysis for key parameters;
- A tool to inform prospective investors of the potential and opportunities for integrated forest management;
- A 3D interactive geographic visualization of the economic, social and environmental outcomes to facilitate direct understanding, also by non-experts.

10.2 Material and Methods

10.2.1 *Methodology*

The system is intended to be a spatially explicit integrative assessment tool for agroforestry projects. It allows the comparison of the economic, social and environmental performance of different management options, by combining spatial data on biophysical features, population, infrastructure and transportation networks with data on economic and technical factors. It also functions as an exploratory tool, being deployed in an interactive environment that enables sensitivity analyses of system performance for the main key factors (e.g. cost of production factors, market prices of commodities) and different spatial options (designs) in forest plantations. The conceptual model behind the spatially explicit cost benefit analysis is illustrated in Fig. 10.1.

The system aims to determine the local performance of agroforestry recipes, defined as a mix of crops that are sequentially cultivated in a certain area. Different recipes have different environmental requirements, timing and economic values in terms of field operations, field inputs, commodities, labour needs and costs. The considered recipes are defined according to the opportunities and constraints set by the biophysical features of specific regions. Examples of recipes are given on the case study description (Sect. 10.3).

The performance of each recipe depends to a large extent on the local biophysical suitability, which affects the attainable yield/productivity per year for different species, which in turn is determined by the combination of local biophysical parameters such as soil, altitude, temperature, precipitation and slope. Suitability and recipes are assessed by local expert knowledge and map analysis, combining different biophysical layers and parameters. Recipes are specially selected to enhance synergies between different crops and other spillover effects such as avoiding soil erosion, protecting watersheds and providing regular jobs and other products.

The produced commodities entail spatially explicit field operations costs and inputs including labour force, planting, maintenance, harvesting and tapping, which are integrated in the total costs of the system. Commodities are dependent on the area suitability and yields per year, hectare and recipe. Commodities include timber and crops (such as cassava and pineapple), as well as by-products (such as broom, roof covers and furniture). Conversion efficiency factors are used to determine the final products, depending on the type of mechanism applied for the transformation. The user is able to introduce and change the commodities produced as needed.

Besides the revenues derived from selling the commodities in the markets and the production costs, there are also highly variable costs that have to be considered, such as transportation and storage. Transportation is one of the economic factors with more expression as access to the production sites is often difficult. Transportation costs are calculated on a combination of geographic data of the roads, ports and markets, simulating the price depending on the distance to transport the commodities via advanced network analysis. This network analysis takes into account the type of roads and fuel costs.

Field operations are defined in terms of the number of each operation per year, per recipe and per unit area. The field operations initially taken into account in the system are land clearing (suppression and removal of existing trees and weeds), seeding and planting, maintenance (e.g. weeding), harvesting and tapping.

The field inputs are the inputs needed for each field operation, such as labour force, fuel consumption, number of seeds and fertilizer per hectare and recipe. These were the initial inputs considered, but the system can support more complexity as needed during the project. Additional field inputs or operations can be added for different progress or scenarios.

10 Interactive Spatial Decision Support for Agroforestry Management

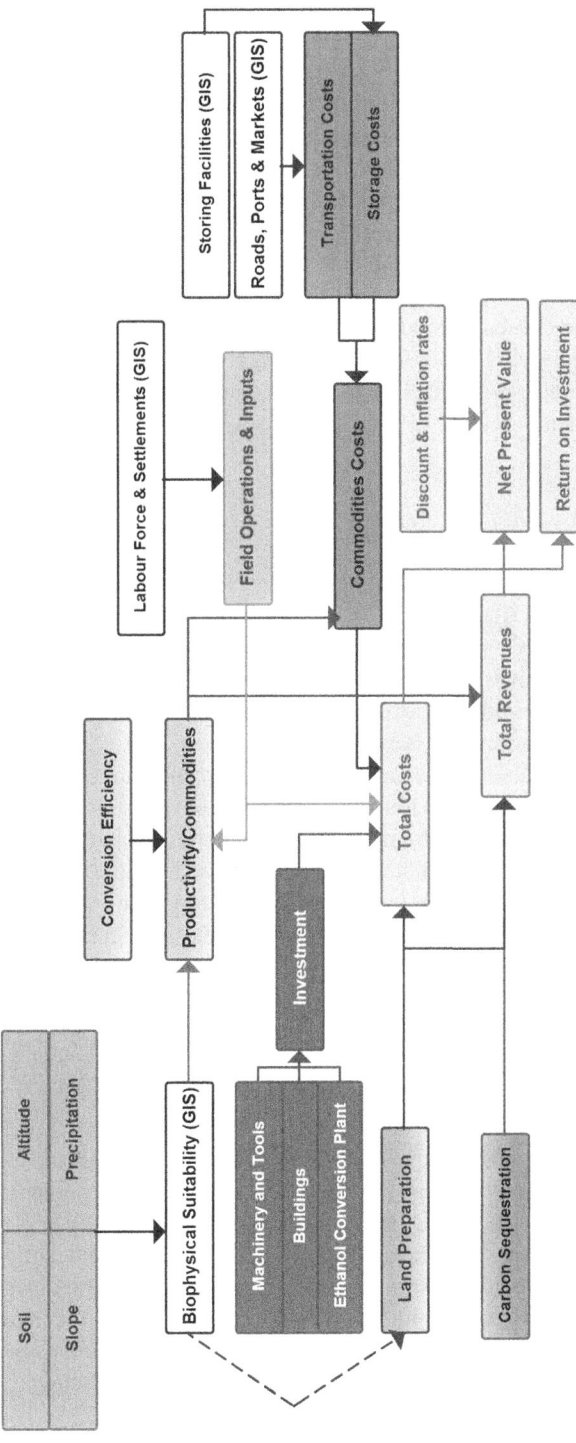

Fig. 10.1 Spatial distribution and economic performance of an agroforestry project

In a first phase approach, all field inputs are combined to give a final value ($/ha) for each field operation described. The total cost of field operations per recipe is calculated by the multiplication of each field operation input with the according number of field operations. Depending on the biophysical suitability of the area and the recipe to apply, a land preparation might also be required in order to reforest it. This land preparation, might include land clearance which is translated into a cost and revenue, as some materials can be sold. Since labour force is required for the realization of the project, it is important to take into account the location of the settlements to choose the most suitable location. The labour force will be determined by the hours of work needed for certain recipe and unit area. Therefore, the recipes can also be chosen or modified to match the amount of labour available or job opportunities needed. Labour has two important perspectives, as a financial cost for the system but also as a social output via the increase of employment rate and welfare improvement.

Economic Aspect

The economic performance is determined by the Net Present Value (NPV) and return on investment (ROI), calculated based on the investment, total costs and revenues. The NPV and ROI are essential economic key factors for investors' decision. Specification of the investment needed for the project is also considered, being determined by the required number of units and costs of machinery, tools, buildings and conversion plants units (e.g. sawmill).

The total costs include field operations, investment and commodities costs (transportation and storage). The total revenues represent the sum of the cash inflow of the entire project, such as cash inflow from commodities, land clearance, as well as carbon emission permits. Both total cost and total cash inflow, are calculated per year, as management decisions are made yearly in the specific case study.

To assess the long-term benefits of different recipes and therefore different land use practices, we calculate the NPV per hectare using the following equation:

$$\mathrm{NPV} = \sum_{t=1}^{N} \frac{\mathrm{Cash\ Flow}}{(1+i)^t} - \mathrm{Investment}(\$/\mathrm{ha}) \qquad (10.1)$$

where NPV is the net present value cumulated to year n; i is the discount rate (%); NPV=Net Present Value of recipe per ha ($/ha); Cash flow=Revenues−Costs ($/ha), t=annuity period (y), N=lifetime of the project.

The discount rate, can also take into account inflation and depreciation rate. The annuity time period considered was 20 years, which is in line with an agroforestry project lifetime. When comparing investments, the one with the highest NPV, assuming the same discount rate, is considered the most desirable on the economic perspective.

ROI is the internal annual rate of return of an investment. It is the compound interest rate that equates the present value of future incomes with the present value of future costs.

$$ROI = \frac{\text{Total Revenues} - \text{Total Costs}}{\text{Total Costs}} \qquad (10.2)$$

Environmental and Social Aspect

So far the system is intended as a "quick scan", easy to use and understandable, therefore instead of complex indices (composite indicators), we represent the social performance by the number of jobs created and the environmental performance by the amount of carbon sequestration (CO_2e tons), per recipe.

Carbon sequestration (CS) of the recipe accessed by the following equation:

$$CS_{recipe} = \sum_{j=1}^{t} n_{T_j} \times \text{Sequestration}_{T_j} (CO_2 e) \qquad (10.3)$$

n_{T_j}—number of trees of a specific tree species; Tj—specific tree species; Sequestration$_{T_j}$—tonnes of carbon stored of a specific tree species; t—number of different tree species of the recipe.

Afforestation and reforestation are included in trading schemes for carbon sequestration offsets, and therefore through credits generates revenue (Eq. 3) (Saundry 2009).

$$CS_{revenues} = \sum_{i=1}^{x} CS_{recipe} \times Ccredits_{price} (\$) \qquad (10.4)$$

Ccredits$_{price}$—CO_2 emission permits market prices; x—number of recipes on the agroforestry project.

However, the system is prepared to receive data from more complex indices, such as biodiversity or soil erosion for environmental performance. Kosonen et al. (1997) developed soil erosion and biodiversity indices for different vegetation covers for a case study on South Kalimantan, where the slope and the richness of bird and tree species were the principal components considered, respectively.

10.2.2 Implementation

Microsoft Excel™ was chosen as implementation environment for the modelling system, due to its flexibility to add and edit different parameters or values, integrating all the system parameters in different spreadsheets. Each parameter has a sheet, in order to ease comprehension and changes for the final user. A dashboard was created to gather the essential controls for the end-user where different parameter values can be simulated.

Reforestation and agroforestry investments can be complex due to the uncertain future conditions. Therefore investors are often sceptical about investing on agroforestry projects. To address this problem an interactive tool with a sensitivity analysis was built so that different parameters could be simulated. For example, price fluctuations can be analysed and simulated in order to evaluate its impact.

The simulation of different parameters through interactive sliders can be instantly visualized spatially in an interactive 3D geographic visualization interface. In this interface, combining geographic location (latitude, longitude) and outcomes data, the user can navigate on the map, helping the comprehension of the different locations benefits.

This way it is possible to simulate spatial and non-spatial variations in a geodesign framework that can help better informed decisions, while exploring different possible scenarios. Besides that, this implementation provides an easy and flexible environment to become aware of the sensitivity to different parameters, allowing a combination of different alternatives and scenarios that wouldn't be possible in a hard copy consulting report.

10.3 Case Study

The methodology is generic and can be applied anywhere in the globe. A model application has been recently developed for a specific study area in Indonesia. Indonesia has the third largest area of tropical forest in the world, 68 % of its landmass, and its impressive biodiversity is contained in those forests. Wood manufacturing paper and printing industry is also an economically significant sector, 3–4 % of the country GDP (Josef et al. 2009). According to 1998 data, almost 24 % of 69.4 million ha under logging concessions were degraded (Kartodihardjo and Supriono 2000).

The study area is located in East Kalimantan, Indonesia, where a local company manages a forest concession of around 200,000 ha. The concession aims to implement a sustainable forestry management strategy, profitable but also fostering development in local communities and promoting the conservation of the surrounding environment. This way, economic, human development and environmental goals can be jointly pursued.

Sustainable use of the forest relies upon a multi-crop reforestation scheme, in which different trees species and crops benefit together from mutual synergies, being therefore more efficient than monoculture schemes for environmental goals (Gamfeldt et al. 2013). Species vary in their nutrients, sunlight and soil moisture requirements to establish and grow successfully (Stringer 2001). Integrating many different species in one unit of land with different spacing, with optimal sunlight utilization through a succession of species will also reduce losses of nutrients. It relies on an integration of growing cycles with different lengths in one total longer rotation of the system. The total success of an ecosystem depends on how the complex processes are adapted to local conditions, and the evaluation of the recipes by a local expert. Everything depends upon competition driven utilization of light and nutrients, as well as strategies in the process of succession during development of a locally stable ecosystem. Matching site-species is a necessity to promote growth and maintain long-term sustainability (Chokkalingam et al. 2006).

In order to maximize the productivity, a recipe has specific timing and biophysical conditions. The following recipe is an example for a wet tropical climate condition on terrain with less than 30 % slopes, well-draining soil, reasonable good access from roads and with enough local labour and local needs for food and energy.

- Start: Land preparation, Planting, Fertilizing
  - Clearing planting spots, digging planting holes, mobilizing compost;
  - Transporting plants to field, planting trees (nitrogen fixer and sugar palms) and cassava mixed;
- Year 1: Harvest and Maintenance, new Planting
  - Harvest of the cassava for food, animal feed and production of ethanol;
  - Maintenance of the planted trees;
  - Planting of banana in between the trees;
- Year 2: Harvest and Maintenance
  - Harvest of the bananas;
  - Maintenance of the trees;
- Year 3: Harvesting and Maintenance
  - Fuel wood from thinning;
  - Harvest of palm fibres;
  - Last maintenance of trees;
- Year 4–6: Harvest of palm fibres and Fuel Wood removal
  - Regular harvesting of palm fibres;
  - In year 6 removal of the remaining fuel wood;
- Year 7–9: Start tapping of sugar palms
- Year 10: Harvesting of sugar palms
  - Last tapping of sugar palms;
  - Harvest of sugar palm fruits and sugar palm wood;
- Restarting the Recipe.

One of the species in the case study is the sugar palm (*Arenga pinnata*). Besides yielding sugar, this palm also provides a great number of other products and benefits to its users, such as bioethanol from the sugar palm juice, after fermentation and distillation. It has a positive contribution to small households (e.g. opportunities for additional sources of income, clean fuel for cooking, transport, electricity, etc.) and requires little maintenance (Mogea et al. 1991; van de Staaij et al. 2011). The bioethanol produced from the sugar palm can then be used to replace gasoline in motorcycles, small vehicles, small machines and generators, and can also be used as cooking fuel in special burners (Smits 2010). A mixed production system can therefore provide food security, energy, regulate water, support biodiversity, sequester more carbon, as well as create jobs year-round, because each culture has its harvesting period.

10.4 Results

10.4.1 Interactive System

A sensitivity analysis is an important tool as an investor or manager can easily see the impact of parameters prices fluctuations on the project.

The goal was to produce an easy to use system, incorporated with sliders that can control different field operational costs. An excel sheet was created for each parameter, example of the field operations sheet on Fig. 10.2, which represents the number of each field operations per year, recipe and hectare.

The system is prepared to easily analyse or edit each recipe, whereas each column represents a year on the 20-year project lifetime considered and the user can change the number of each field operation per year and observe the impact on economic aspects. For example, when labour becomes a limiting factor at a certain moment in time (e.g. because a new industry nearby offers higher paid jobs) the absence of maintenance can directly be translated in terms of income and less carbon sequestered.

Commodities (Fig. 10.3) are organized in a list where the user can input the quantity (tons or m^3) produced per year and price ($) of the crops and raw materials, final products or by-products. Options to control the price through sensitivity analysis sliders is also provided on the dashboard.

A dashboard has been developed (Fig. 10.4) where the user is able to see the content of the recipe, the map suitability of the recipe, as well as the possibility to change the field operations and commodity value prices and instantly see the impact in terms of Total Revenues, Total Costs, Net Present Value and Return on Investment.

The objective of the sensitivity analysis sliders is to explore the critical factors of the agroforestry project. In the present case study, tapping and transports are the most critical ones. The user can also decide the area of cultivation to be calculated, having an instant result of that change on the dashboard. For other changes, for instance, when a disease wipes out certain seedling planting stock in the nursery, other recipes can be chosen to make up for the loss. This might mean planting fewer recipes but larger areas of each of them.

The distance map (Fig. 10.5) is a first approach for the decision algorithm that will give the most suitable locations depending on the roads and settlements available. In the Fig. 10.5, the settlements are represented as dots, and the colours depending on the distance and roads available to the settlements. Green surfaces are the closest areas to the villages and red the most distant and inaccessible. For labour intensive recipes, it is more suitable to be close to the labour force.

This type of network analysis can also be applied to determine the best cost-effective way to transport the commodities to the markets and ports.

As the project consists of a large geographical area with heterogeneous characteristics, visualization can help to support planning and management. Each geographical unit has unique geographical coordinates, it is then possible to combine the model outputs and visualize them in an interactive 3D geographic visualization. In this way, the user can see which areas and recipes are more profitable (Fig. 10.6) or the ones that have a higher carbon sequestration or higher employment.

10 Interactive Spatial Decision Support for Agroforestry Management

Field operations per year for each crop

☐ Recipe A   ☑ Recipe B
☐ Recipe C   ☐ Recipe D

Number of field operations per year	Recipe B (years)																				
	0	1	2	3	4	5	6	7	8	9	10	11	12	13	14	15	16	17	18	19	20
Land Clearing/Site preparation	1	1	0	0	0	0	0	0	0	0	0	1	0	0	0	0	0	0	0	0	0
Seeding/Planting (nr)																					
Planting (seedlings and nursery incl.)	1	1	0	0	0	0	0	0	0	0	1	1	0	0	0	0	0	0	0	0	0
Maintenance (eg. weeding)																					
Maintenance	0	1	1	1	0	0	0	0	0	0	0	1	1	1	0	0	0	0	0	0	0
Harvest (nr)																					
Harvesting	0	1	1	1	1	1	1	0	0	0	0	1	1	1	1	1	1	0	0	0	0
Tapping	0	0	0	0	0	0	0	360	360	360	360	0	0	0	0	0	0	360	360	360	360
Other	0	0	0	0	0	0	0	0	0	0	0	0	0	0	0	0	0	0	0	0	0

Fig. 10.2 Field operations tab, recipe B example. (N.B: Fictional data due to company confidentially)

List of commodities produced each year per ha & Market Prices

☐ Recipe A ☑ RecipeB
☐ Recipe C ☐ RecipeD

											Recipe B (years)										
	1	2	3	4	5	6	7	8	9	10	11	12	13	14	15	16	17	18	19	20	Lifetime
Crops & Raw materials (ton)																					
Sugar Palm Juice (m3)	0	0	0	0	0	0	612	612	612	612	612	612	612	612	612	612	612	612	612	612	8568
Cassava	0	8	8	8	8	8	8	8	8	8	8	8	8	8	8	8	8	8	8	8	152
Albicia	0	0	0	0	0	0	0	0	0	0	0	0	0	0	0	0	0	0	0	80	80
Agathis	0	0	0	0	0	0	0	0	0	0	0	0	0	0	0	0	0	0	0	0	0
pineapple	0	0	0	0	0	0	0	0	0	0	0	0	0	0	0	0	0	0	0	0	0
Acacia	0	0	0	0	0	0	0	0	0	0	0	0	0	0	0	0	0	0	0	0	0
beans	0	0	0	0	0	0	0	0	0	0	0	0	0	0	0	0	0	0	0	0	0
corn	0	0	0	0	0	0	0	0	0	0	0	0	0	0	0	0	0	0	0	0	0
papaya	0	0	0	0	0	0	0	0	0	0	0	0	0	0	0	0	0	0	0	0	0
banana	0	0	0	0	0	0	0	0	0	0	0	0	0	0	0	0	0	0	0	0	0
cacao	0	0	0	0	0	0	0	0	0	0	0	0	0	0	0	0	0	0	0	0	0
Wood (m3)	0	0	0	0	0	0	0	0	0	0	0	0	0	0	0	0	0	0	0	0	0
Final Products																					
ethanol (m3)	0	0	0	0	0	0	48,96	48,96	48,96	48,96	48,96	48,96	48,96	48,96	48,96	48,96	48,96	48,96	48,96	48,96	685,440
other (fill if applicable)	0	0	0	0	0	0	0	0	0	0	0	0	0	0	0	0	0	0	0	0	0
By-products (units)																					
roof cover	0	0	0	0	0	0	0	0	0	0	0	0	0	0	0	0	0	0	0	0	0
handy craft	0	0	0	0	0	0	0	0	0	0	0	0	0	0	0	0	0	0	0	0	0
floorings	0	0	0	0	0	0	0	0	0	0	0	0	0	0	0	0	0	0	0	0	0
furniture	0	0	0	0	0	0	0	0	0	0	0	0	0	0	0	0	0	0	0	0	0
other (fill if applicable)	0	0	0	0	0	0	0	0	0	0	0	0	0	0	0	0	0	0	0	0	0

Fig. 10.3 Commodities tab, recipe B example. (N.B: Fictional data due to company confidentially)

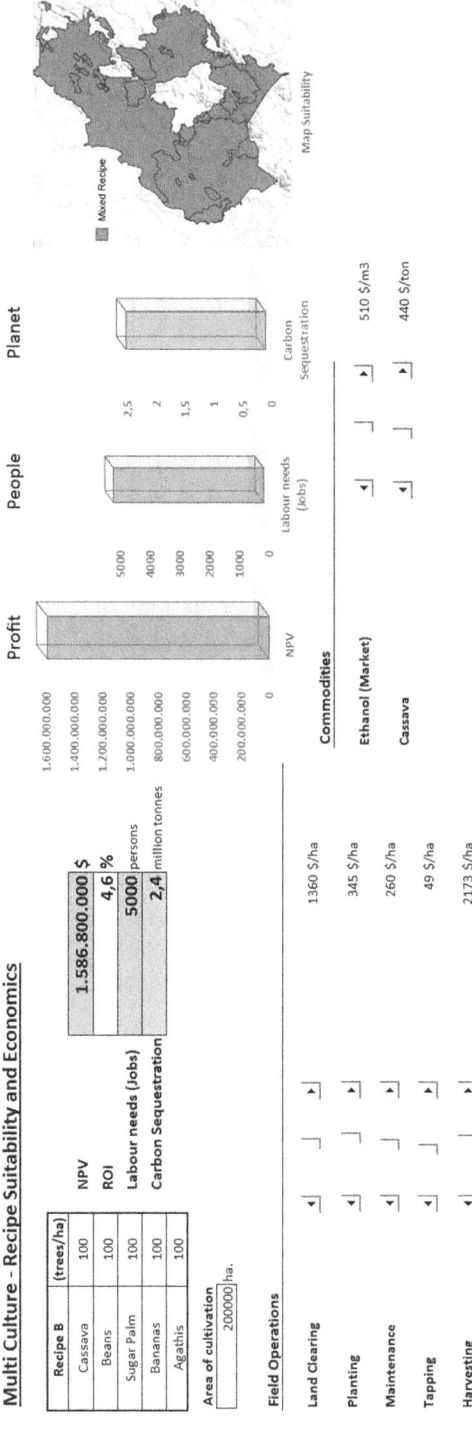

Fig. 10.4 Multi-culture - Suitability and Economics dashboard

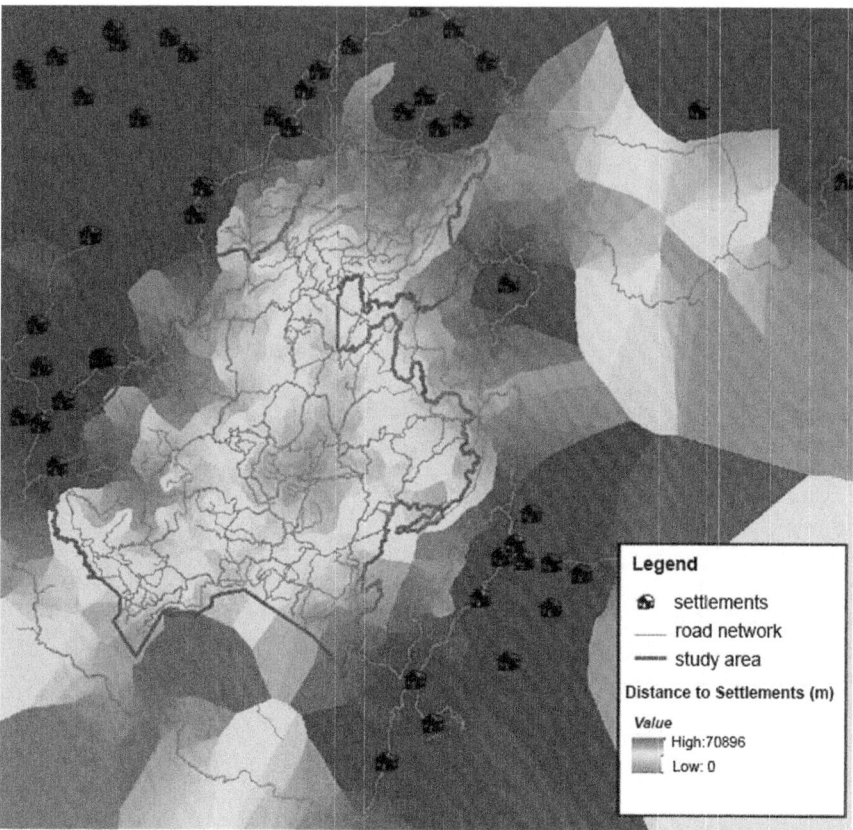

Fig. 10.5 Distance to every point based on settlements and roads available

An agroforestry project can have lifetimes from 15 to 30 years usually, which is a long-term investment, therefore it's important to show geographically, through time how the project will develop and when the economic investment pays off (Fig. 10.7).

The interactive geo-visualization was developed using Microsoft™ Power Map Preview. Each column on the 3D graph, is geographically positioned via the latitude, longitude and represents the value (cash inflow, jobs or carbon sequestration). The user can click on the desirable column and to access additional specific information, such as the exact value.

10.4.2 System Application

The proposed system is hereby exemplified with a hypothetical[1] case study in Indonesia. A system application was developed to assess and compare the economic, en-

---

[1] Due to company confidentiality the data presented is fictional.

10 Interactive Spatial Decision Support for Agroforestry Management

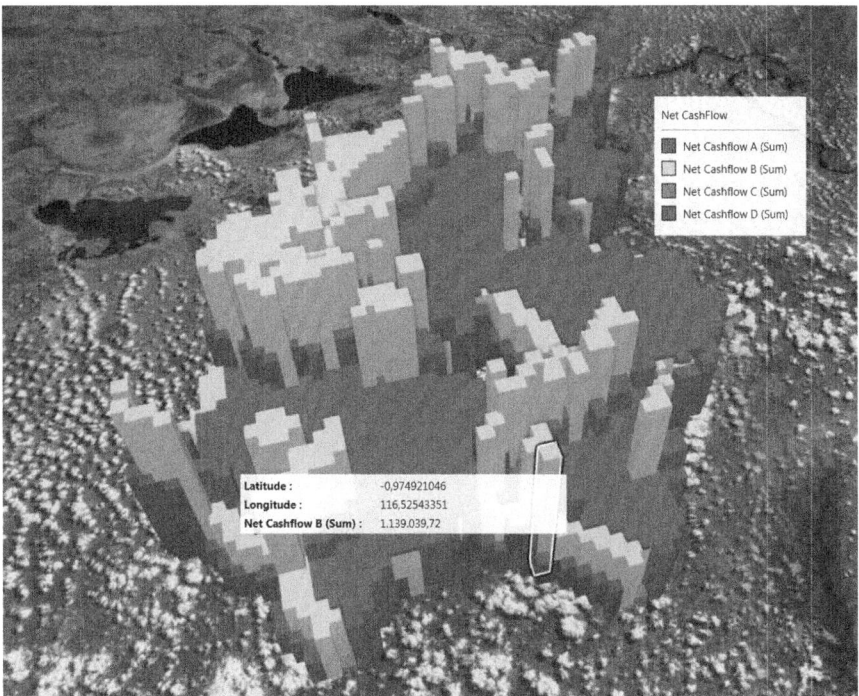

Fig. 10.6 Interactive geo-visualization of the net cashflow per recipe

vironmental and social impact of three different possible agroforestry approaches: a recipe of a monoculture scheme of timber production, a mixed recipe and a mixed design approach with different recipes for all the area.

As different recipes have different biophysical suitability, it's important to maximize the use of the area depending on the suitability for each recipe. In the mixed recipes approach (Fig. 10.7), 5 different recipes were implemented for the entire area, according to their best suitability.

Analysing the three plantation schemes (Fig. 10.8 and Table 10.1), we can evaluate the result of the different implementations and see which is more profitable or which provides more jobs or carbon sequestration.

The sliders on the recipes dashboards (Fig. 10.4) give the possibility of an interactive sensitivity analysis of each commodity. A sensitivity analysis of the timber price is illustrated in Fig. 10.9. It provides important information on the variation of the overall economic performance of the system due to volatility of market prices. In the present example, we can see that the plantation scheme with the five mixed recipes is less vulnerable to fluctuation of timber prices. Furthermore, NPV remains positive even if timber market prices are much lower than initially assumed. Therefore, it can be concluded that financial risks are distributed over different crops in this plantation scheme. On the other hand, monoculture schemes appear to be much more vulnerable to sudden changes in commodity prices.

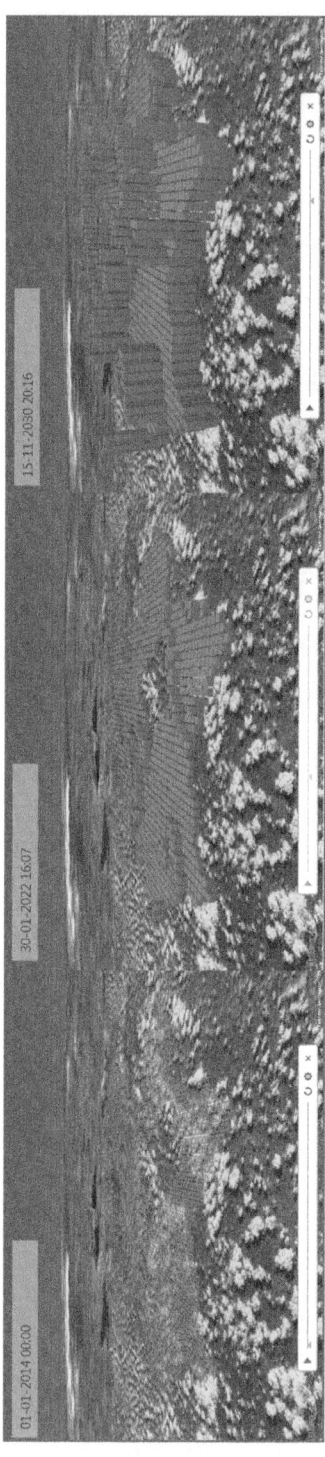

Fig. 10.7 Interactive geo-visualization of the net cashflow per recipe, through time

10 Interactive Spatial Decision Support for Agroforestry Management

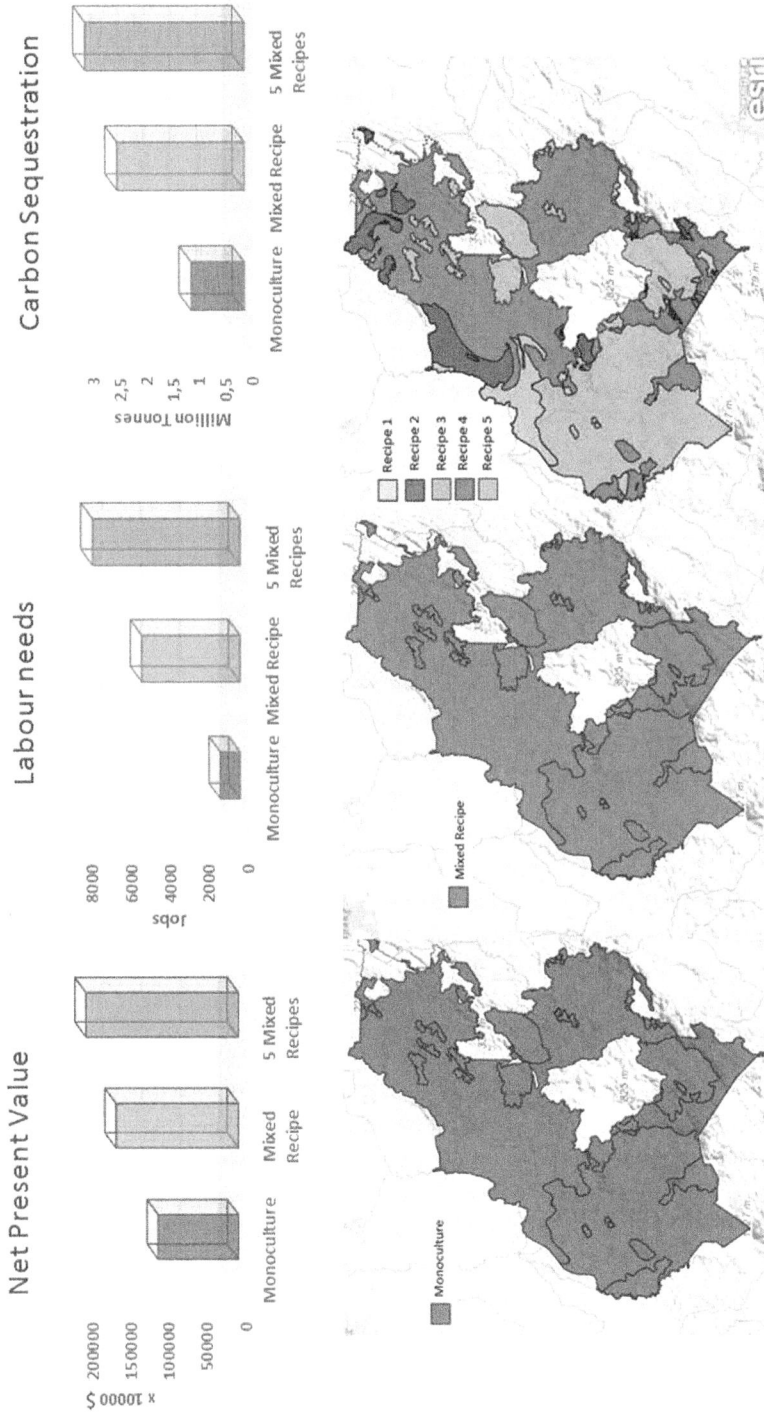

Fig. 10.8 Recipes' perfomance

Table 10.1 Recipes' performance. (N.B: Fictional data due to company confidentially)

	Monoculture	Mixed recipe	Five mixed recipes
Net present value ($)	1,042,300,000	1,586,800,000	1,972,200,000
ROI (%)	2.5	4.6	6.1
Labour needs (Jobs)	1000	5000	7500
Carbon Sequestration (Million tonnes)	1.1	2.4	3.0

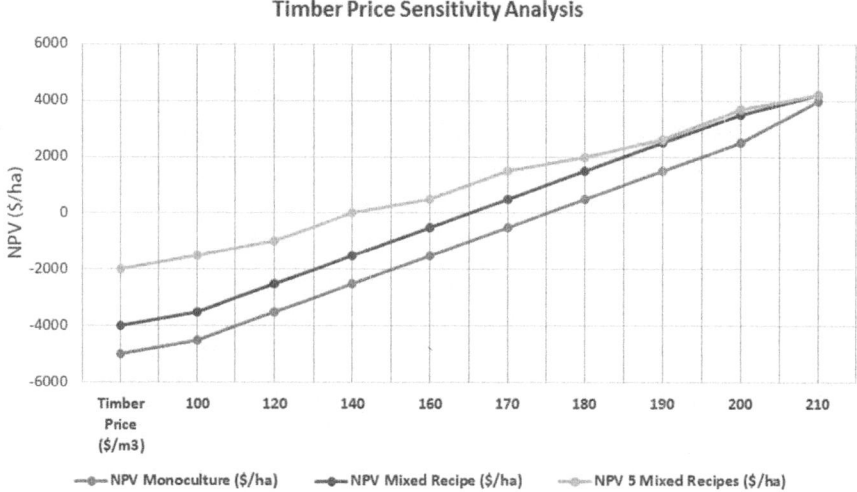

Fig. 10.9 Timber price sensitivity analysis example. (N.B: Fictional data due to company confidentially)

10.5 Conclusions

This paper describes the methodological framework of a spatially explicit decision support system being developed for sustainable forest management, integrating economic, social and environmental performance.

The system could also be used as a tool to analyse beforehand the performance of agroforestry projects, taking into account regional-specific environmental challenges in terms of climate change and soil and forest degradation.

Reforestation projects can benefit and gain efficiency through decision support systems that help to evaluate the feasibility and the overall security of the project. The geographical visualization is also an important decision and communication tool, especially in large area projects with spatial variability of biophysical conditions.

From an economic point of view, a higher NPV is generally desired, but other non-economic factors also need to be taken into account when determining project feasibility such as the carbon sequestration and number of jobs created. This system can provide insights and explore possible win–win solutions for reforestation projects between local residents, the environment and the economy while enhancing transparency and fairness.

This system, integrating spatial and non-spatial information for a better decision, has enormous potential for geodesign in agroforestry projects as provides powerful information of the most beneficial location for a sustainable forest management. Moreover, it can inform stakeholders providing them a tool to better understand the impacts of the project and its exact location.

The use of suitability maps allowed assessing expected productivity and the economic performance of growing different agro-forestry commodities. An underlying assumption of this approach is that the maps are static and maximum production yields are always attained. Negative impact from short-term events (e.g. heat waves or excess of rainfall) and long-term dynamic processes (changes in climate conditions or soil erosion) are not explicitly incorporated in the current model. Therefore, the tool may over-optimize the real capacity for delivering commodities and, as a result, the determined economic performance can actually be lower than what is being determined by the model. Nevertheless, the present model is able to provide an indication on what could be attained under optimal biophysical circumstances, as well as exploring the sensitivity to changing conditions. In addition, it should be noted that environmental spatial externalities (e.g. resulting from the movement of materials such as water, soil, plants, pests and contaminants) and factors related to economies of scale (e.g. clustering of production systems) were not explicitly taken into account. Our system is nevertheless able to inform the main areas where the production of different commodities could become economically attractive and thus provide an indication for decision-makers on the areas where positive externalities and increasing returns to scale are worth being explored while minimizing ecological risks.

Future developments on the system will emphasize user friendliness and spatial design capabilities on the interactive map, powering it up as a geodesign tool. In addition, the development and incorporation of more complex indices for the social and environmental performance should be pursued, as well as methods to estimate the benefits resulting from economies of scale.

References

Chertov, O., Komarov, A., Mikhailov, A., Andrienko, G., Andrienko, N., & Gatalsky, P. (2005). Geovisualization of forest simulation modelling results: A case study of carbon sequestration and biodiversity. *Computers and Electronics in Agriculture, 49*(1), 175–191. doi:10.1016/j.compag.2005.02.010.

Dias, E. S., Linde, M., Rafiee, A., Koomen, E., & Scholten, H. J. (2013). Beauty and brains: Integrating easy spatial design and advanced urban sustainability models. Chapter 27. In: Geertman, S., Toppen, F., & Stillwell, J. C. H. (Eds.), *Planning support systems for sustainable urban development* (pp. 469–484). Berlin: Springer.

Ervin, S. (2011). A system for geodesign. In E. Buhmann, S. Ervin, D. Tomlin, & S. Pietsch (Eds.), Teaching landscape architecture—Preliminary proceedings (pp. 145–154), Bernburg.

FAO. (2005). *Global forest resources assessment 2005*. Rome: Food and Agriculture Organization of the United Nations.
FAO. (2010). Global forest resources assessment 2010. http://www.fao.org/docrep/013/i1757e/i1757e.pdf. Accessed 10 March 2014.
FAO. (2013). Advancing agroforestry on the policy agenda: a guide for decision-makers. http://www.fao.org/docrep/017/i3182e/i3182e00.pdf. Accessed 10 Feb 2014.
Gamfeldt, L., Snäll, T., Bagchi, R., Jonsson, M., Gustafsson, L., Kjellander, P., & Bengtsson, J. (2013). Higher levels of multiple ecosystem services are found in forests with more tree species. *Nature Communications, 4,* 1340. doi:10.1038/ncomms2328.
Gustafson, E. J., Shifley, S. R., Mladenoff, D. J., Nimerfro, K. K., He, H. S. (2000). Spatial simulation of forest succession and harvesting using LANDIS. *Canadian Journal of Forest Research* 30:32–43.
He, H. S., & Mladenoff, D. J. (1999). Spatially explicit and stochastic simulation of forest landscape fire disturbance and succession. *Ecology, 80,* 81–99.
Hinssen, P. J. W., & Rukmantara (1996). The cost comparison model: A tool for financial budgeting of reforestation projects. *IBN Research Report, 96*(2), 57. Wageningen: Instituut voor Bos en Natuuronderzoek.
Josef et al. (2009). *Investing in a more sustainable Indonesia: Country environmental analysis. CEA series, East Asia and Pacific region.* Washington, DC: World Bank
Kartodihardjo, H., & Supriono, A., (2000). *Dampak Pembangunan Sektoral terhadap Konversi dan Degradasi Hutan Alam: Kasus Pembangunan HTI dan Perkebunan di Indonesia*. Bogor: Center for International Forestry Research (CIFOR).
Kosonen, M., Otsamo, A., & Kuusipalo, J. (1997). Financial, economic and environmental profitability of reforestation of Imperata grasslands in Indonesia. *Forest Ecology and Management, 99*(1–2), 247–259. doi:10.1016/S0378-1127(97)00210-7.
Mogea, J., Seibert, B., & Smits, W. (1991). Multipurpose palms: the sugar palm (*Arenga pinnata* (Wurmb) Merr.). *Agroforestry Systems, 13*(2), 111–129. Kluwer. doi:10.1007/BF00140236.
Mönkkönen, M., Juutinen, A., Mazziotta, A., Miettinen, K., Podkopaev, D., Reunanen, P., & Tikkanen, O.-P. (2014). Spatially dynamic forest management to sustain biodiversity and economic returns. *Journal of Environmental Management, 134,* 80–89. doi:10.1016/j.jenvman.2013.12.021.
Paletto, A., De Meo, I., Di Salvatore, U., & Ferretti, F. (2013). Stakeholders' perceptions on sustainable forest management. Socio-economic analyses of sustainable forest management.
Phillips, P. D., Brash, T. E., Yasman, I., Subagyo, P., & van Gardingen, P. R. (2003). An individual-based spatially explicit tree growth model for forests in East Kalimantan (Indonesian Borneo). *Ecological Modelling, 159*(1), 1–26. doi:10.1016/S0304-3800(02)00126-6.
Reynolds, K. M. (2005). Integrated decision support for sustainable forest management in the United States: Fact or fiction? *Computers and Electronics in Agriculture, 49*(1), 6–23. doi:10.1016/j.compag.2005.02.002
Sampson, N. (1998). Farm and Forest: Which Way to Sustainability. *American Farmland Trust* Center for Agriculture in the Environment. DeKalb, Illinois. Working Paper CAE/WP98-7. http://aftresearch.org/research/resource/publications/wp/wp98-7.html. Accessed 10 Feb 2014.
Schoeneberger, M., & Ruark, G. (2003). Agroforestry—Helping to achieve sustainable forest management. UNFF intersessional experts meeting on the role of planted forests in sustainable forest management, 24–30 March 2003, New Zealand.
Segura, M., Ray, D., & Maroto, C. (2014). Decision support systems for forest management: A comparative analysis and assessment. *Computers and Electronics in Agriculture, 101,* 55–67. doi:10.1016/j.compag.2013.12.005.
Smits, W. (2010). Ethanol production unit and method for the production of ethanol, U.S. Patent No. 20100267101 A1.
Steinitz, C. (2012). *A framework for geodesign: changing geography by design*. Redlands: ESRI.
Stringer, J. W., (2001). Kentucky Forest Practice Guidelines for Water Quality Management. Cooperative Extension Service, College of Agriculture, University of Kentucky UK.

Varma, V. K., Ferguson, I., & Wild, I. (2000). Decision support system for the sustainable forest management. *Forest Ecology and Management, 128*(1–2), 49–55. doi:10.1016/S0378-1127(99)00271-6.

Van der Hilst, F., Dornburg, V., Sanders, J. P. M., Elbersen, B., Graves, A., Turkenburg, W. C., Faaij, A. P. C. (2010). Potential, spatial distribution and economic performance of regional biomass chains: The North of the Netherlands as example. *Agricultural Systems, 103*(7), 403–417.

van de Staaij, J., van den Bos, A., Hamelinck, C., Martini, E., Roshetko, J., & Walden, D. (2011). Sugar palm ethanol: analysis of economic feasibility and sustainability. Working Papers b17084, World Agroforestry Centre, Library Department. http://www.worldagroforestry.org/downloads/publications/PDFs/RP17084.PDF. Accessed 10 Feb 2014.

Vierikko, K., Vehkamäki, S., Niemelä, J., Pellikka, J., & Lindén, H. (2008). Meeting the ecological, social and economic needs of sustainable forest management at a regional scale. *Scandinavian Journal of Forest Research, 23*(5), 431–444.

Wang, J., Chen, J., Ju, W., & Li, M. (2010). IA-SDSS: A GIS-based land use decision support system with consideration of carbon sequestration. *Environmental Modelling & Software, 25*(4), 539–553. doi:10.1016/j.envsoft.2009.09.010.

Zaichi, Z., & Chokkalingam, U. (2006). Stakeholder perspectives on constraints and lessons learned from Guangdong Province. Center for International Forestry Research (CIFOR).

Part III
Heritage and Placemaking

Chapter 11
History Matters: The Temporal and Social Dimension of Geodesign

Jan Kolen, Niels van Manen and Maurice de Kleijn

11.1 Introduction: Geodesign and Landscape History

In this chapter, we would like to elaborate on the geodesign framework as it has been developed by Steinitz (2012). We will do so from a somewhat different angle, that is from a perspective of the historical development and the heritage of regions and landscapes and see an important role for geospatial technologies to make this operational. History and heritage do appear in several of the use cases presented by Steinitz in his recent synthesis (Steinitz 2012). The planning history is relevant to the change models, or, as Steinitz puts it, "every plan has a past" (Steinitz 2013). Furthermore, "heritage", "cultural identity", "historically significance" appear as constrains and/or goals that inform the decision models. Yet, in theory and practice, economic performance and other quantifiable criteria often take precedent. Steinitz acknowledges that this is a limitation: "not everything that can be counted counts, and not everything that counts can be counted" (Steinitz 2012, pp. 183). We argue that this is a particularly costly omission with regards to the historical dimensions of (attitudes to) the place subject to spatial intervention. Current processes in the landscape can only be understood properly if their long-term origins and characteristics are taken into account. Impacts of new interventions can only be foreseen once we have grasped their likely interactions with decisions and interventions taken a long time ago. And, the feasibility of decisions is to an important degree determined by attitudes and values—what Steinitz refers to as "cultural knowledge"—that are taking shape with reference to the past. Therefore, we argue that there should be explicit emphasis on the long-term perspective in all the analytical models in Stein-

---

J. Kolen (✉)
Faculteit Archeologie & Centre for Global Heritage and Development (CGHD),
Universiteit Leiden, Van Steenisgebouw, Einsteinweg 2, 2333 CC Leiden, The Netherlands
e-mail: j.c.a.kolen@arch.leidenuniv.nl

N. van Manen · M. de Kleijn
Faculty of Economics and Business Administration, Spatial Economics,
SPINlab, VU University Amsterdam, De Boelelaan 1105,
1081 HV, Amsterdam, The Netherlands

itz's framework and that the historical character of the values that influence the change and decision models should be considered.

We will first introduce the concept of historical and heritage landscape and explore its historical relationship with planning and design practices. This is followed by an explanation of the potential value of historical and heritage information to geodesign. This potential is then illustrated through the example of the Dutch river delta. Finally, concluding remarks will summarise the practical steps that we feel should be taken in order to incorporate the temporal and heritage dimension in geodesign, with an important role for a Spatial Data Infrastructure and associated geospatial technologies.

11.2 Landscape History and Heritage

Until recently, spatial planning, landscape history and heritage practices as fields of study and practice operated and evolved in relative isolation from one another. Given their historical roots, this should not come as a surprise. Spatial planning and historical research as professions both find their roots in Europe and North American in the early nineteenth century with the birth of the nation state. But whereas the former was strictly future-oriented, spatial interventions as a means to construct the nation state, historical research was embedded in a historicist paradigm, focusing on the distinct qualities of past eras. Therefore, spatial planners did not view the past as a blueprint or even point of reference for their designs and historians reconstructed the past without reference the present. Similarly, the first initiatives to study and manage built heritage in the early decades of the twentieth century focused on protection of objects, buildings and sites as if they were part of a museum collection, thus shielding them from new uses or spatial interventions (Janssen et al. 2014). This emphasis on protection persisted well into the second half of the twentieth century, when the definition of heritage was widened to include landscapes and regions. Despite the odd dissenting voice, notably Jane Jacobs who famously argued that "new ideas require old buildings", the modernist tendencies in urban planning and design from the 1920s reinforced planners' rejection of the past.

Since the 1980s, a more constructive relationship can be observed between landscape history and heritage on the one hand, and spatial planning and design on the other. Postmodernism has made spatial planners more receptive to past designs and practices and helped to anchor history and heritage more firmly in the present and future. This has resulted in a number of successful initiatives to integrate historical knowledge and material heritage in (urban) landscape interventions, including the Internationale Baauastellung Emscher Park in Germany's Ruhrgebiet, aimed at strengthening the identity of the place and strengthening its social and economic vitality while at the same time finding new purposes for remnants from the past (Raines 2011). In several countries, such examples inspired research programmes and planning policies explicitly aimed at the fruitful symbiosis of heritage and planning practice.

The Belvedere Memorandum in the Netherlands resulted between 1999 and 2009 in funding to 400 planning-with-heritage projects, a major academic research programme as well as an interuniversity teaching network. A recent evaluation of Belvedere highlights the tremendous enthusiasm invoked among planning agencies, landscape designers and citizens and the major impact of the programme on definitions and practices in heritage management (Janssen et al. 2014). Yet, the long-term impact of Belvedere on landscape planning and design in the Netherlands, and of similar initiatives elsewhere, is highly uncertain. The recent financial crisis and economic downturn and new social and demographic challenges have rapidly changed the field in which the planners and designers operate. Furthermore, a new emphasis on innovation in content (concepts such as "smart cities", "resilient cities", "sustainable cities" and "transit oriented development") and process (new public-private partnerships, bottom up or citizen-led design) could well result in a renewed distancing between design and the past.

Through this article, we seek to help prevent such a trend and indeed inspire those participating in planning and design tasks, especially those applying the geodesign framework, to continue to seek collaboration with historians and heritage practitioners and to consider the history and heritage landscape as a foundation layer for their practices.

11.3 Landscape History, Heritage Values and Geodesign

The geodesign framework consists of a sound and logical series of analytical and interpretative steps, which should be taken in an iterative way. The first step entails a description of the area in the form of representation models. This step is followed by an analysis of the processes that operate in the area today, and, subsequently, by evaluation models, change models, impact models and decision models. In the course of this process, scenarios for change are developed, tested and adapted, a cycle which can be repeated to improve the quality of the decision making process. Geodesign provides an analytical framework for creating a better future world based on a deep understanding of the complexities of the current landscape and of the impact of potential interventions envisioned by the creative minds of landscape designers, urban planners and engineers. Yet, the long-term character of processes, the often prolonged impact of decisions taken in the distant past and the influence of the past on social values and thus on decisions, do have an obviously place in the models in Steinitz's framework, but are rarely treated systematically by those applying his framework. Therefore, geodesign-informed interventions could profit considerably from a deeper engagement with the past and historical spatial information. Recent technological innovations have matured and as such we are now at a turning point where historical spatial data and information and knowledge about the past generated by historians and heritage scholars can easily be shared. Both approaches require digitization of historical data, making large bodies of data available. However, due to disciplinary unawareness and ontological issues the main difficulties are to access the information and understand the value of the data. Recent technological innovations of user centric Spatial Data Infrastructures (SDI) have produced tools which we expect to enable aid in bridging the gap between different disciplines (De Kleijn et al. 2013). This is also the explicit goal of the Rediscovering Landscape Programme that has recently been established between the VU University Amsterdam and the University of Leiden: to come to better designs and better decisions by incorporating historical insights in every step of the geodesign framework aided by applying innovative SDI technologies.

We can illustrate the value of adding a temporal dimension to geodesign by reflecting on the historical development, heritage and current dynamics of a particular landscape or region. For this purpose, we will zoom in on the landscape in which the first European Geodesign Summit was held, the river landscape of the central Netherlands, which in turn is part of the densely built-up Dutch delta. Based on this case study, some preliminary observations will be made about why history should matter to geodesign and how the application of a user-centric SDI can help in the process of designing with history.

11.4 A Long-term Perspective on Landscape Change

By making a representation model (the first step in the geodesign framework) of the Dutch river landscape, we could list thousands of characteristics and their interrelationships, even if we focus on a single environmental issue or problem. The same is true for a process model, although it would not be difficult to present an overview of processes currently operating in the area that would influence the landscape the most in the near future. One of the major issues is the fact that climate change and changing hydrological conditions ask for a reorganization of the landscape system, for which measures are taking place already. Yet, to understand these processes better, and in order to make sounder decisions for the future, we have to take into account their long-term nature and their historical background.

Placing recent transformations in the river landscape in a long-term context reveals that these transformations are of a very complex nature, that they are the result of path-dependent developments, and that they are guided by human decisions and interventions that were often taken a very long time ago (Van de Ven 1993; Harten 2000; Renes 2005). This becomes evident only if we consider the long chain of cause and effect, of natural events and human interventions, which eventually produced today's landscape of canalized rivers, dikes, floodplains and polders. The human control over the rivers began rather cautiously with the construction of local dams and dikes in the tenth and eleventh century AD. From the thirteenth century onwards, virtually everywhere in the Dutch delta, these local dikes were incorporated into closed dike systems, forcing the rivers into a straightjacket that later turned out to have problematic effects, especially in seasons when the river's capacity to transport water was pushed to the limits. An additional problem was that the river now showed a tendency to raise its bed at rapid pace as a result of sedimentation within the narrow corridors between the dikes. In contrast, the inhabited area behind the dikes sank as a result of the effectiveness of the drainage system (Renes 2005). This created an increased difference in altitude between the river and its inhabited environment, producing periodical floods with disastrous effects. This problem worsened time and again as the solution was sought in strengthening and raising the closed dike systems, resulting in a continuous elevation of the riverbed and flood plain.

The history of the river landscape thus tells a story about the long-term, the accumulated and often delayed effects of human interventions in the past. We should first understand the nature of this long-term development in order to predict the possible implications and impact of new interventions that may significantly change or continue the course of this development in the future. Yet, for this purpose, archaeological, historical and ecological data and information have to be analyzed much more thoroughly and systematically than has been done so far in geodesign. For this purpose use can be made, for example, of geomorphogenetic datasets, archival information on the history of water management and catastrophic events, archaeological site distribution and on-site stratigraphic information, historical maps and plans, etc. It is these kinds of both "hard" and "soft" data that should be combined in agent-based dynamic modelling.

11.5 Landscape Change and Social Time

Careful consideration of the past also reveals that the landscape is not simply a huge technology that enables people to use or exploit their environment in the best possible way, or a technology that enables people to adapt themselves optimally to changing environmental, climatic and hydrological conditions, but that it should be seen as a social landscape at one and the same time. This explains why significant changes in the landscape also need *social* time to take place, which means that time is needed for people to accept and appreciate fundamental transformations of their living environment (Minc 1986; Connerton 1989; Gell 1992).

A historical case from the Dutch river delta that particularly illustrates this fact is the shift from farming to the early industrial use of the river landscape, which took place in the nineteenth century (Stadhouders 2010). In the farmers' perspective, the landscape was not really owned by them, but the reverse was true: they were owned by the landscape (cf. Akkermans 1991). This means that they perceived the farmland as an inalienable possession of a whole family, or community, which had to be transmitted carefully from one generation to another. Yet, from the perspective of the early industrial entrepreneurs of the area, people obviously had a moral and social right to exploit the landscape and its resources to their own end, and to transform the floodplains accordingly (Elings 2007).

It should be realized that this shift was also made in the social context of single families. We have some detailed accounts of members of the farming community who shifted from farming practices to the systematic exploitation of clay for the industrial production of stone bricks (see Elings 2007 for an example from De

Loowaard), for which the demand increased considerably with the growth of the Dutch cities at the end of the nineteenth century. We know that the position of these innovators was highly contested from a social point of view (Stadhouders 2010; Elings 2007). In fact, it often took two or three generations to turn a family into a successful company in this new industrial sector, indicating that it took social time to accept this fundamental transformation of the river landscape as well. It will be clear that this shift brought about not only a fundamental change in people's historical and genealogical relationship with the land, but also a fundamental transformation of the landscape and its natural biotopes as well. Therefore, in order to better understand socially and economically driven landscape changes, history certainly matters.

11.6 The Role of Heritage Values

The social acceptability of transformations, of course, also determines the success of spatial transformations in the short and longer term. Therefore, the effect of human interventions in the landscape, and in fact of all processes operating in the landscape, is not only a matter of utility and functionality, but of social values and appreciations as well. We shall elaborate somewhat on this particular argument, as it obviously plays a significant role in the public acceptance of measures to make the river landscape "climate-change-proof".

In the early 1990s, when the Dutch river delta was threatened rather unexpectedly by extensive floods, a debate was initiated within Dutch society about the future safety of the area. This debate was dominated by three different visions about the landscape's future and about the question of how to deal with the rapidly changing hydrological situation (De Bruin et al. 1987; Landschap als geheugen 1993; Van de Cammen and De Klerk 2003; Ruimte voor de Rivier 2007; Van Toorn 2011). The first vision was a plea for entirely resetting the landscape, the second for drastically repairing the existing landscape system, whereas the third was a plea for *conserving* the landscape, introducing only minor revisions.

Propagating the first mentioned option, several engineers, landscape designers and ecologists in a sense proposed to reinvent the landscape by replacing the established system of canalized rivers, closed dikes, floodplains and polders entirely. They proposed to make the Dutch river landscape "climate-change-proof" by removing the dikes from parts of the landscape, thereby creating enough space for the rivers in the densely built-up delta, and, by combining this strategy with the development of wetland nature, also restoring biodiversity in the area.

The second scenario was based upon the conviction that we could stick to the existing system, but that the dikes have to be strengthened and raised considerably in order to be prepared for new and more severe floods in the future. In fact, this solution ignored the problem of riverbed sedimentation, resulting in a constant need to further strengthen and raise the dikes. Furthermore, this would mean that the new dikes would become very dominant elements in the landscape, which turned out to be a highly contested issue.

These scenarios were opposed by a local group of inhabitants and particularly of artists from the region, who argued that resetting the landscape or drastically repairing the landscape would inevitably destroy the scenic and historic qualities of the Dutch river landscape, which is why they proposed to adopt a conservative attitude. To keep the landscape as-is does not seem to be a favourable solution in the light of recent floods, but somewhat surprisingly this local opposition was supported by at least some radical adherents of participatory planning, who argued that people have the right to choose for living in a risky environment, and that planners, engineers and landscape designers should not decide for them. But that is beside the point that we wish to make here.

What is important in the context of geodesign, is that all scenarios were based on an appreciation of the river landscape as *heritage*, that is: basic social values, with implicit or explicit reference to current attitudes to the past, were underlying the different concepts of future processes. In the third scenario, local groups adopted a rather essentialist notion of the landscape as a huge collection of material heritage. In their view, the most characteristic land forms and physical structures in the river landscape were created by past communities in the area—a set of *lieux de mémoire* with considerable authentic and scenic value. This view, in fact, corresponds to a much more common attitude towards the river landscape in Dutch society as a whole. This explains why so many people in the Netherlands, including those acting on behalf of local, regional and national government agencies and organizations for heritage management, strive at preserving this typically Dutch landscape product for our descendants, as an icon and symbol of our collective mentality (Schoonderbeek et al. 2006). Hands off our landscape of rivers, floodplains and dikes, for that landscape is *the* reflection of our spatial culture. The Netherlands' contribution to the list of UNESCO's World Heritage is the solidified pride of our long history of successful water management (Van Gorp and Renes 2003). Such an appreciation of the river landscape *as* heritage, in fact, was also recognizable in the second scenario that proposed to repair the existing system. Yet, this scenario stressed the idea that the river landscape is the *changeable* result of the Dutch national character. Our ability to shape the landscape out of an indefinable mixture of mud and water, and so constantly resisting the whims of nature, implicitly formed the basis for the engineers' confidence in the prudent transformability of our country (Kolen 2007). Therefore, in the second scenario, the engineers and landscape designers involved adopted a developmental and certainly less essentialist approach to heritage.

In the first scenario, which proposed to drastically reset the landscape system as a totality, a quite different appreciation and valuation of the past was explicitly referred to. In this case, engineers and landscape designers emphasized that it is that same valuable tradition of engineering, entrepreneurship and adaptability that has now revealed the need to put that history behind us. According to them, the solutions of the past have been stretched to the limit, but can no longer guarantee our safety. From such a perspective, the old spatial technology of canalized rivers, dikes and controlled floodplains have ultimately become a hindrance. In fact, by allowing the river to flow more freely, engineers will create space for an "unspoiled" wetland nature to dominate the landscape again. In this way a much older, pre-modern (or even pre-cultural) landscape picture could be reintroduced successfully in the hypermodern living environment of the Dutch (De Bruin et al. 1987).

This example makes clear that exploring scenarios and solutions for our densely built-up river delta is much more than a technological or spatial task. It is also a social and cultural one, which touches on the social and heritage values that "surround" the physical organization of our landscape and, in part, have also *created* it. The building of scenarios for the Dutch river area reveals the operational nature of very different but fundamental social appreciations of the landscape, which always involve heritage values, and which guide both the creative *thinking* and political *debate* about the most favourable processes in the future, be it explicitly or implicitly.

To operationalise this approach, information and knowledge form a vital component. We believe that geodesign provides the tools and technologies for different stakeholders to interact and debate. The location is where the different scholars interact and where they can exchange their disciplinary knowledge. Attempting to enforce the different landscape approaches to express them spatially, based on spatial information, creates a ground for discussion and decision.

11.7 Concluding Remarks

In this contribution, we intended to observe challenges of taking history into account in landscape interventions and illustrate methods to address the issues. Firstly, that the processes operating in landscape are always multi-scale, not only in a spatial but also in a temporal sense. Secondly, that these processes are often the result of long-term and path-dependent developments that have been initiated by human decisions that were taken a long time ago. Thirdly, that these processes not only involve the functionality and utility of place and space, but also the social appreciation and heritage value of landscapes (cf. Bazelmans 2013). And, finally, that we are currently at a point where interdisciplinary cross fertilization, through using innovative technologies, will aid the processes given above (De Kleijn et al. 2013).

To our opinion, all these aspects of the region, of the processes operating in it today and of the debates regarding its future, could and should be taken into account to ensure the quality of the landscape intervention. The case of the Dutch river delta illustrates that it is indeed feasible to think about the processes of landscape change in this way. We hope that it will encourage landscape geographers, archaeologists, historians and heritage scholars to join projects in which the geodesign framework is applied, and that landscape designers and environmental scientists will incorporate historical insights into their work. This will reveal whether geodesign, by explicitly incorporating a historical dimension, truly facilitates fitter designs and more robust decisions. Finally, we expect that using innovative geospatial tools and technologies will aid in this process. One of the main goals of the Rediscovering Landscape programme (De Kleijn et al. 2013) is therefore to generate theoretical and technology driven methodologies and empirically test them in order to generate best practices for more generic tools enabling to incorporate the historical dimension.

References

Akkermans, A. J. M. (Ed.). (1991). *Herinneringen uit mijn leven*. Zevenaar: Jan Willem Conrad Koch, een Liemerse autobiografie.

Bazelmans, J. (2013). Waarde in meervoud. Naar een nieuwe vormgeving van de waardering van erfgoed. In: S. Van Dommelen & C.-J. Pen (Eds.), *Cultureel erfgoed op waarde geschat. Economische waardering, verevening en erfgoedbeleid* (pp. 13–23). Amsterdam: Platform 31.

Connerton, P. (1989). *How societies remember*. Cambridge: Cambridge University Press.

De Bruin, D., Hamhuis, D., Van Nieuwenhuize, L., Overmars, W., Sijmons, D., Vera, F. (1987). *Ooievaar. De toekomst van het rivierengebied*. Arnhem: Stichting Gelderse Milieufederatie.

De Kleijn, M.,Van Manen, N., Kolen, J., & Scholten, H. J. (2013). *User-centric SDI framework applied to historical and heritage European landscape research*. Research Memorandum 2013.

Elings, A. (2007). *De aanraking met het wezen der dingen. Tijdservaringen als nieuw erfgoedperspectief* (Ma Thesis). Utrecht: Universiteit Utrecht.

Gell, A. (1992). *The anthropology of time. Cultural constructions of temporal maps and images*. Oxford: Berg.

Harten, J. D. H. (2000). Rivierkleilandschap. In: S. Barends et al. (Eds.), *Het Nederlandse landschap. Een historisch-geografische benadering* (pp. 92–103). Utrecht: Matrijs.

Janssen, J., Luiten, E., Renes, J., & Rouwendal, J. (2014). Heritage planning and spatial development in the Netherlands: Changing policies and perspectives. *International Journal of Heritage Studies, 20*(1), 1–21.

Kolen, J. (2007). The Dutch cultural landscape and the cultural dialogue about spatial identity. In: A. Evelein (Ed.), *The artificial land* (pp. 21–31). Utrecht: Centrum Beeldende Kunst Utrecht.

Landschap als geheugen. (1993). *Landschap als geheugen. Opstellen tegen dijkverzwaring*. Utrecht: Cadans.

Minc, L. D. (1986). Scarcity and survival: The role of oral tradition in mediating subsistence crises. *Journal of Anthropological Archaeology, 5*, 38–113.

Raines, A. B. (2011). Wandel durch (Industrie) Kultur [Change through (industrial) culture]: conservation and renewal in the Ruhrgebiet. *Planning Perspectives, 26*(2), 183–207.

Renes, J. (2005). Water management and cultural landscapes in The Netherlands. In H. S. Danner, J. Renes, B. Toussaint, G. P. Van de Ven, & F. D. Zeiler (Eds.), *Polder pioneers* (pp. 13–32). The influence of Dutch engineers on water management in Europe. Netherlands Geographical Studies 338, Utrecht.

Ruimte voor de Rivier. (2007). *Planologische Kernbeslissing Ruimte voor de Rivier*. The Hague: Ministerie van V&W, Ministerie van VROM en Ministerie van LNV.

Schoonderbeek, M., Kolen, J., & Sijmons, D. (2006). The landscape as historical montage. A dialogue between Jan Kolen and Dirk Sijmons. *OASE (NAi), 69*, 80–93.

Stadhouders, K. (2010). *Steenfabrieken. Beelden van een veranderend landschap*. Amsterdam: Stokerkade.

Steinitz, C. (2012). *A framework for geodesign: Changing geography by design*. Redlands: ESRI.

Steinitz, C. (2013, 20 September). Lecture at the First European Geodesign Summit, Herwijnen, The Netherlands.

Van de Cammen, H., & De Klerk, L. (2003). *Ruimtelijke ordening. Van grachtengordel tot Vinexwijk*. Houten: Het Spectrum.

Van de Ven, G. P. (Eds.). (1993). *Man-made lowlands. History of water management and land reclamation in The Netherlands*. Utrecht: Matrijs.

Van Gorp, B., Renes, J. (2003). The Dutch landscape: A way of seeing. In B. Van Gorp, M. Hoff, & J. Renes (Eds.), *Dutch Windows. Cultural geographical essays on The Netherlands* (pp. 55–75). Utrecht: FRW.

Van Toorn, W. (2011). *Het grote landschapsboek*. Amsterdam: Querido.

Chapter 12
Urban Landscape archaeology, geodesign and the city of rome

Gert-Jan Burgers, Maurice de Kleijn and Niels van Manen

12.1 Introduction

Throughout the history of Western society, the classical Greek and Roman world has figured as a model for architectural and spatial design and planning, with regard to both individual buildings and entire cities and landscapes (Fig. 12.1). It might therefore come as a surprise that classicists—those scholars who actually study the classical world—have refrained from employing classical models to further modern needs. They investigate ancient art and architecture and even ancient town planning, but usually only as examples of the unique character of classical civilization. In this paper, we observe that in recent decades approaches have changed in this regard and that classical archaeologists have much to offer to modern spatial planning and design. In regard to these changes, we will demonstrate how archaeological and heritage information can be integrated in geodesign-informed spatial planning and the role that geospatial technologies can play in this integration. A Spatial Data Infrastructure (SDI) has been developed and deployed for an urban design task in Testaccio, a neighbourhood in Rome, the very heart of Classical society. Our evaluation of this case study will conclude with suggestions for further promoting the mutually beneficial interchange between archaeology and gedesign.

G.-J. Burgers (✉)
Faculty of Arts, VU University Amsterdam,
De Boelelaan 1105, 1081 HV, Amsterdam, The Netherlands
e-mail: g.l.m.burgers@vu.nl

M. de Kleijn · N. van Manen
Faculty of Economics and Business Administration, Spatial Economics,
SPINlab, VU University Amsterdam,
De Boelelaan 1105, 1081 HV, Amsterdam, The Netherlands
e-mail: mtm.de.kleijn@vu.nl

N. van Manen
e-mail: n.van.manen@vu.nl

Fig. 12.1 Classical inspiration: Capitol Hill, Washington, D.C. Photo by Geoglance

12.2 Recent Trends in Classical Archaeology and Heritage Studies

The changes that we refer to are of a contextual nature. First, in spatial terms, classical archaeologists have traditionally focused mainly on the study of ancient cities and, within them, of classical monuments in relative isolation, that is, as highlights in the history of European art and architecture. However, as a result of successive, post-WWII stages of theoretical thinking, ranging from processual and Marxist to contextual approaches, the city and its monuments are now analysed in their widest spatial and social contexts (Bernard Knapp 1992; Bintliff 1991; Morris 1994; Millett 2007; Osborne 1987). Studies focus not only on emperors and aristocrats, but also on common people and slaves—the 'people without history', to quote from a famous book by anthropologist Eric Wolf (1982)—and on the world beyond the city, from the fertile countryside to marginal landscapes such as marshes and mountains. Many excavations are now carried out on the periphery of modern Rome, to investigate ancient harbour and industrial areas, rural villas, vineyards and drove roads. Likewise, spatial analyses are performed to relate all these elements to each other.

A related shift can be observed in archaeological heritage management. Nowadays, following European treaties such as the European Convention on the Protection of the Archaeological Heritage and the European Landscape Convention, when we talk about the archaeological heritage of the city of Rome we refer to the entire assemblage of physical traces of the ancient landscape within the modern territory of Rome, from monuments to simple cottages, and from marble statues to the pits of ancient vineyards.

A second factor that is important to understand is that changing attitudes towards the Roman heritage is of a chronological nature. Most of the archaeological monuments in the modern city date to the Imperial period. On the one hand this is because in that period construction techniques had reached their apogee, which of course influences the lifespan of monuments and buildings. On the other hand, sustainability is also heavily influenced by later perceptions of the very same objects. From the latter perspective it has to be pointed out that throughout the later Roman history, particularly Imperial heritage has been cherished and preserved, excavated and restored, with the aim of glorifying the Rome of the emperors (e.g. Manacorda and Tamassa 1985; Painter 2005). The monuments have become symbols of that glory.

But there is of course much more than the Imperial monuments: the Roman landscape was and still is highly dynamic. Following the same theoretical approaches mentioned above, the chronological context has widened significantly in both archaeological narratives and heritage management. Attention is also being paid to the Iron Age hut compounds on the Roman hills that preceded the classical period (e.g. Carandini 2007), to the periods of disintegration of the city and to the landscape of ruins of the Middle Ages (Menighini and Santangeli Valenzani 2007). Some even specialize in the archaeology of industrial buildings of the last century, or in the modern Roman landscape (Bjur and Santillo Frizell 2009). In fact, attention is slowly shifting to a completely diachronic history of the entire urban landscape of Rome from prehistory to modern times. In this context, our colleagues at the Swedish Institute in Rome, who are working on the Roman Via Tiburtina, speak of 'urban landscape archaeology', namely the study of the present-day urban texture in all its historical diversity, as a palimpsest of closely interwoven histories (Bjur et al. 2009). Central to this are the continuous transformation processes that have been responsible for this palimpsest.

A third change that can be observed in archaeological heritage management, also in Rome, concerns the very concept of heritage. This is a complicated concept, one that arouses many different associations and that is used by many scholars in different ways. Archaeologists dealing with heritage have commonly emphasized its relation with the past. Archaeological heritage is considered primarily a source to reconstruct the past; being able to read, to interpret this heritage is a premise for understanding the past. This approach can be generalized under the label of historicism (Kolen 2005, pp. 70–76).

Other scholars, however, emphasize that heritage is related primarily to the present. In this perspective, heritage is argued to relate to an idea, that is, the idea of having inherited something, of being the rightful heir to an object, a building or a specific past. In this case, heritage is perceived as a modern construction; it belongs to the domain of the present and is often related to political, ideological or economic use of the past. This approach can be defined as constructivism (Kolen 2005, pp. 70–76). It became popular in the 1980s through major works such as those of Hobsbawn and Ranger (1983), Lowenthal (1985) and Nora (1984–92). In this perspective, the past is not perceived of as an objective historical reality. The approach to the past and its interpretation at any given time are rather seen as mirrors of contemporary society; monuments and museums are not neutral transmitters of the past but *lieux de memoire*, or places and modes through which communities, whether nations or other communities, create and safeguard their collective memory.

The strength of the more recent, constructivist approaches lies in promoting critical thinking on how the classical ideal has been constructed and used. These approaches encourage us to consider how the classical monuments have been idealized, and to accept that they have never stood there the way they are presented nowadays: they have been consciously excavated, reused or restored by popes, princes, politicians and even local peoples throughout the centuries. That is the case especially with the historical centre of Rome, which does not represent a homogeneous and stable state of affairs, reflecting the past 'as it really was'; rather, it is a heterogeneous selection of monuments, buildings and artefacts of all ages, preserved and spatially arranged in a museum context through successive interventions.

The same critical approaches also encourage archaeologists to become aware of their being part of this continuous process of re-appropriation and redesign as well as of their responsibility towards contemporary society; again with regard to the Roman 'open-air museum', it is present-day society that maintains it, integrates it into the modern urban tissue and organizes it according to its own needs, whether these needs are related to identity formation, tourism, or urban design and planning.

12.3 The Challenging Testaccio Project

It is this awareness that increasingly opens up the work of classical archaeologists to the needs of present-day society (Burgers 2007; 2009). It is also central to the Roman research that we are involved in, that is, the 'Challenging Testaccio. Urban Landscape History of a Roman Rione' project (Fig. 12.2). This is a joint project between the Soprintendenza Speciale per I Beni Archeologici di Roma, the Royal Netherlands institute in Rome and VU University Amsterdam. The project is focused on the neighbourhood of Testaccio, in the sub-Aventine plain, immediately west of the Aventine hill (Fig. 12.3). In ancient times, the city's river harbour was located here. The modern neighbourhood was built on top of it in the late nineteenth and early twentieth century, to provide accommodation for working-class citizens. The neighbourhood is now being restyled and we have been invited to collaborate in a study of the history and archaeology of the Testaccio area, which is to inform the urban redevelopment process. The Challenging Testaccio project has three major aims that are very much in line with the recent trends in Roman archaeology and heritage management discussed above.

The first aim is to carry out a comprehensive landscape study of the ancient harbour area. This means an in-depth investigation of the spatial organization and use of the harbour. To that end, we have for example carried out excavations of the standing remains of the Porticus Aemilia (Burgers et al. in press), one of the largest buildings of ancient Rome and a central element of the new harbour (*Emporium*). The excavations have been especially informative on the ancient phases of the building, revealing for instance the remains of a *cella* of a *horreum*, a large warehouse for the storage of grain.

However, the excavations have also enabled us to study significant layers of the post-antique phases of abandonment and reuse of the building and of the wider area around it. In the late Roman period, the area loses its original purpose and over the

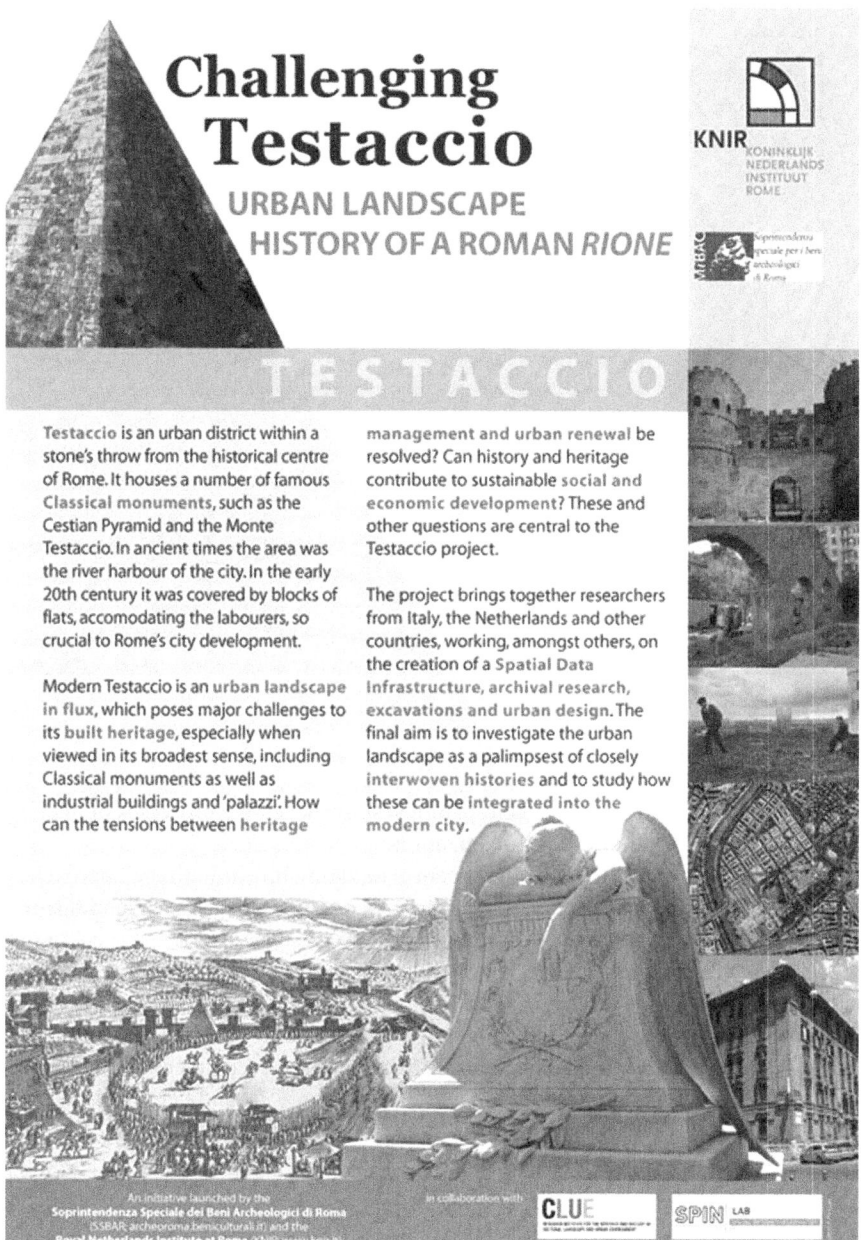

Fig. 12.2 The Challenging Testaccio Project

course of centuries transforms progressively into a suburban countryside, maintaining this character until the threshold of the twentieth century. This leads us to the second aim of the project: to do diachronic urban landscape archaeology as outlined

Fig. 12.3 Location of the Roman neighbourhood of Testaccio

above, namely to study the Testaccio palimpsest of successive processes of ancient urbanization, ruralization and re-urbanization.

The third aim is in line with the recent paradigm shift in heritage approaches discussed in the previous section; that is, to present our archaeological and historical analyses in such a way as to inform and inspire the ongoing urban regeneration process. We are doing so through public outreach events and urban design projects, in close collaboration with the local authorities, citizen groups, architects and urban planners. Moreover, we have developed geospatial tools to facilitate such collaborations. The remainder of the chapter will focus on the development of these tools and their deployment in a first experiment with archaeology and heritage informed planning and design in Testaccio.

12.4 The Spatial Data Infrastructure

All three components of the Challenging Testaccio project require varied and extensive data sets and new methodological frameworks informed by multiple disciplines. The foundation for the research conducted in the project is therefore an integrated platform that gives access to data and information used by the various disciplines, for both disciplinary and interdisciplinary research, related to this study area. Given the strong interest in place and space of the different disciplines, a Spatial Data Infrastructure (SDI) is created as a vehicle to develop innovative research strategies enabling cross disciplinary knowledge exchange and providing tools for interaction.

Spatial Data Infrastructures have been discussed and constructed since the early 1990s. The aims and purposes for SDIs have evolved significantly. Whereas the first SDIs were static repositories with Spatial Information, nowadays SDIs are more focussed on providing a platform where users can efficiently cooperate to handle spatial data (Hennig and Belgui 2013; de Kleijn et al. 2013; Rajabifard et al. 2006). Within the Challenging Testaccio project, we approach an SDI as a collaboration platform in which geospatial tools and services enables users to interact and make use of each other's disciplinary knowledge. Therefore, the user is placed at the heart of the development process, aimed at generating a useful and user-friendly SDI.

Developing the User-Centric SDI for Challenging Testaccio is done iteratively, through successive "waves" of design, deployment and evaluation. Together these different waves produce building blocks with information and functionality with which the Challenging Testaccio SDI is built. Focussing on particular sub projects enables us to get a clear view on the user requirements for data and functionalities, making the development process steered by users thus stimulating high usability.

So far, only the first wave has been fulfilled. This first wave was led by the SDI developers, as many of the participants in Challenging Testaccio had little to no knowledge of GIS and were therefore unable at the outset to accurately formulate their data and functionality needs. It included basic GIS training to programme participants, the collecting of all available historical and heritage data related to Testaccio, and the development and deployment of an app in a planning and design competition.

Starting point for developing the app is the idea that archaeological, historical and heritage information presented in an interactive mapping interface can support the dialogue between future oriented disciplines like spatial planning and architecture and past oriented disciplines like archaeology and history, enabling both to make better informed decisions in understanding, using and discovering historical and heritage features. The app was developed as a digital version of the biography of the landscape research strategy described and discussed by Roymans (Roymans et al. 2009). Based on the biography of the landscape research methodology, the information in the app has been elaborated to be usable for designers and spatial planners. Besides carefully selected historical maps, geotagged historical images and points of interest representing the *museo diffuso* developed by SSBAR (Sebastiani and Serlorenzi, 2008), a retrospective cartographic reconstruction was inserted into the app (de Kleijn et al. 2013). This cartographic reconstruction is obtained by analyzing historical maps in combination with an analysis of archaeological excavations. It contains thematic maps representing all significant transformations of Testaccio. The spatial information available in the app is also accessible through services for other purposes. However, since the app is meant to be used outdoors, a direct connection between the SDI services and the app is not in place. All spatial information are integrated in the app and stored locally on the device.

To test the hypothesis that the app stimulates interdisciplinary interaction resulting in better historic and heritage informed decisions an experiment was organised. Thirty-one architecture students, divided in groups of three, were given the task to develop a design for a particular square in Testaccio. Half the groups worked with the app, the other half with conventional sources including books, maps and photographs, enabling to measure differences between the different groups by analysing their results and by questionnaire surveys before and after the design task.

The experiment was done on a relatively small sample and the participants were presumably slightly biased since they were educated to design with heritage and informed of the organisers' objectives. Nevertheless, the results of the experiment do suggest that, overall, the architects considered the app useful to the task (De Kleijn et al. forthcoming). Participants stated that they used 40% (n=17) of their time in consulting the app during the design task. The percentage they were willing to reserve budget to develop the app is 20% (n=31). A more detailed examination of the questionnaires that were completed by the participants will need to reveal which data and functionalities they found particularly useful.

This analysis will be the starting point for involving designers and planners closely in the further development of the SDI and tools such as the app. The app *was* developed based on insights from previous attempts at heritage and planning integration (Bosma and Kolen et al. 2010; Elerie and Spek 2010; de Kleijn et al. 2013). Yet, the data and functionality of the app, including the cartographic reconstruction and the interface, were developed by non-architecture past-oriented researchers. Furthermore, to promote truly transdisciplinary methods, follow-up experiments in subsequent waves should also involve participants from disciplines other than planning and design. Nevertheless, the GIS training, data inventory and experiment have produced valuable first building blocks for the SDI.

12.5 Conclusions

The geodesign framework has always had the scope to involve archaeological and historical information in analysis, design and evaluation tasks. Yet, the changing character of archaeology, as outlined in this chapter, provides new opportunities to fulfil this potential. Archaeologists have widened their spatial, chronologic and thematic scope and become more inclined to connect their research to the present and the future. As a result, the information they produce is more attuned to the needs of planners and designers. As highlighted by the Testaccio design competition, when presented in attractive and interactive formats, this information can inform spatial interventions. Yet, our evaluation of this competition also highlights that more is required in order to facilitate a creative exchange between the past and future oriented disciplines and develop truly transdisciplinary methods. In particular, representatives of these disciplines should be given joint task, to become acquainted with each other's methods and to stimulate the formulation of new, joint lines of research. As highlighted by the Testaccio SDI, geospatial technologies can play a facilitating role in this process, but only if the users are closely involved in their development.

References

Bernard Knapp, A. (Ed.). (1992). *Archaeology, annales, and ethnohistory*. Cambridge: Cambridge University Press.

Bintliff, J. (Ed.). (1991). *The annales school and archaeology*. New York: New York University Press.

Millett, M. (2007). What is classical archaeology? Roman archaeology. In S. E. Alcock & R. Osborne (Eds.), *Classical archaeology* (pp. 30–50). Oxford: Oxford University Press.

Bjur, H., & Santillo Frizell, B. (Eds.). (2009). *Via Tiburtina. Space, movement and artefacts in the urban landscape*. Stockholm: Acta Instituti Romani Regni Sueciae.

Bosma, K., & Kolen, J. (2010). *Geschiedenis En Ontwerp: Handboek Voor de Omgang Met Cultureel Erfgoed*. Nederlands: Vantilt.

Burgers, G.-J. (2007). Classical archaeology and local communities. Setting up archaeological parks in the Italian region of Apulia. In *Interpreting the past. Who owns the past? Heritage rights and responsibilities in a multicultural world*. Proceedings of the Second Annual Ename International Colloquium, Ghent 22–25 march 2006, Brussels, pp. 107–118.

Burgers, G.-J. (2009). L'Archeologia classica tra dimensione internazionale e realtà locali, In A. L. D'Agata, S. Alaura (a cura di), *Quale futuro per l'archeologia? Atti del workshop internazionale, Roma, 4–5 dicembre 2008*. Roma, pp. 27–39. Gangemi Editore.

Burgers, G-J., De Leonardis, V., Della Ricca, S., Kok-Merlino, R-A., Merlino, M., Sebastiani, R., & Tella, F. (in press). *Porticus una extra portam Trigeminam: nuove considerazioni sulla Porticus Aemilia*. In Proceedings of the XVIII International Congress of Classical Archaeology, Mérida Spain.

Carandini, A. (2007). *Roma. Il primo giorno*. Laterza: Roma-Bari.

De Kleijn, M., Van Manen, N., Kolen, J., & Scholten, H. J. (2013). *User-centric SDI framework applied to historical and heritage European landscape research*. Research Memorandum 2013.

De Kleijn, M., Dias, E., & Burgers, G.-J. (forthcoming). *Testaccio, a geospatial heritage instrument*. Results of an Experiment.

De Kleijn, M., Van Aart, C., Van Manen, N., Burgers, G.-J., & Scholten, H. J. (2013). *Testaccio, a digital cultural biography app*. PATCH Conference Proceedings.

Elerie, H., & Spek, T. (2010). *The cultural biography of landscape as a tool for action research in the Drentsche Aa national landscape (Northern Netherlands). The Cultural Landscape Heritage Paradox. Protection and Development of the Dutch Archaeological-historical Landscape and its European Dimension* (pp. 83–113). Amsterdam: Amsterdam University Press.

Hennig, S., & Belgui, M. (2013). User-centric SDI: Addressing users requirements in third-generation SDI. The Example of Nature-SDIplus. *Geoforum Perspektiv, 10*. pp. 30–42.

Hobsbawn, E., & Ranger, T. (Eds.). (1983). *The invention of tradition*. Cambridge: Past and Present Publications.

Kolen, J. (2005). *De Biografie van het Landschap. Drie Essays over Landschap, Geschiedenis en Erfgoed*. Dissertation, VU University Amsterdam.

Lowenthal, D. (1985). *The past is a foreign country*. Cambridge: Cambridge University Press.

Manacorda, D., & Tamassa, R. (1985). *Il Piccone del Regime*. Rome: Curcio.

Meneghini, R., & Santangeli Valenzani, R. (Eds) (2007). *I Fori Imperiali: Gli Scavi del Comune di Roma (1991–2007)*. Rome: Viviani.

Morris, I. (Ed.) (1994). *Classical Greece: Ancient histories and modern archaeologies*. Cambridge: Cambridge University Press.

Nora, P. (Ed.). (1984–1992). *Les Lieux de Mémoire*, Paris: Gallimard.

Osborne, R. (1987). *Classical landscape with figures: The ancient Greek city and its countryside*. London: George Philip.

Painter, B. W. (2005). *Mussolini's Rome. Rebuilding the eternal city*, New York: Palgrave MacMillan.

Rajabifard, A., Binns, A., Masser, & Williamson, I. (2006). The role of sub-national government and the private sector in future spatial data infrastructures. *International Journal of Geographical Information Science, 20,* 727–741.

Roymans, N., Gerritsen, F., Van der Heijden, C., Bosma, K., & Kolen, J. (2009). Landscape biography as research strategy: The case of the South Netherlands project. *Landscape Research, 34,* 337–359.

Sebastiani, R., & Serlorenzi, M. (Eds.). (2008). Il progetto del Nuovo Mercato di Testaccio. *Workshop di Archeologia Classica, 5,* 137–171.

Wolf, E. R. (1982). *Europe and the people without history.* Berkeley: University of California Press.

Chapter 13
GIS-based Landscape Design Research: Exploring Aspects of Visibility in Landscape Architectonic Compositions

Steffen Nijhuis

13.1 Introduction

Knowledge of spatial design is at the core of landscape architecture. This implies that the development of skills for exploring and defining landscape designs as architectonic compositions is a necessity for landscape architects (Steenbergen et al. 2002; Dee 2001). The concept 'composition' refers to a conceivable arrangement, an architectural expression of a mental construct that is legible and open to interpretation. In that respect the landscape design is regarded as an 'architectonic system' by which rules of design common to all styles are established (Colquhoun 1991; Steenbergen et al. 2002). Landscape design research is a vehicle to acquire knowledge of spatial composition via architectonic plan analysis. It is a matter of developing and deploying spatial intelligence, the architectural capacity or skill to think and design in space and time (Gardner 1999). This includes the ability to understand, represent and construct landscape architectonic compositions. Because the fundamental importance of spatial intelligence, designers have always been eager to employ manual and digital media which can support thinking and communicating about spatial compositions (Zube et al. 1987; Bishop and Lange 2005; Nijhuis 2013). These tools are extensions of the designers' perception and help to measure *what* we see and determine also *how* we see (e.g. Horrigan 1995). Here seeing is equated with knowledge acquisition. In fact this dialectic between research and the involved technology (the tools) and the representation and interpretation of reality has been at the core of science and art for centuries (Kemp 1990).

S. Nijhuis (✉)
Faculty of Architecture and the Built Environment, Delft University of Technology, Julianalaan 134, 2628 BL Delft, The Netherlands
e-mail: s.nijhuis@tudelft.nl

13.1.1 GIS In Landscape Design Research

Although GIS is potentially a powerful tool for landscape design research, in landscape architecture GIS is often regarded as a landscape planning tool or 'map machine' to document, visualize and present geographic realities. But not as a tool to increase spatial intelligence for landscape architectonic research and design. However, Jack Dangermond—founder of ESRI and landscape architect—points out ('Foreword' in: Longley and Batty 2003):

> The real heart of GIS is the analytical part, where you explore on a scientific level the spatial relationships, patterns and processes of geographic, cultural, biological and physical phenomena.

Although landscape architectonic phenomena are not mentioned, this definition implies a wide range of possible applications of GIS in landscape design research, since geographic (contextual) relationships and spatial patterns and structures are key concepts for understanding landscape architectonic compositions. Via advanced spatial analysis GIS generates specific information which can reveal knowledge contained in landscape architectonic compositions. It facilitates researchers in landscape architecture to get a grip on spatial (future) realities by offering a vehicle for capturing, analyzing, manipulating ideas, forms and relationships in a geographic context while using visual representations.

13.1.2 Objective and Structure

This study explores the application of GIS in landscape design research in order to get a grip on the 'DNA' of landscape architectonic compositions. The presumption is that GIS offers a tool for measurement of relevant and new aspects of the composition—influencing what knowledge we acquire—as well as provides an alternative way of understanding compositions—influencing how we acquire knowledge. But how can landscape architectonic compositions be studied in a transparent and systematic way? What GIS-based methods and techniques are suitable for modelling, analyzing and representing them? How can these GIS-based methods and techniques be applied in landscape design research and what are the results? What are the possibilities and limitations of GIS for the development of landscape design research methods and techniques, and what is the contribution to the design-knowledge apparatus of landscape architecture?

Although these questions cannot be addressed fully here, this chapter aims to showcase some applications of GIS for the analysis, simulation and evaluation of the landscape architectonic composition 'from the inside out', exemplifying the potential of GIS in landscape design research. Although other aspects of the composition are also very important, this chapter focusses mainly on GIS-based analysis of visibility (cf. Ervin and Steinitz 2003). A more complete account on the possibilities of GIS in landscape design research can be found in Nijhuis (2015).

The next paragraph elaborates on a framework for landscape design research, followed by a GIS-based visibility analysis, exemplified by case studies of Piazza San Marco and Stourhead landscape garden. The chapter closes with conclusions and prospects for GIS-based landscape design research.

13.2 Understanding Landscape Architectonic Compositions

Landscape architectonic compositions embody a great wealth of design knowledge as objects of our material culture. They carry knowledge about how to satisfy certain requirements, how to perform tasks, and it is a form of knowledge that is available to everyone (Cross 2006). In particular, by studying landscape architectonic compositions knowledge can be acquired of the possible relationships between conceptual thinking and the three-dimensional aspect (Steenbergen and Reh 2003). In this respect the landscape design, as expressed by its spatial composition, is a container of design knowledge and serves as a basis for new designs (Nijhuis and Bobbink 2012). Therefore landscape architectonic compositions can be considered as an important source of design knowledge.

13.2.1 Analytical Framework for Landscape Design Research

Grounded in the notion of precise geographic and geometric delineation of landscape architectonic compositions Steenbergen *cum suis* proposed a framework for landscape design research (Steenbergen and Reh 2003; Steenbergen et al. 2008). This analytical framework consists of four general categories that lay out the relation between the various aspects of the architectonic form and its perception in a systematic way. By application of this analytical framework the design principles that constitute the design can be understood. It provides the basis for a deeper understanding of the landscape architectonic composition via analysis of the:

- Basic form: the way in which the topography of the natural landscape or the man-made landscape is reduced, rationalized and activated in the ground plan of the design;
- Visible form: the form and functioning of three-dimensional landscape space, which creates a spatial dynamic. This might be, for example, the framing of a landscape or urban panorama, or the construction of a spatial series along a route, making a pictorial landscape composition;
- Metaphorical or symbolic form: the way in which symbolic, iconographic and mythological images and architectonic structural forms are connected with one another and with elements from nature, such as water, the relief and vegetation. Routes are important operative structures accommodating narratives of design and reception;

- Programmatic form: the spatial program leads to a functional zoning or layout in relation to logistics and functional patterns of movement (path structure).

This study utilizes GIS as a tool for landscape design research in order to acquire knowledge from landscape architectonic compositions, offering possibilities for measurement of relevant and new aspects (what) as well as offering an alternative way of understanding (how) (Fig. 13.1). However, the remainder of this chapter will elaborate on the application of GIS in exploring the visible form of the landscape architectonic composition using an architectural landscape and urban landscape as example. The application of GIS in exploring the basic form, symbolic form and programmatic form, as well as an elaboration on GIS in visible form research can be found in Nijhuis (2015).

13.2.2 GIS-based Analysis of Visible Form

Visible form is about the visual manifestation of three-dimensional forms and their relationship in outdoor space, expressed by its structural organization (e.g. balance, tension, rhythm, proportion, scale) and ordering principles (e.g. axis, symmetry, hierarchy, datum, transformation) (Hubbard and Kimball 1935; Bell 1993). Visible form refers to the appearance of objects; it is about the 'face' of the composition. However, the meaning attached to it is referred to as semantic information, and is dependent on the receiver (Haken and Portugali 2003; Blake and Sekuler 2006).

Here landscape architectonic compositions are regarded as visibility fields and address those parameters that are observable by a viewer located within space (a horizontal perspective), and those configuration properties that can be discovered by visual experience evoked by optical axes, visibility fields and sequences of visual information (Psarra 2009; Tzortzi 2004). The concepts of visual perception such as organization of visual logic, space-making, composing views and the control of movement are important properties to be explored (Nijhuis 2011). In order to convey the composition from an observers point of view and to enable visual analysis of the landscape Tandy (1967) introduced the concept of isovists or viewsheds. An isovist or viewshed is an area that is visible from a specific location, also called "limit-of-vision plottings" or "visual watersheds". Early applications can be found in Higuchi (1975), Lynch (1976) and Benedikt (1979, 1981).

Due to advances in computer science visibility-analysis is nowadays a widespread phenomenon with a broad palette of applications (e.g. Nijhuis et al. 2011). Moreover, advances in GIS offer researchers in landscape design interesting clues to engage in the field of visual research, particularly the GIS-based isovists (sight field polygons) (e.g. Batty 2001; Rana 2002) and viewsheds (e.g. Gaffney and Stančič 1991; Wheatley 1995; Llobera 2003). Both concepts address the physiognomy of space with visibility as a key element. The potential of 'being able to see' is mapped out and addresses plausible and/or probable visible space (Fisher 1995, 1996; Weitkamp 2010). It exposes spatial patterns composed of open spaces, surfaces, screens and volumes as it could be experienced by an observer moving through

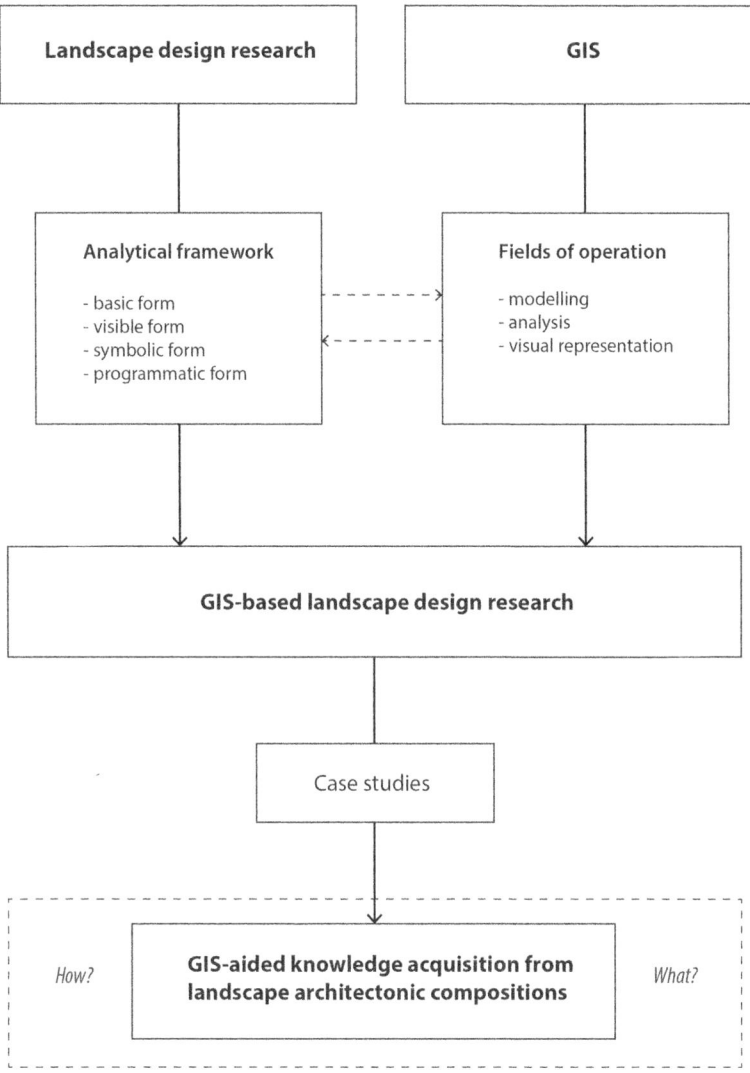

Fig. 13.1 Framework for GIS-based landscape design research. This study utilizes GIS as a tool for landscape design research in order to acquire knowledge from landscape architectonic compositions, offering possibilities for measurement of relevant and new aspects (what) as well as offering an alternative way of understanding (how). In this chapter the focus is on the analysis of the visible form. (Source: Nijhuis)

a virtual space, making use of GIS-based isovists and viewsheds. The technical difference between the two concepts is that the raster-based viewsheds represent parts of space that are visible, taking into account vertical viewing angle and elevation, while vector-based isovists consider visible space in the horizontal plane. The result is a closed polygon that can be characterized with different numerical parameters (Batty 2001; Turner et al. 2001).

13.3 Two Examples of GIS-based Visibility Analysis in Landscape Design Research

This section explores the use of GIS-based viewsheds and isovists in landscape design research in order to reveal some important aspects of the visible form of two well-documented examples: Piazza San Marco (Venice, Italy) and Stourhead landscape garden (Wiltshire, UK). Both sites are widely acknowledged for their designed spatial qualities and articulated visual system, which have the potential to be tested and verified by means of GIS. The GIS-based analysis offers an actual (non- or a-historical) and formal reading of the sites and showcases that GIS-based isovists and viewsheds have the potential of measuring visual phenomena which are often subject of intuitive and experimental design. The description of the examples is based on Nijhuis (2011, 2015), here also more backgrounds can be found on the theory and methods, as well as the modeling, analysis and visual representation with GIS.

13.3.1 Digital Landscape Models

As a basis for the GIS-based landscape design research both examples are digitized and abstracted into highly accurate geo-referenced Digital Landscape Models (DLMs) by means of GIS, CAD and 3D-modeling software. The DLMs consist of a terrain layer, a Digital Elevation Model (DEM), supplemented with a volume layer of 2D and 3D referenced objects, like buildings, trees and other artefacts (Li et al. 2005; Van Lammeren 2011). This DLM can be represented via a surface, vector or raster definition in a 2D or 3D geo-referenced setting. In this respect the DLM is merely a representation of things that exist and represent the landscape architectonic composition in the formal system of the digital world (Van Lammeren 2011).

The DLM of Piazza San Marco is based on data derived from field surveys by the University of Venice (1:100; 1:500) and research by Samonà et al. (1970) and Morresi (1999). The DLM of Stourhead landscape garden is mainly based on recent digital maps (1:2,000; 1:10,000; 1:25,000) provided by The National Trust and the British Ordnance Survey (2010). For the location and nature of the planting recent aerial photographs (orthographic) are used, as well as inventories of Woodbridge (1976, 1970, 1996) and field visits (2009, 2011). The reconstruction of the route and path-structure is based on research by Woodbridge (1976) and Reh (1995). The models, as well as the results of the analysis were verified and corrected via observations in the field (2009, 2011) and recent photographs.

Fig. 13.2 Piazza San Marco, 2010. (Photo: Nijhuis)

13.3.2 *Piazza San Marco: GIS-based Analysis of Space Relationships and Articulation of Space*

The Piazza San Marco is one of the quintessential parts of Venice and is highly appreciated by inhabitants as well as thousands of tourists. The square is a symbol that represents the city of Venice, its history, politics, religion and social and ethical values. The vicissitudes of the piazza's transformation are slow and far-reaching and have occurred over a long period of time (e.g. Samonà et al. 1970; Morresi 1999; Schulz 1991). The piazza is divided into two parts that form an L-shape: the actual piazza and the *piazzetta* (little square). The L-shape is one of the most challenging designs for a square, and the least liable to succeed. This shape has a distinct disadvantage as each branch, the piazza and the piazzetta, has a hidden counterpart (Figs. 13.2, 13.3 and 13.4). Nevertheless, the architectonic composition is very successful and is acknowledged for its spatial qualities such as the articulation of space and space relationships (e.g. Janson and Bürklin 2002; Newton 1971; Samonà 1970). Particularly the entrances to the square and the space-turning role of the campanile are of great importance to the visible form of the square, which can be explored by means of GIS-based isovists (Nijhuis 2011).

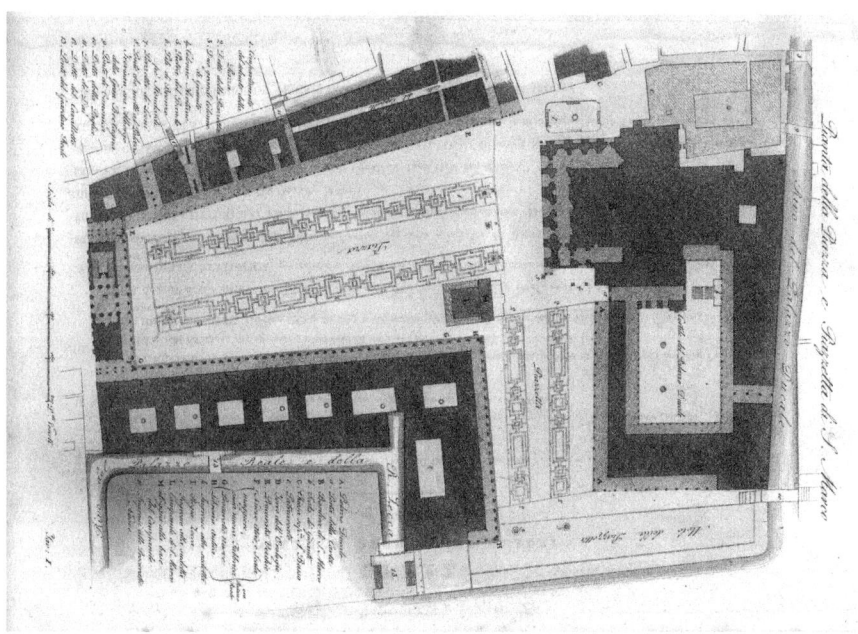

Fig. 13.3 Plan of the Piazza San Marco. Dionisio Moretti, 1828. (Source: Supernova Edizioni)

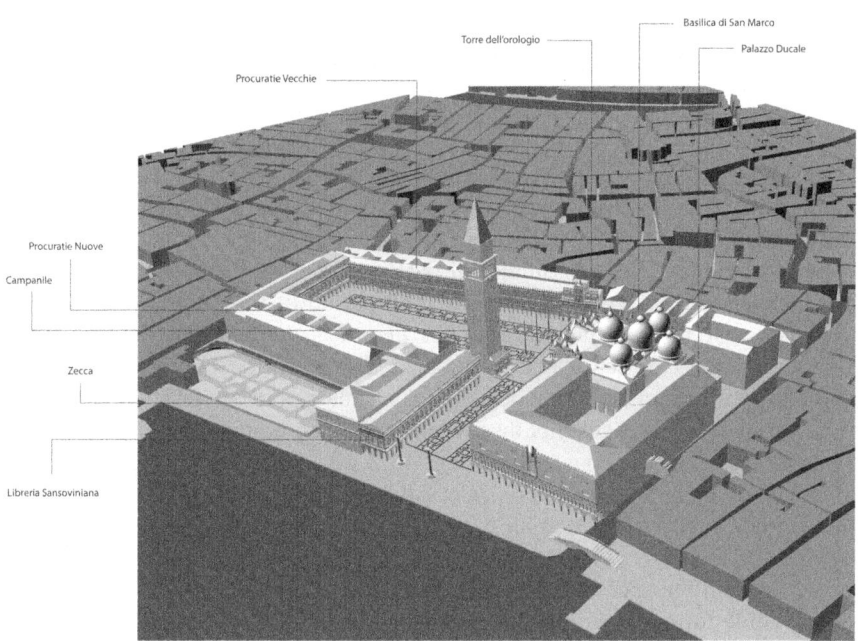

Fig. 13.4 Three-dimensional model of the square. (Model by Nijhuis and Wiers)

Fig. 13.5 Sequence of views entering the square at the Torre dell'orologio. (Source: Nijhuis)

Entrance from the Clock Tower

As Piazza San Marco is a square, the experience of visible form is not directed by paths or routes, but by the entrances to the square and the visual effect of the architecture and space relationships.

Here the visible form of the approach to the square via the *Torre dell'orologio* (Clock Tower) is analyzed. This clock tower is one of the most important links between the piazza and the rest of the city. GIS-based isovists (at eye-level; 1.60 m) are employed in a sequence of viewpoints to study the perceptual order of the entrance. The sequence of isovists shows the framed views into the piazza, across the façade of the Basilica, straight out through the piazzetta, until *San Giorgio Maggiore*. On the opposite side, it provides visual reference, taking the eye past the piazza and on in the direction of Rialto. However, towards the square the optical axis points towards the piazzetta, to gradually open out over the whole piazza. This slow sequence of frontal views can also be represented as a *Minkowski-model* (Benedikt 1979) showing the relation between visible form and time (movement). The model is a sequential stacking of individual isovists and shows the gradual change of visible space by moving forward entering the square (Figs. 13.5 and 13.6). In this respect the landscape architectonic composition affords movement by its openings, offers a sense of direction by its spatial orientation and offers arousal/attraction by its visual composition.

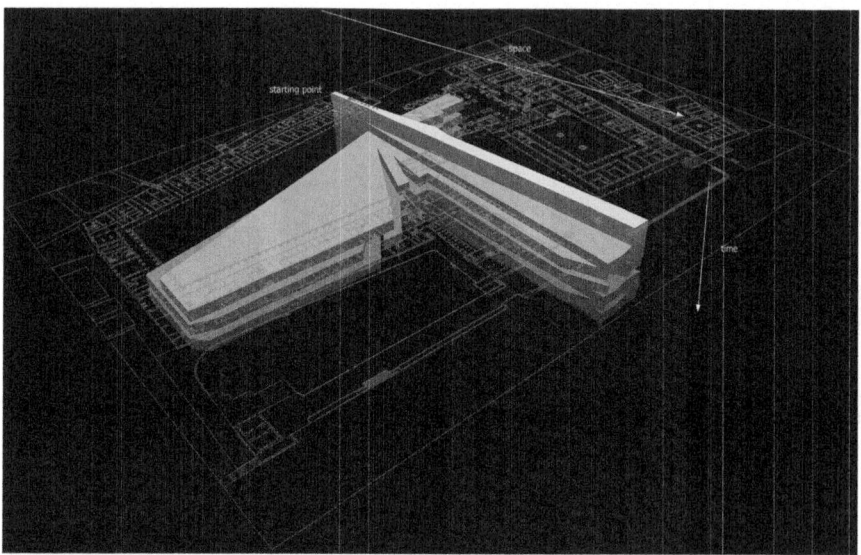

Fig. 13.6 Minkowski-model from Piazza San Marco approached via the Torre dell'orologio. The *top layer* of the model represents the first isovist at the starting point; the *bottom layer* represents the last isovist. (Source: Nijhuis)

Space-turning Role of the Campanile

The campanile acts as pivotal point or hinge on which the two spaces turn; the relatively greater height of the tower, compared to the *Libreria* and *Procuratie Nuove*, undoubtedly enhances its space-turning role (Janson and Bürklin 2002; Newton 1971; Von Meiss 1991). This architectural phenomenon is here further investigated using isovist fields.

The shape and size of the isovists is liable to change with position and therefore generate specific characteristics. Because of the geometrical nature of these sight field polygons we can characterize them mathematically. Numerical measurements can quantify salient size and shape features such as perimeter, area, diameter, radius, circularity, etc. (e.g. Batty 2001; Rana 2002). These measurements are turned into a set of scalar or isovist fields. These isovist fields provide an overview of the visual properties of the architectonic space analyzed. They show syntactical relations between isovists and can generate parameters such as a clustering coefficient, complexity or drift (Turner et al. 2001).

As visible in Fig. 13.7. the tower as occluding element gives the piazza and the piazzetta relative autonomy, yet at the same time they announce each other's presence. The position of the campanile articulates the connection between piazza and piazzetta as an intermediate member, blocking a direct transition between the two areas of the piazza. As regards the movement of passers-by, this translates into a pause and a change in direction or division of space. This initiates an interesting shift of scenery, a constant change of visibility at eye-level.

13 GIS-based Landscape Design Research

Fig. 13.7 Serial vision from the west-end of the piazza to the south-end of the piazzetta showing the crucial role of the campanile in the changing visibility (degree of shifting scenery) of the spatial transition from the piazza to the piazzetta. (Model by Nijhuis and Wiers)

The shift of scenery can be calculated by using the clustering coefficient parameter in an isovist field at eye level. The clustering coefficient gives a measurement of the proportion of intervisible space within the visibility neighborhood of a point. It indicates how much of an observer's visual field will be retained or lost as the individual moves away from that point (Turner et al. 2001). In order to show the impact of the campanile, a comparison of the piazza with and without the bell-tower can be seen. The analysis points out that the campanile has a great impact on the variation of visibility, and influences large parts of both squares (Fig. 13.8.).

The results from the analysis show that crucial aspects of the visible form can be explored using GIS. Here space relationships are studied with isovists and isovist fields, such as the sequential unfolding of visual space at the entrance of the square and the hinge-effect of the bell-tower introducing a high degree of shifting scenery. This example showcases that GIS enables measurement and alternative visualization, which revealed some particularities of the perceived architectonic space which are often regarded as subjective and hard to represent. Here GIS made it possible to study spatial relationships and articulation of space in a precise, systematic/transparent, and quantified manner.

13.3.3 *Stourhead Landscape Garden: GIS-based Analysis of Composed Views and their Sequence*

A fine example of a landscape architectonic composition that provides individuals with composed views or 'pictures' is the pictorial circuit of Stourhead landscape garden, especially the valley garden (Moore et al. 2000; Grandell 1993; Watkin 1982) (Figs. 13.9, 13.10 and 13.11.). The landscape garden has a long history and was designed and developed by the owners themselves, unassisted by landscape architects (Woodbridge 1976, 1996). The valley garden has a double visual structure, with axial views and circuitous, serial views with a lake as the reflecting pool mirroring the scenes. The first is about stationary vision and framed views directed across a lake, providing scenes with Classical and Gothic emblems dramatically juxtaposed in a larger valley landscape. In fact, these strategic foci are goals, as a stage in a circuit walk and thus initiate movement. The counter-clockwise defined route directs the observer through slow-motion vision and tactile experience (going up and down) through a series of shifting views, offering sequential and gradual discovery of the various features involved. Multiple single viewpoints and their sequential/specific organization are crucial for understanding the visible form of Stourhead landscape garden, and are analyzed by means of GIS-based viewsheds (Nijhuis 2011, 2015). Since the valley garden is a designed space mainly of relief and vegetation, viewsheds are here more suitable for analysis then isovists, because it takes into account vertical viewing angle and elevation, both with wide implications for visibility.

13 GIS-based Landscape Design Research 205

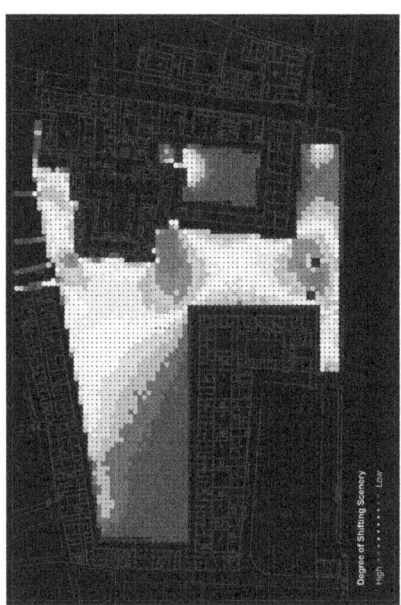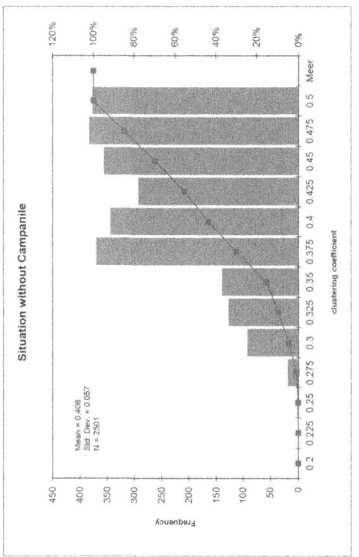
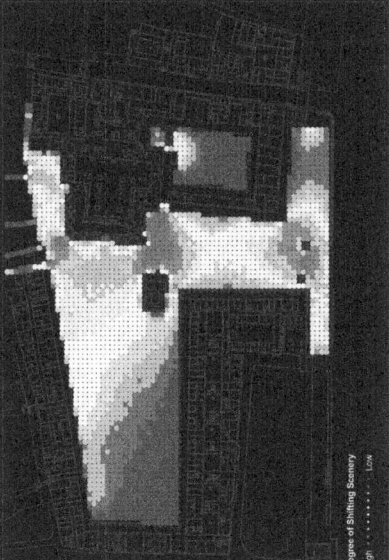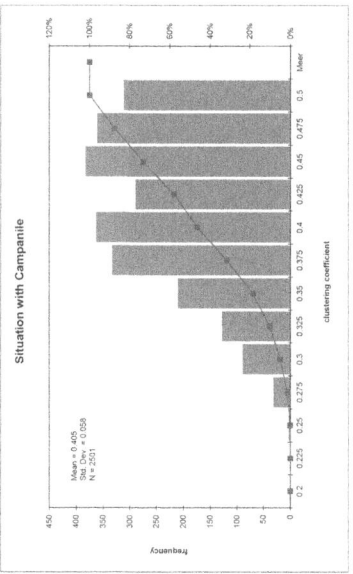

Fig. 13.8 Degree of shifting scenery with and without the campanile. The bell-tower articulates the visual transition between the two spaces by occlusion offering a wide variation in (inter)visibility and influences both spaces (gradual transition). In the situation without the bell-tower the variation concentrates at the corner (sudden transition). The differences are also visible in the graphs which explain the numerical distribution (*left axis*) of the clustering coefficient (*bottom axis*) showcasing a gradual distribution of the values versus sudden changes in the distribution. *Left*: Degree of shifting scenery WITH campanile. *Right*: Degree of shifting scenery WITHOUT campanile. (Source: Nijhuis)

Fig. 13.9 Stourhead landscape garden, 2011. (Photo: Nijhuis)

Framed Views and Focal Points

The path structure facilitates the stroll starting at the Temple of Flora and ending at Bristol High Cross and directs the movement through the three-dimensional composition (Fig. 13.12). By following the counter-clockwise circuitous route the visual form becomes cinematic, because of the sequence of staged views. The axial views are framed by contrasting masses of light- and dark-toned trees and under planting. As a result several composed picture-like views with a foreground, middle ground and background can be seen, reflected by the lake. Occlusion is the most powerful depth-cue involved, exaggerating the perceived distance. By utilizing GIS-based viewsheds it was possible to analyze the visible area from the major viewpoints, measurement of the (angular) extent of the views and object count within the view (Figs. 13.13 and 13.14).

The viewshed-analysis points out that the optimum angular extent of the composed views corresponds with the center of the field of vision in the range of 20–30° binocular view (Table 13.1). This corresponds with the zone with the highest degree of acuity in our field of vision (Snowden et al. 2006; Ware 2004). The analysis suggests that this is the decisive factor for framing the view and (visual) grouping of the focal points in the scene. It is designed 'by eye' as a three-dimensional painting or theatre, rather than using rulers and a compass. This perceptual order is also

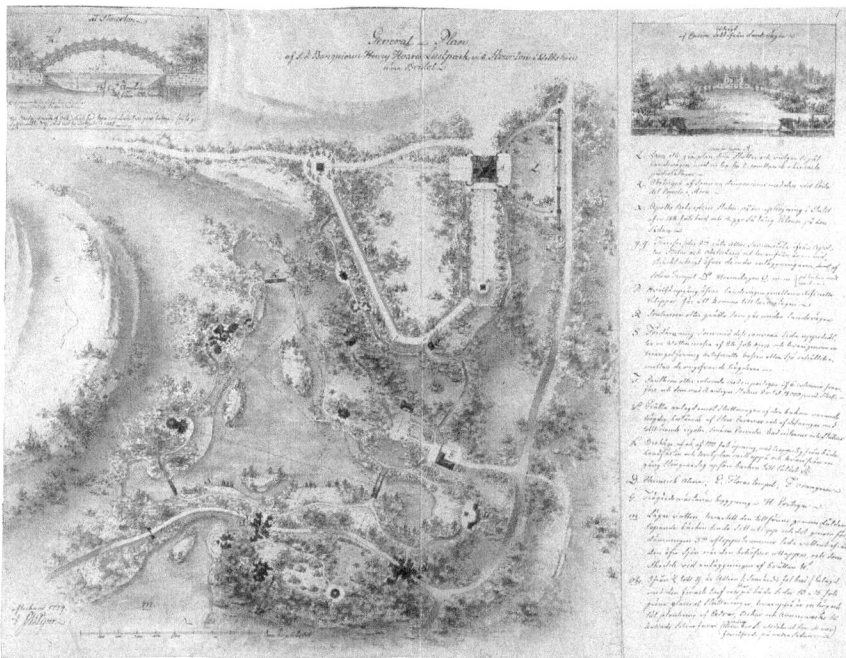

Fig. 13.10 Plan of the valley garden at Stourhead. F.M. Piper, 1779. (Source: Royal Academy of Fine Arts, Stockholm)

expressed in the metric length of the lines of sight between the focal points across the lake establishing the axial relationships. The average distance is about 431 m making sure that that the artefacts and their characteristics can be recognized. The maximum distance for recognition of characteristic elements in a landscape is about 500 m (Van der Ham and Iding 1971).

It would be interesting to compare these findings with other landscape architectonic compositions in order to find other angular and distance relationships in composed views. Also other aspects such as the topographical height of the visible objects in the view, or relationships in occlusion and visibility per view, remain interesting clues to be investigated by means of viewsheds.

Sequence of the Views

The slow-motion vision through following the path, offers sequential frontal and/or lateral perception of scenes and gradual discovery of the various features involved. This gradual change offers a sense of scenic intricacy that arouses and sustains curiosity. Upon arrival, the focal points (i.e. the temple) are used for enjoyment and repose for those walking through the valley garden and become viewpoints for other scenes as stages in the circuit walk.

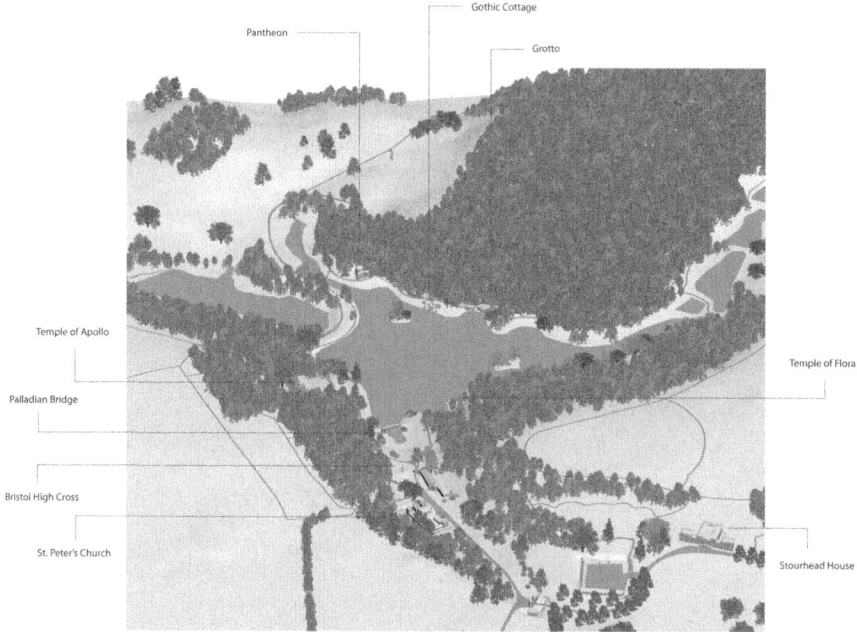

Fig. 13.11 Three-dimensional model of the valley garden. (Model by Nijhuis and Wiers)

The resulting cinematic experience is a reflection of the visual story being told; and the storyline becomes a physical construction, starting originally at Stourhead House and ending in Stourhead's Inn (Woodbridge 1976). Usually only citing the extant features, scholars have imposed divergent allegories upon the garden (see Magleby 2009 for an overview). Whether or not a specific iconographic program was in the mind of the gardens creators, they surely created a dream world inhabited by the gods, goddesses, and heroes of classical antiquity and England's history.

With regard to the allegorical nature of the pictorial sequence organized by the circuitous route, the views are analyzed via counting and characterizing the elements within the views (Fig. 13.15). The analysis show that almost every view (in the present state) contains juxtaposed Classical and Gothic architecture suggesting an allegorical dialogue between historical events. Note that there is a balanced amount of artefacts within the view counting an even number of emblems. In other words, every Classical element is counterbalanced by a Gothic iconographic object. It also interesting to consider the relation of the viewpoints and the course of the path. In a horizontal direction there is a certain timing, with varying intervals, between the major views. In vertical direction the relation is in going upward and downward e.g. descending to the Grotto, ascending to the Pantheon and the steep climb to the Temple of Apollo. Whether this tactile experience and the related staging of views reflects a story with a deeper meaning, or is a kind of memory system

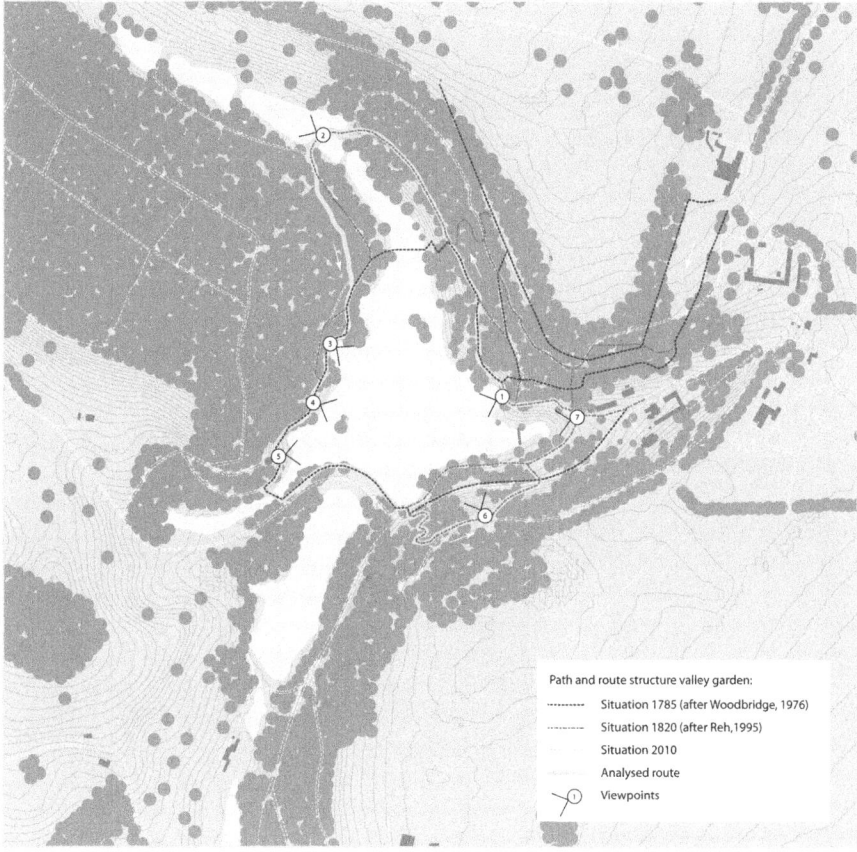

Fig. 13.12 Path structure and related major viewpoints in the valley-garden. (Source: Nijhuis)

facilitating pleasure and relaxation, it is a rich site which promotes and provokes a wide range of emotions, ideas and stories.

At Stourhead landscape garden the analysis of the angular extent, the visual coverage of (composed) framed views and counting focal points by means of viewshed analysis, especially their angular extent in relation to the physiology of vision and the balanced amount of emblematic focal points within these views, gives an interesting result. Here GIS enabled the measurement and visualization of their sequential relationship in time, based on slow-motion vision by walking, taking into account tactile properties such as differences in heights along the course of the path.

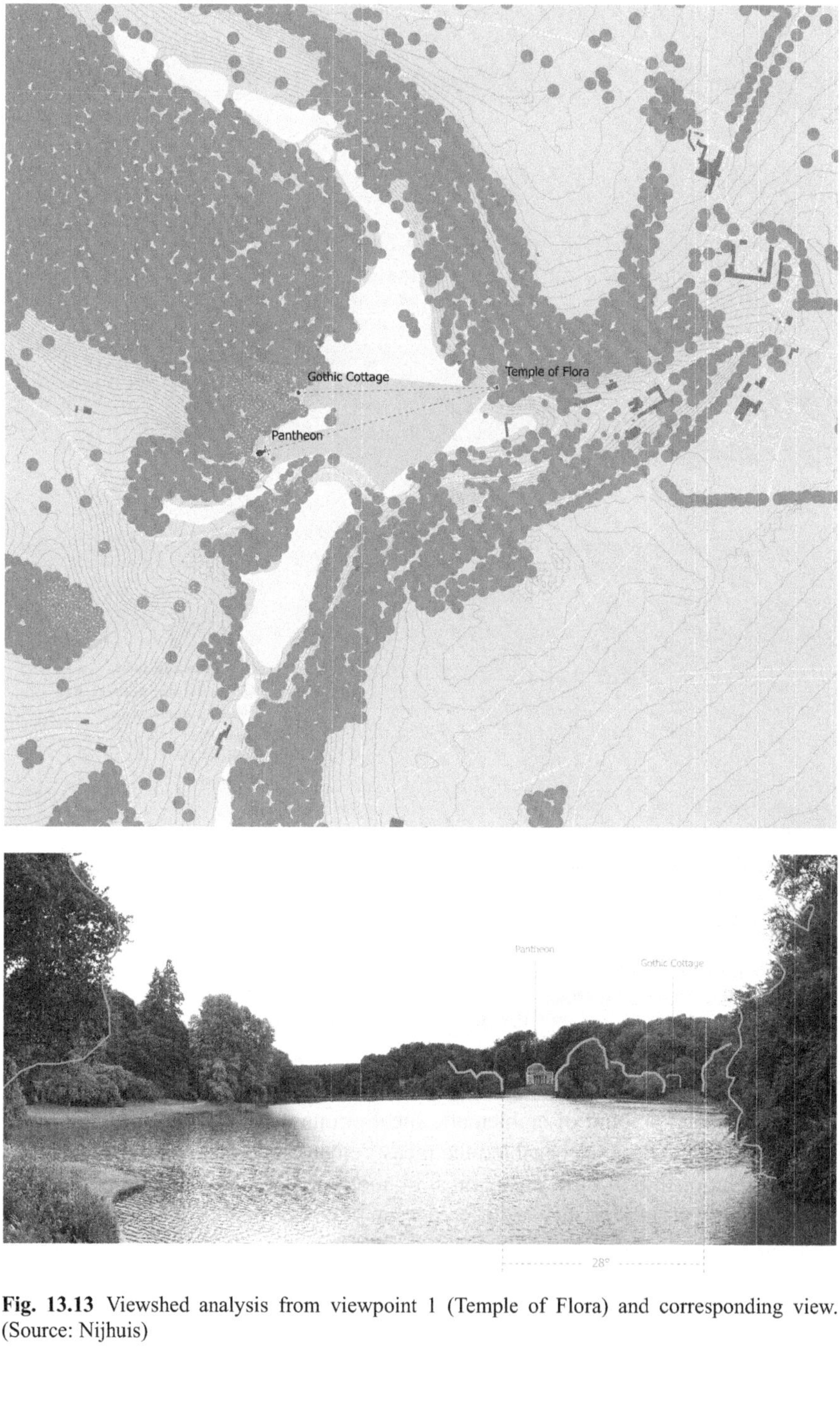

Fig. 13.13 Viewshed analysis from viewpoint 1 (Temple of Flora) and corresponding view. (Source: Nijhuis)

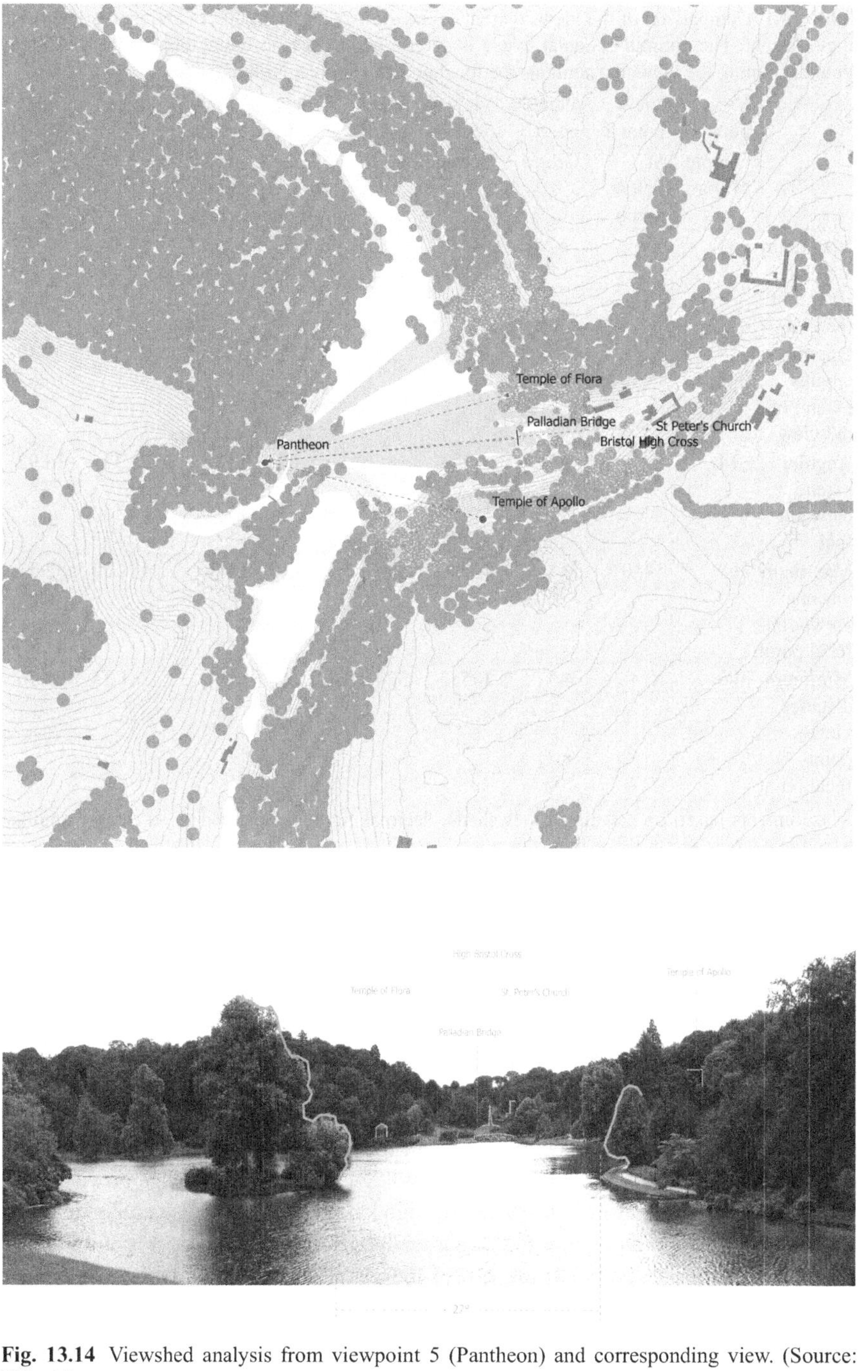

Fig. 13.14 Viewshed analysis from viewpoint 5 (Pantheon) and corresponding view. (Source: Nijhuis)

Table 13.1 Comparison of the views; extent of the view in angular degrees and metric length of lines of sight. The optimum angular extent is determined by the occluding objects in the middle ground, framing the view that contains the focal points. (Source: Nijhuis)

	View point 1	View point 2	View point 3	View point 4	View point 5	View point 6	View point 7	Mean	Std. dev.
	Temple of flora	St. Peters pump	Grotto	Cottage	Pantheon	Temple of Apollo	Bristol high cross		
Maximum angular extend of the view	53	X	67	43	62	86	36	57.83	17.97
Optimum angular extend of the view	28	X	31	28	22	32	24	27.50	3.89
Angular extend between foci	14	X	23	12	13(30[a])	60	12	22.33	18.92
Maximum distance viewpoint-focal point	368	1440[b]	318	497	494	3120[b]	478	431.00	82.57
Minimum distance viewpoint—focal point	306	X	343	305	324	320	90	281.33	94.76

Measurements based on calculated viewsheds, decimal figures converted to an integer angular extend in degrees and distance in meters
[a] Incl. Temple of Apollo
[b] Outside the valley garden

13.4 Conclusion and Outlook

Although there is a lot left to be explored in the examples, this chapter exemplified that GIS-based landscape design research can offer clues for deeper understanding of particular spatial phenomena that constitute visible form. By exploring the physiognomy of the composition, as it is encountered by an individual within it, moving through it, it is possible to acquire object related and typological design knowledge on visual aspects. This is important for acquisition of design knowledge, but is also crucial in management and restoration of sites like Stourhead. GIS turned out to be a useful vehicle for systematic and transparent research of the visible form. Moreover the discussed examples showcase that GIS-based isovists and viewsheds have the potential of measuring visual phenomena which are often subject of intuitive and experimental design, taking into account physiological, psychological, and anthropometric aspects of space. It offers the possibility to combine general scientific knowledge of visual perception and wayfinding with the examination of site-specific design applications.

13 GIS-based Landscape Design Research

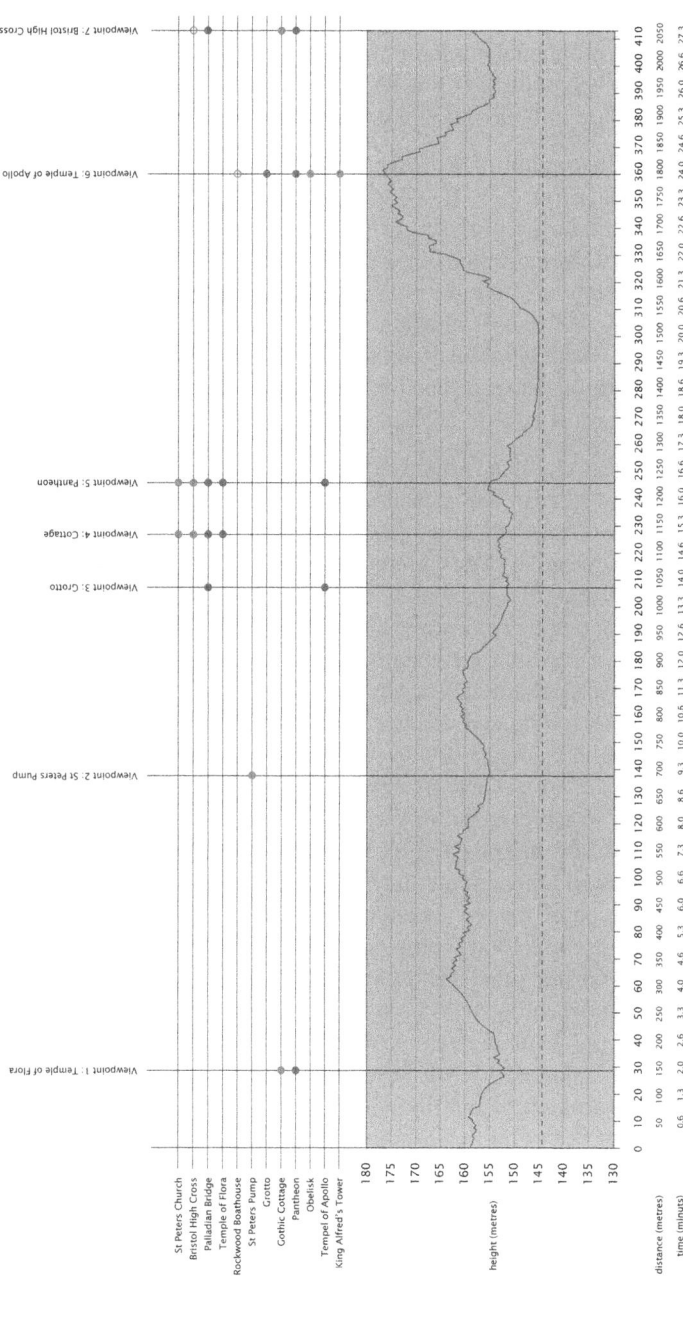

Fig. 13.15 The stroll at Stourhead explored. The sequence of the views in relation to distance, time and height of the path via GIS-based measurements. (Source: Nijhuis)

This chapter exemplifies the potential of GIS as a tool for landscape design research while enriching the analytical framework of Steenbergen *cum suis*, here particularly regarding analysis of the visible form. Although other aspects of the landscape architectonic compositions are not addressed here, such as aspects of the basic, symbolic and programmatic form, it seems that GIS could be useful for landscape design research exploiting GIS in its powerful integrating, analytical and graphical capacities in at least three fields of operation:

- Spatial modelling: description of existing and future landscape architectonic compositions in digital form;
- Computer-aided architectonic analysis: exploration, analysis and synthesis of landscape architectonic compositions in order to reveal new or latent architectonic relationships, while utilizing the processing capacities and possibilities of computers for *ex-ante* and *ex-post* simulation and evaluation;
- Computer generated visual representation: representation of (virtual) landscape architectonic compositions in space and time in order to retrieve and communicate information and knowledge of landscape design.

In order to develop GIS-based landscape research it is important to deepen and broaden the body of knowledge of GIS in landscape architecture in two ways. Firstly by following the discipline and developing specific aspects of it via applications of GIS in addressing the 'same types of design-knowledge', but in a more precise, systematic/transparent, and quantified manner. It makes for precise delineation and alternative ways of representation of landscape architectonic compositions. GIS helps to reproduce and transfer research methodology and offers an integrative, transparent and systematic approach for advanced spatial analysis. It also comprises of measurement (quantities), testing and verification of expert knowledge, or known architectonic phenomena in landscape architecture.

Secondly by expanding the field by setting in motion fundamental new developments via GIS in generating 'new types of design-knowledge' by advanced spatial analysis and the possibility of linking up/integrating other information layers, fields of science and data sources. In that respect the field of landscape design research can be extended by offering alternative ways to understand landscape architectonic compositions. GIS offers the possibility of integrating and exploring other fields of science (e.g. visual perception, way-finding studies) and dealing with complexity (more variables). Also the availability of other types of data such as Web 2.0, terrestrial LiDAR, Location Based Services (LBS), and Crowd Sourcing is important in this respect. GIS-based landscape research offers the possibility to enrich formal reading by revealing tactile and sensorial potentialities of a design, which was hardly possible before, and also expands the analysis with data derived from psychological and phenomenological approaches addressing matters of reception of a design.

References

Batty, M. (2001). Exploring isovist fields. Space and shape in architectonic and urban morphology. *Environment and Planning B: Planning and Design, 28,* 123–150.
Bell, S. (1993). *Elements of visual design in the landscape.* London: E&FN Spon.
Benedikt, M. L. (1979). To take hold of space: Isovists and isovist fields. *Environment and Planning, B, 6,* 47–65.
Benedikt, M. L., Burnham, C. A. (1981). Perceiving architectural space: from optic arrays to isovists. In W. H. Warren, R. E. Shaw, & N. J. Hillsdale, (Eds.). *Persistence and Change* (pp. 103–114). Lawrence Erlbaum: Connecticut.
Bishop, I., Lange, E. (Eds.). (2005). *Visualization in landscape and environmental planning. Technology and applications.* New York: Taylor and Francis.
Blake, R., Sekuler, R. (2006). *Perception.* McGraw-Hill: New York.
Colquhoun, A. (1991). Composition versus the project. In idem, *Modernity and the Classical Tradition. Architectural Essays, 1980–87.* Cambridge: MIT Press.
Cross, N. (2006). *Designerly ways of knowing.* London: Birkhauser.
Dee, C. (2001). *Form and fabric in landscape architecture.* London: Routledge.
Ervin, S., Steinitz., C. (2003). Landscape visibility computation: Necessary, but not sufficient. *Environment and Planning B: Planning and Design, 30*(5), 757–766.
Fischer, P. F. (1995). An exploration of probable viewsheds in landscape planning. *Environment and Planning B: Planning and design, 22,* 527–546.
Fisher, P. F. (1996). Extending the applicability of viewsheds in landscape planning. *Programmetric Engineering and Remote Sensing, 62*(11), 1297–1302.
Gaffney, V., Stančič, Z. (1991). *GIS approaches to regional analysis. A case study of the island of Hvar.* Ljubljana: University of Ljubljana.
Gardner, H. (1999). *Multiple intelligences. The theory in practice.* New York: Basic Books.
Grandell, G. (1993). *Nature pictorialized.* Baltimore: The Johns Hopkins University Press.
Haken, H., Portugali, J. (2003). The face of the city is its information. *Journal of Environmental Psychology, 23,* 385–408.
Higuchi, T. (1975). *The visual and spatial structure of landscapes.* Cambridge: The MIT Press.
Horrigan, P. (1995). Visual books: Representing landscapes. *Proceedings of the Council of Educators in Landscape Architecture, VII,* 35–48.
Hubbard, H. V., Kimball, T. (1935). *An introduction to the study of landscape design.* New York: Macmillan.
Janson, A., Bürklin., T. (2002). *Scenes. Interaction with Architectural Space: the Campi of Venice.* Basel: Birkhäuser.
Kemp, M. (1990). *The science of art.* New Haven: Yale University Press.
Li, Z., Zhu, C., Gold, C. (2005). *Digital terrain modeling: Principles and methodology.* Boca Raton: CRC press.
Longley, P. A., & Batty, M. (Eds.). (2003) *Advanced spatial analysis. The CASA book of GIS.* Redlands: ESRI.
Llobera, M. (2003). Extending GIS-based visual analysis: the concept of visualscapes. *International Journal for Geographical Information Science, 17*(1), 25–48.
Lynch, K. (1976). *Managing the sense of a region.* Cambridge: MIT Press.
Magleby, M. A. (2009). *Reviewing the mount of Diana: Henry Hoare's Turkish tent at Stourhead.* PhD-thesis, Ohio State University.
Moore, C. W., Mitchell, W. J., et al. (2000). *The poetics of gardens.* Cambridge: The MIT Press.
Morresi, M. (1999). *Piazza San Marco. Istituzioni, Poteri e Architettura a Venezia nel primo Cinquencento.* Milano: Electa.
Newton, N. T. (1971). *Design on the land. The development of landscape architecture.* Cambridge: Harvard University Press.
Nijhuis, S. (2011). Visual research in landscape architecture. *Research in Urbanism Series, 2,* 103–145.

Nijhuis, S. (2013). New tools. Digital media in landscape architecture. In J. Vlug, et al. (Eds.), *The need for design. Exploring Dutch landscape architecture* (pp. 86–97). Velp: Van Hall Larenstein, University of Applied Sciences.

Nijhuis, S. (2015). *GIS-based landscape design research. Stourhead landscape garden as a casestudy*. Delft: Delft University of Technology (in preparation).

Nijhuis, S., & Bobbink, I. (2012). Design-related research in landscape architecture. *Journal of Design Research, 10*(4), 239–257.

Nijhuis, S., Lammeren, R. van, & Hoeven, F. D., van der. (Eds.). (2011). *Exploring the visual landscape. Advances in physiognomic landscape research in the Netherlands*. Amsterdam: IOS Press.

Psarra, S. (2009). *Architecture and narrative. The formation of space and cultural space*. Abingdon: Routledge.

Rana, S. (2002). *Isovist analyst extension*. Version 1.1. Computer software program produced by the author at the Centre for Advanced Spatial Analysis, University College London.

Reh, W. (1995). *Arcadia en metropolis. Het landschapsexperiment van de verlichting*. Delft: Delft University of Technology.

Samonà, G. (1970). Caratteri morfologici del sistema architettonico di Piazza San Marco. In G. Samonà, et al. (Eds.), *Piazza San Marco. L'architettura la storia le funzioni* (pp. 9–42). Padova: Marsilio Editori.

Samonà, G. et al. (Eds.). (1970). *Piazza San Marco. L'architettura la storia le funzioni*. Padova: Marsilio Editori.

Schubert, O. (1965). *Optik in Architektur und Städtebau*. Berlin: Verlag Gebr. Mann.

Schulz, J. (1991). Urbanism in Medieval Venice. In Molho, A., Raaflaub, K., & Emlen, J. (Eds.), *City–states in classical antiquity and Medieval Italy. Athens, Rome, Florence and Venice* (pp. 419–445). Stuttgart: Franz Steiner.

Snowden, R., Thompson, P., & Troscianko, T. (2006). *Basic vision. An introduction to visual perception*. Oxford: Oxford University Press.

Steenbergen, C. M., Meeks, S., & Nijhuis, S. (2008). *Composing landscapes. Analysis, typology and experiments for design*. Basel: Birkhäuser.

Steenbergen, C. M., Mihl, H., & Reh, W. (2002). Introduction; design research, research by design. In Steenbergen, C. M., et al. (Eds), *Architectural design and composition* (pp. 12–25). Bussum: Thoth.

Steenbergen, C. M., Reh, W. (2003). *Architecture and landscape. The Design experiment of the Great European gardens and landscapes*. Basel: Birkhäuser.

Tandy, C. R. (1967). The isovist method of landscape survey. In C. R. Murray (Ed), *Methods of landscape analysis* (pp. 9–10). London: Landscape Research Group.

Turner, A., Doxa, M., O'Sullivan, D., & Penn, A. (2001). From isovists to visibility graphs: A methodology for the analysis of architectural space. *Environment and Planning B: Planning and design, 28*, 103–121.

Tzortzi, K. (2004). Building and exhibition layout. Sainsbury wing compared with castelvecchio. *Architectural Research Quaterly, 8*(2), 128–140.

Van der, Ham, R. J. M., & Iding, JA. (1971). *De landschapstypologie naar visuele kenmerken. Methodiek en gebruik*. Wageningen: Wageningen University.

Van, Lammeren, R. (2011). Geomatics in physiognomic landscape research—A dutch view. *Research in Urbanism Series, 2*, 73–97.

Von, Meiss, P. (1991). *Elements of Architecture*. London: E&FN Spon.

Ware, C. (2004). *Information visualization. Perception for design*. Burlington: Morgan Kaufmann.

Watkin, D. (1982). *The English vision. The picturesque in architecture, landscape, and garden design*. Londen: Harper & Row.

Weitkamp, G. (2010). *Capturing the view. A GIS based procedure to assess perceived landscape openness*. Wageningen: Wageningen University.

Wheatley, D. (1995). Cumulative viewshed analysis: A GIS-based method for investigating intervisibility, and its archaeological application. In G. Lock & Z. Stančič (Eds.), *Archaeology and GIS: A European perspective*. London: Routledge.

Woodbridge, K. (1970). *Landscape and antiquity. Aspects of English culture at Stourhead 1718–1838.* Oxford: Clarendon.

Woodbridge, K. (1976). The planting of ornamental schrubs at Stourhead. A history, 1746–1946. *Garden History, 4*(1), 88–109.

Woodbridge, K. (1996). *The Stourhead landscape.* The National Trust (Reprint edition 2002).

Zube, E., Simcox, D., & Law, C. (1987). Perceptual landscape simulations: History and prospect. *Landscape Journal, 6*(1), 62–80.

Chapter 14
3D LOS Visibility Analysis Model: Incorporating Quantitative/Qualitative Aspects in Urban Environments

Dafna Fisher-Gewirtzman

14.1 Introduction

The well-being of urban dwellers is influenced by a variety components reflected in their perception of space. The perceived density is one of the most influencing response to the stimulus sensed from the environment. Many parameters influence perceived density: subjective and objective. Previous research has confirmed that the comparative volume of visible space has indicated the comparative perceived density (Fisher-Gewirtzman and Wagner 2003a; Fisher-Gewirtzman et al. 2006). Hence, a larger volume of free space would indicate a lower perceived density. This was the first attempt to conduct an objective index to assign a geometrical attribute that can give a strong initial indication to the perceived density. Visual privacy is one of the parameters influences the perceived density and the well-being of dwellers in their dwelling units. Visual privacy can be easily indicated by the measured distance between view-point and visual target. Four ranges of distances from dwelling-units windows were defined (Shach-Pinsley et al. 2011) for level of privacy. In both works the geometry and measurements of the environment are in focus. Nevertheless, not only objective measures influence perception but also what is observed and its significance and interpretation to the observer; personal preferences, type of activity, etc. A follow-up recent work, focusing on a local neighborhood was conducted. A hundred local residents participated in the study. Participants were asked to fill-in a questionnaire containing some general questions regarding their satisfaction from their apartments and several multiple choice questions regarding their perceived density, visual privacy and preference of view in every room in their apartments. At the same time their apartments were accurately modeled and inserted into a virtual model of the neighborhood to enable objective analysis calculating the visible volume of space from every window in each function in all apartments taking part in the study.

D. Fisher-Gewirtzman (✉)
Faculty of Architecture and Town Planning,
Technion - Israel Institute of Technology, Haifa, Israel
e-mail: ardafna@tx.technion.ac.il

Voxel-based visibility analysis was used to measure the visible volume of space from each window in each apartment. This model is presented in Fisher-Gewirtzman et al. (2013). The realistic environment is virtually represented in a 3D virtual model. This model subdivides a three dimensional virtual environment by a three dimensional grid where the basic component is a 3D pixel, i.e. the voxel. The size of this voxel and extent of environment for analysis is defined by the user. The model automatically counts the number of visible voxels from an indicated viewpoint. The sum of visible voxels constitutes the visible volume for each view point. The outcomes of the visible volume of space for each window, in every room of all apartments were compared to the dwellers response regarding their perceived density and visual privacy. A high correlation was found between participants' response to their perceived density and privacy and the measured volume of space from each relevant window. The full report on this study is in progress. Since every participant lives in a different apartment with a different view, analyzing the outcomes has become a very complex process.

For example, one of the participants ranked his perceived density in a room as very low while the measured average volume of visible space was supposed to indicate on average perceived density. Since all the rest of the participants' answers had high correlation with the measurements, the reason for the difference was not clear. The complete file including the questionnaire and the documentation of the apartment was revisited. His comment to the open question in the questionnaire indicated that this person likes very much the view of some nice green tall trees to be seen outside that specific window. Therefore we acknowledged that his preference of the view compensated on the quantity of the visible space with the quality of view. Previous work already reported on willingness to pay for a view (Bishop et al. 2004) and market value of a room with a view (Lang and Schaeffer 2001). The assumption is that an analytical model integrating the quality of the view with the quantity would be much closer to explaining human perception than the subjective ranking of participants and objective analysis.

Another interesting and more complex example, which will be discussed and explored later in this chapter, is the case of one of the participants that his answers had high correlation with the measured volume of space. The only comment that drew our attention was his answer to one of the background questions: The question focused on the participants' satisfaction with his apartment and apartment house. He was expected to be very happy and satisfied since his questionnaire indicated on very low perceived density from all windows, visual privacy and that he likes the distant and close by view. Surprisingly his answer was that he is not so satisfied since the next door school is a great disturbance. This participant is a university student without any children; his reaction to the adjacent school is understandable. The analytical model in its present state could not refer to such information and produce the relevant outcome, indicating his well-being in his apartment. The 3D voxel based visibility model we used for calculating the visible volume of space has no indicators as to the sense and quality of elements contained in space. Identifying the presence of the near-by school or the profile of a possible dweller and his needs and preferences would result in a much more comprehensive analysis.

This was the trigger to challenge us to try and develop a quantitative measure that will encompass also qualitative aspects utilizing geospatial data. The aim was to develop a 3D visibility model that would quantify the volume of visible space and distances along with identifying the type of objects that are contained within the visible space, integrating their studied influence on human perception. The valuable information used is drawn from GIS. In the future, such analytical outcomes may serve as a valuable input into the system.

A novel 3D Line of Sight (LOS) visibility analysis model is introduced. The model analyzes the visibility from any internal or external viewpoints regarding the volume of visible space and the distances of lines of sight from identified elements in the environment; Hence, buildings (building types), roads (road types), trees, topography, the sea view or any other physical entities being part of the scenery. This model provides the framework for integrated qualitative and quantitative parameters of the visible environment. Its great potential in a geodesign processes is presented.

14.2 Objectives

The main objective of this work was to develop an automated 3D visibility analysis model integrating quantitative (volume and distances) and qualitative (type and quality) visual analysis reflecting human perception of the environment. This model utilizes GIS layers of information. This was subdivided into several sub-goals:

The 3D visibility model will analyze a large extent of any environment and would be compatible to off the shelf drawing tools. The quantitative module of the model would enable measurements of the visible volume of space and will indicate distances from every view obstruction.

Variable view components influence the observer in different ways. GIS layers of information would be integrated into the geometry of the virtual model. This way the geometry of the virtual model would inherit the attributes from the relevant GIS layer and would mark the different view components. Within its power, the analytical model will inform what view components are in the scenery and in what distance and to compile their integrated influence on the observer. Lines of sight projected from any view-point will identify the target (as being topography, trees or a building, etc.) in accordance to GIS layer of information inserted to the system. This way the type of element and the distance from view-point would be integrated in the analysis. The great potential of this model will be demonstrated on a case study.

14.3 Background

14.3.1 Geodesign

In the last decade many disciplines began using GIS for applied and theoretic research in the areas of architecture and urban design. The ability to visualize a multi-variable physical condition, to analyze it in diverse ways and at the same time to cross information, enables a wide knowledge base for quantitative research. It enables a 2D and 3D visualization of the layers of information. This opened up extensive research opportunities in architecture and urban design. Geodesign unites the art and creativity of design with the power and science of geospatial technology. Geodesign can produce more informed, data-based design options and decisions (Dangermond 2010). Geodesign may support research development. It may support urban and architecture design and analyses through spatial tools. It enables visualizing a large amount of data and quick analysis to be used in a design process.

This work utilized GIS layers of information that are provided by the local municipality. The information was inserted into the system for a more comprehensive visual analysis. This is aiming at providing an analysis that would be very close to the human perception meaning this model can recognize and respond to the geometry of the three dimension representation of the built and natural environment together with additional characteristics on top of the geometry; such as type of building, its esthetics, the type of scenery, type of roads, etc. We know all of the above have different implications on human perception. Human perception is influenced by many visual attributes of the environment and also the status of the observer in a specific timing. If the same view would be observed from two different functions or activities it may be perceived differently.

14.3.2 Visual Analyses

A variety of analyses were developed in the past decade. Some graph approaches to street network make use of navigational and visibility principles. A visibility graph may be applied to the urban environment by imposing a regular or an irregular grid on the top of urban space. Such a graph can be created in terms of how each point of the grid is visible to others. Visibility graph application differs in respect to implementation of visibility graph analysis; some of them analyze the open space between the built forms, while others (Kruger 1979), (De Floriani et al. 1994) map the intervisibility of built components. Another graph representation of the city is so called Space Syntax axial lines (Hillier 1984). This graph representation of the urban environment is constructed in the following way: the nodes are straight street segments tracing over every longest line of sight, which are then linked into a network via their intersections and analyzed as a network of movement choices.

Turner et al. (2001) and Turner and Penn (1999) applied a visibility graph to the analysis of architectural space taking a grid of all existing points across the space rather than selecting a few key locations. Jiang and Claramunt (2002) adopted the visibility graph relying on a set of characteristic points in the urban layout and further used the graph to analyze the structure of the street network. The characteristic points are defined as the nodes of an urban structure and visibility connections between them as edges schematized as a graph. This method has several advantages over the axial line representation: it is computable and cognitively meaningful. It is easier to understand and makes more sense visually, and enables a shift from the primal problem to the dual and back, thus providing a much richer interpretation of the space syntax (Batty 2004). It also allows incorporating metric distance in one framework with topological properties. An Integrative Visibility Graph (IVG), a quantitative method, based on visibility analysis of urban structure and its functioning was proposed by Natapov et al. (2013).

Benedikt (1979) was the first to introduce the isovist and to develop a set of analytic measurements of isovist properties to be applied in order to achieve quantitative descriptions of spatial environment. A number of researchers have developed measurement methods and tools for automated 'Isovist' analysis, amongst them Turner (2003) that showed how a set of Isovist can be used to generate a graph of mutual visibility between locations and developed the 'Depthmap' for visibility graph analysis. Batty (2001, 2004) describes how a set of 'Isovist' forms a visual field whose extent defines different 'Isovist' fields of different geometric properties. He suggested a feasible computational scheme for measuring "Isovist' fields and illustrated how they can visualize their spatial and statistical properties by using maps and frequency distributions. Several models have been developed to examine the Isovist in different ways; The 'Spatialist' by (Peponis et al. 1998) and the 'Axwoman' by (Jiang et al. 1999). Several methods show that visibility is connected to accessibility. For example, the space syntax method (Hillier 1999; Hillier and Hanson 1984) examines the relation between spatial configurations and movement, and connects them with social, cultural and economic-functional aspects.

The Spatial Openness Index (Fisher-Gewirtzman and Wagner 2003a 2006) can explore the 3D visibility of spatial configurations. It was the first real attempt to simulate human three dimensional visual perceptions (Fisher-Gewirtzman and Wagner 2006, 2003a). It can also be described as a 3D Isovist. SOI measurements of alternative spatial configurations were correlated with comparative perceived density, thus, the objective measurements indicated the subjective response. A voxel based 3D visibility analysis is the current attempt to progress the SOI measurements (Fisher-Gewirtzman et al. 2013). This visibility analysis model enables accurate representation of the three dimensional geometry of the built environment, focusing on building structure and the terrain. Large trees can be represented similarly to buildings. A bounding box is defined by the user regarding the area and height of space within which the analysis can take place. The defined space is subdivided into basic 3D pixels, i.e. voxels. The user may define a view-point or multiple view-points on a selected building for analysis and the radius (distance for visibility analysis to all directions). The analysis results can be visualized by a colored hemisphere, where

Fig. 14.1 Voxel based analysis. The analysis results visualized by a colored hemisphere

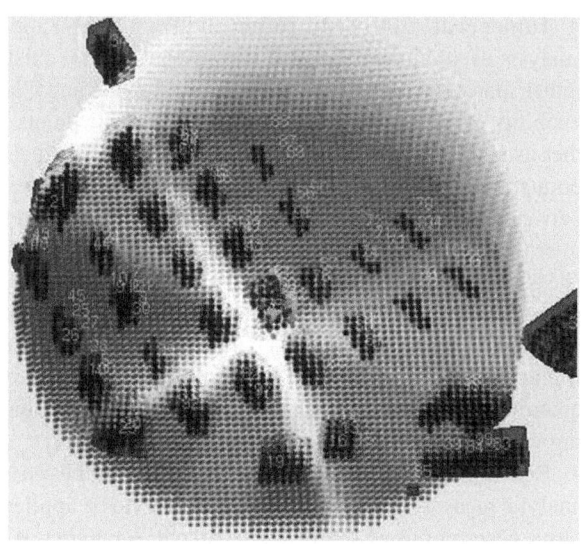

the view-point is marked in red and the visual volume of space is indicated by white voxels and the hidden view (shaded view) is colored in dark brown. See Fig. 14.1. Detailed description is provided in Fisher-Gewirtzman et al. 2013.

3D visual analysis comes closer to human perception of space since humans experience the 3D environment where features such as buildings, topography, and vegetation has a 3D presence. The impact of the three dimensional environment cannot be fully represented in a 2D map projection. Voxel based analysis considers the complexity of the 3D geometry of the realistic environment. The size of the voxel is determined by the user. The smaller the voxel is, the more precise the visibility calculation will be.

Some methods were developed to dealing with exact visibility in 3D scenes, without considering environmental constraints. Plantinga and Dyer (1990) used the aspect graph—a graph with all the different views of an object. A shadow-determination algorithm that uses a data structure, called a backprojection, to represent the visible portion of a light source from any point in the scene was presented by (Drettakis and Fium 1994). Similar approaches were also suggested by Stewart and Ghali (1994) and Teller (1992). The applicability of these works to a large scene is limited, due to computational complexity. Automatic generation or modeling of complex 3D environments, such as the urban case, can be a very complicated task dealing with fast computational analysis. Visibility computation in 3D environments is a very complicated task, which can hardly be performed in a very short time using traditional well-known visibility methods (Gal and Doytsher 2012). Most of them do not support off the shelf drawing tools.

In this current work we developed a 3D visibility analysis based on Line Of Sight. This model can analyze large scale environments with complex detailing which is valuable for urban environments. The lines of sight are launched from individual view-points indicated in advance or selected later by the user. Each LOS exports information regarding the distance and the type of object blocking its view. LOS visibility analysis is introduced in the next section.

14.4 Introducing the LOS Visibility Analysis Model

The LOS model analyzes the sum and segmentation of LOS, measuring the distances and visible volumes of space in each viewpoint selected in a virtual built environment. The extent of detailing is reliant upon the user. View-points may be indicated in advance or selected ad-hoc by the user. This model was developed in collaboration between architecture and computer science researchers. The platform for the model is the GUIRIT (Elber 2011). IRIT is a solid modeling environment that allows one to model basic, primitive based, models using Boolean operations as well as freeform surface's based models. The IRIT modeling package was originally developed by Elber (Elber 2011). IRIT has several unique features such as strong symbolic computation, support of trivariate spline volumes, multivariate spline functions and triangular patches. A rich set of computational geometry tools for freeform curves and surface is offered. The solid modeler is highly portable across different hardware platforms. The system is designed for simplicity and is geared toward research. As such, a graphical user interface (GUI) is not part of IRIT but is considered an extension package (See GuIrit). Version 11.0 of the IRIT solid modeling system contains tools that can aid in research and development in the areas of computer aided geometric design and computer graphics (Elber 2011). The IRIT modeling package was exploited for a whole variety of research activities.

Preparing a virtual model for analysis Relevant data for the analysis is drawn out of GIS. This information is converted into a Computer Aided Design (CAD) format. This preliminary conversion is essential stage for building a digital model in a common format and for running analyses in various platforms. GIS includes a load of characteristics that needs to be simplified. In the preliminary stage the focus is on two layers: buildings and topography. In the GIS layers of information the topography is represented as height lines and the buildings as polygons with height characteristics. These were transformed into a CAD format saving the height characteristics. In the second stage, the CAD files were transformed into the Google SketchUp software: 3D common simple modeling software used by potential benefiters of such an analysis. The topography can be created by using the triangulation command and the buildings can be represented by a simple extrude command. Figure 14.2 presents the SketchUp model used for this study. The next step was to transform the SketchUp model into a standard 3D format (OBJ). This was exported into the GuIrit platform with its visibility extension, as illustrated in Fig. 14.3. This analysis model can calculate extensive visibility volumes in large scale models in a very short time. In case of a tilted topographic condition where distant views are

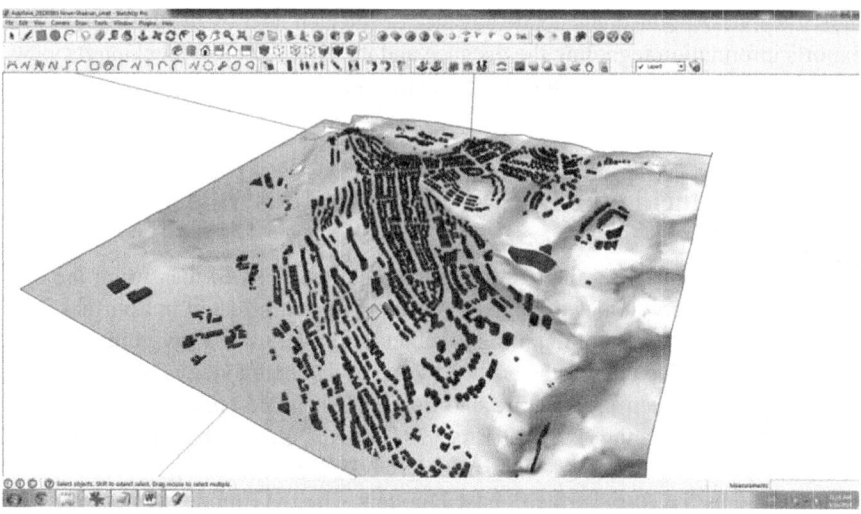

Fig. 14.2 The SketchUp model used as a case study for this study. Two by three km' in size

Fig. 14.3 The model exported into the GuIrirt platform with its visibility analysis extension

common, it is very useful to be able to analyze large scale modeled environments. This way the actual distant views viewed by people can be represented in the virtual model.

This model was originally developed to enable precise visualization analysis in complex three dimensional environments enabling measuring the visibility of

14 3D LOS Visibility Analysis Model

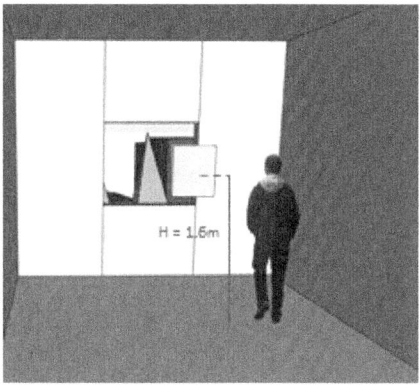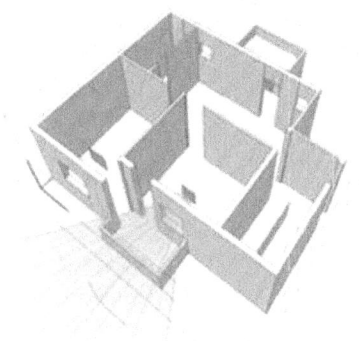

Fig. 14.4 On the *left* illustrating visibility of a person standing in a specific point in a room and looking out-side. On the *right* an example of an apartment with view-points spread in some of the rooms

Fig. 14.5 Depth-view as visual outcomes for the LOS analysis. Considering topography and buildings. The *red color* represent the shortest distance (the wall in the room) and the *green* represent the most distant views

a person standing or sitting in a specific point in a room and looking out-side as illustrated in Figs. 14.4 and 14.5 presents a depth-view as visual outcomes of the LOS analysis for a window. Distances from the view point are indicated by colors: The red color represents the close distance from the inner wall and the bright-green represent the distant sky and all LOS that were not blocked by any object. Outcomes can be represented by graphs illustrating the distribution of distances. Numerical results are available as well.

This model enables analyzing extensive large built environments simultaneously in a fairly short time. It is an excellent interface for off the shelf drawing tools used by the potential users such as architects and urban planners. The model can provide

Fig. 14.6 the apartment located in a hilly Haifa neighborhood overlooking the Haifa bay

separate depth-views for each component of the environment. This will be demonstrated further in Sect. 14.6.

14.5 Introducing the Case-Study

The case-study is a small apartment located on the upper story of a small apartment house in a hilly neighborhood overlooking the Haifa bay (Fig. 14.6). The apartment is inhabited by a university student. Documentation of the case-study includes plans and photographs of the view from each window. The visible volume of space was measured using the voxel based analysis and documented for each window in the apartment. At the same time, the tenants' response to the questionnaire regarding perceived density, visual privacy and preference of the view was explored. Figure 14.7 is presenting the apartment plan and the view from each significant window in the apartment. From each window the tenant can view ether a nice distant sea and mountains view or a close by green-trees view. The trees also contribute to his visual privacy from close by buildings by blocking the view to the neighbours.

14.5.1 Preliminary Analysis Using Voxel Based 3D Visibility Analysis

The visible volume of space was calculated from each window in the apartment using the voxel-based visibility analysis (Fisher-Gewirtzman et al. 2013). The outcomes were compared to the participants' response. The virtual model used for the calcula-

14 3D LOS Visibility Analysis Model

Fig. 14.7 : presenting the apartment plan and the view from each window in the apartment

tions simulated the buildings only; no trees or any other component was simulated. Therefore the measurements reflect the maximum potential visible space defined by the surrounding built structures. In this case the view blocked by the trees was not considered. At the same time, the positive impact of the trees was not considered. Table 14.1 presents the visibility analysis results right next to the actual view from each window together with the participants' response to the question regarding his perceived density. Results for the windows in this apartment were amongst the highest (largest visual volume of space) in all of the previous research case studies (~60 apartments) as expected. Correlation between viewshed calculations and participants' responses were very good. The largest volume of visible space was calculated for his study room (windows 1 and 2). The participants' response was the highest (1—very low density) for all of his rooms. In general this apartment is facing a large open distant view. Therefore we expected the dweller to report on low perceived density. In addition, the close-by trees add a nice green screen contributing to the visual privacy from close by buildings, while enabling observing the distant view. This can explain why despite having reduced visible volume of space, the side windows were ranked by our participant as 1—very low density.

Galil 101/3

Room	Study		Bedroom	Kitchen	Apartment
Window no.	1	2	3	4	
Viewshed					
View					
Window area (m²)	1.2	1.7	1.2	3.8	
Viewshed Vol. (10³ m³)	381	615	615	253	
Viewshed × window area	1.2 × 381=457	1.7 × 615=1045	1.2 × 615=738	3.8 × 253=961	
Total Viewshed (10³ m³)	1502		738	961	3201
Participants answers: 1. low density, 5. high density					
1 101.3.1 (m)	1/5		1/5	1/5	1/5

Table 14.1 Presenting the viewshed calculations for each room in the apartment, compared to the actual views from each window and to the participants' response to the perceived density

This participant was expected to be very satisfied with his apartment. Surprisingly, as mentioned in the introduction to this paper, the participants' response to the general open question: "are you satisfied with your apartment?" was unexpected. His answer was: "not so satisfied since a close by school is a big disruption". The voxel based 3D visibility analysis measures the geometry of the environment. It can't consider the meaning of the geometry in reality or any other attributes that have subjective implication on a human view-point. Nothing regarding the qualitative attributes in the visible range of the observer and inhabitants profile indicating on his needs or general known preferences was accumulated in the analysis. Analyzing the geometry of the environment can give only partial explanation to how humans perceive the environment. An integrated analysis regarding geometry together with building uses, such as the close by school could help predict this students' lack of satisfaction regarding his current profile. A close by school is considered as a drawback because of the noise and traffic in the morning and afternoon. Same location may be considered as a great advantage for a different kind of user profile. If the participant had a child, he may relate to the close-by school as an advantage. Integrating additional qualitative information that influence peoples' perception in a quantitative model may be a big step for spatial analysis and planning and design processes. Such information can be easily incorporated from GIS. In the next section we will present the first steps to integrate GIS data to the 3D visibility LOS analysis model.

Fig. 14.8 Haifa municipality usage map of the area. Our case-study is marked in *red*, next to it a *plot* marked as public building (the small school)

Fig. 14.9 Orthophoto of the same area showing the trees and other vegetation. The case study is marked in *red*

14.6 Case-Study Analysis

To prepare the virtual model for analysis we looked at some layers of GIS supplied by the Haifa municipality. We looked at the usage map of the area, where our case-study is marked in red (Fig. 14.8). The plot right next to our case-study is marked as a public building (a small school). Since this map did not contain any vegetation, an orthophoto of the same area was considered as well (Fig. 14.9).

The information from the usage map and orthophoto was inserted in the Sketch-Up model. In this stage the information was inserted manually as color- coded to the geometry of the model. To be more accurate about the trees location detailed photos from the actual place were used. The SketchUp model used for the analysis

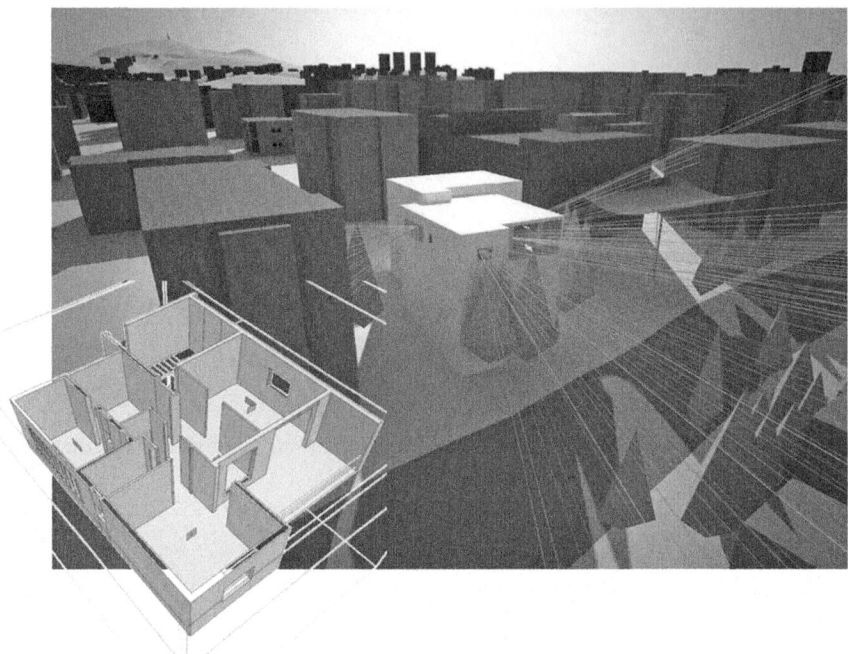

Fig. 14.10 The updated SketchUp model integrating the GIS data regarding building uses, trees, streets etc. illustrating Lines Of Sight aiming from each view-point located in the apartment

Fig. 14.11 General depth-views regarding the geometry of the environment only

is presented in Fig. 14.10: The apartment house with the case-study is colored in white. All other buildings are colored in grey. Roads are colored in a different hue of grey. Trees are colored in green and the school is colored in dark red. The yellowish planes inside the rooms represent the location of the view-points for analysis. The school is not visible from the window of this apartment. Only the shading system of the back-yard is slightly visible. The updated SketchUp model was inserted into the GUIRIT visualization extension and a set of depth-views was provided for each window, as illustrated in Fig. 14.11.

To demonstrate a detailed visual analysis for each component in the environment we focused on window no. 1 and window no. 3. This way it was possible to learn more about the components of the environment that may have influenced the participants' perception of space. The various components were isolated in accordance to their representation in the updated model: the current building-the case study, other

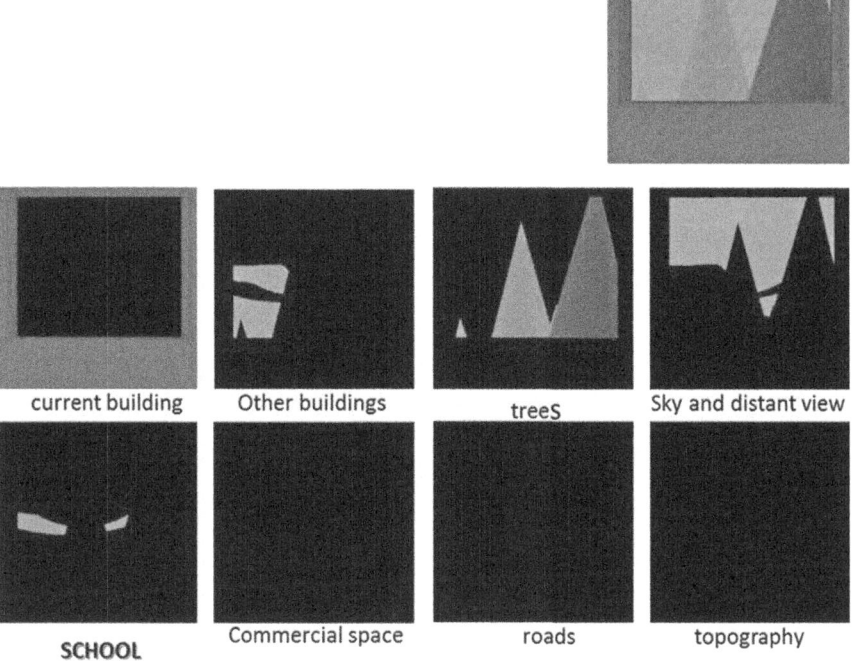

Fig. 14.12 Isolated depth-view for window no.1 in the apartment

apartment buildings, public buildings, trees, the sky and distant view, roads and topography. For each view a separate histogram was processed. Updated information on the distribution of distances for each component was provided (the distribution of distances to visible trees; the distance to other buildings etc.) Fig. 14.12 presents the separate depth view for window no.1 and Fig. 14.13 presents the separate depth view for window no. 3.

The small patches of color in Fig. 14.12—'School' indicate the view to the schools' back yard in between the trees. The numerical results show that the public building is quite close to the participants' apartment. The view of the school from the window is almost unnoticeable, but the implications of noise because of the distance all along the day and traffic jams in the morning and afternoon are very dominant on the perception and satisfaction of the tenant.

14.7 Conclusions and Recommendations

The LOS 3D visibility analysis model is a first attempt to combine quantitative and qualitative attributes of the built environment in one automated model. We show the great potential of using the GIS layers as a natural step. The level of detailing in-

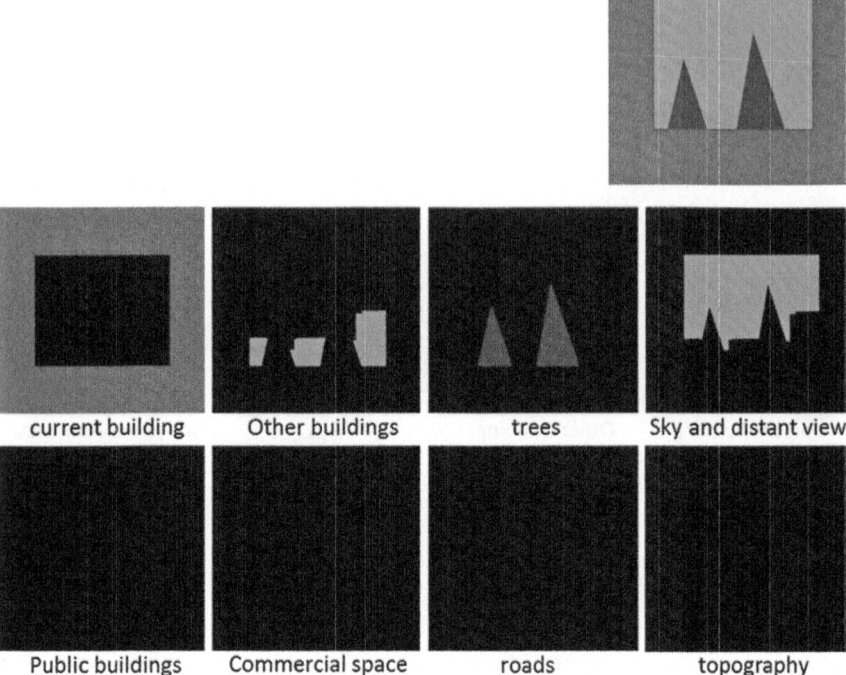

Fig. 14.13 Isolated depth-view for window no. 3 in the apartment

corporated in the representation of the built environment is infinite and can be done very easily. This work demonstrated how incorporating the function of a close-by building can provide an added value to the visual information and result in the potential well-being of a tenant. Esthetic value of built environment, or buildings defined for preservation in an historic sector of a city may influence the economic value, the willingness to live close-by and influence quality of life. This may not suit every inhabitant. A model that could combine a very detailed visual analysis of the environment regarding all visible components, their attributes and their implications on human perception; would enable ranking their value in accordance to the specific profile of existing or potential population. Such a model would help monitor, plan and design more livable environments.

The outcomes of such analysis may be inserted in the GIS system as an additional valuable layer of information. Integration of GIS data in the LOS visibility analysis model is currently operated manually. No doubt an automatic transformation would contribute to the easy use in every planning and design processes. This is one of the next steps of developing the model.

A current study is utilizing a virtual reality laboratory to investigate the relative influence of variant elements in variant distances from a viewpoint on the human perception (Fig. 14.14). This study is expected to provide a relative index to be incorporated in the automated model.

Fig. 14.14 A participant in the Virtual Reality lab. Visualizing an alternative external view from a room. The (HIS) Hybrid Ideation Space was kindly lended to the Technion by Prof. Tomas Dorta

This model can be used as an analysis tool in case of existing built environments to help propose improvements. In addition, it can be used as a comparative analysis tool along the planning and design development process, to compare between design alternatives. The future user will insert as many influencing components as required. Such a model would contribute much to the geodesign of our existing and future environments, for better, safer and healthier cities.

Acknowledgments I would like to thank Prof. Gershon Elbar for his collaboration in developing the LOS analysis model. This model has opened new horizons for our research work.

I would also like to thank my research assistants Natali Polanski, Igal Tartakovsky and Liran Malka.

This research was supported by the ISRAEL SCIENCE FOUNDATION (grant No. 722/11)

References

Batty, M. (2001). Exploring isovist fields: space and shape in architectural and urban morphology. *Environment and Planning B: Planning and Design, 28,* 123–150.
Batty, M. (2004). A new theory of space syntax. *Syntax, 44*(0), 36 (http://discovery.ucl.ac.uk/211/1/paper75.pdf).
Benedikt, M.L. (1979). To take hold of space: Isovist fields. *Environment and Planning B: Planning and Design, 6,* 47–65.
Benson, E. D., Hansen, J. L., Schwartz, JR. A. L., & Smersh, G. T. (1998). Pricing residential amenities: The value of a view. *Journal of Real Estate Finance and Economics, 16*(1), 55–73.
Bishop, I. D., Lang, E., & Mahbubul, A. M.(2004). Estimation of the influence of view components on high-rise apartment pricing using a public survey and GIS modeling. *Environment and Planning B: Planning and Design, 31,* pp 439–452.

Dangermond (2010). *GeoDesign and GIS—Designing our futures*. Proceedings of digital landscape architecture.
De. Floriani, L., Marzano, P., & Puppo, E. (1994). Line-of-sight communication on terrain models. *International Journal of Geographical Information Systems, 8,* 329–342.
Drettakis, G., & Fiume, E. (1994). A fast shadow algorithm for area light sources using backprojection. In Proceedings of SIGGRAPH'94 the 21st Annual Conference on Computer Graphics and Interactive Techniques, pp. 223–230.
Elber, G. (2011). IRIT 11.0 user's manual, Version 11.0, (Technion-IIT, Haifa, Israel. http://www.cs.technion.ac.il/~irit/)
Fisher-Gewirtzman, D., & Elber, G. (2013). *LOS based 3D visibility analysis—from housing interior to the surrounding environment*. Working paper 2013.
Fisher-Gewirtzman, D., & Wagner, I.A. (2003a). Spatial openness as a practical metric for evaluating built-up environments. *Environment and Planning B: Planning and Design, 30*(1), 37–49.
Fisher-Gewirtzman, D., & Wagner, I.A. (2006). The 'Spatial Openness Index': An automated model for 3-D visual analysis of urban environments. *Journal of Architecture and Planning Research, 23*(1), 77–89.
Fisher-Gewirtzman, D., Shashkov, A., & Doytsher, Y. (2013). Voxel based volumetric visibility analysis of urban environments. *Survey Review, 45*(333), 451–461.
Gal, O., & Doytsher, Y. (2012). Spatial 3D analysis of built-up areas. In Proceedings of FIG Annual Conference FIGWW'2012, pp. 1–18, 2012.
Hillier, B. (1999). The hidden geometry of deformed grids: or, why space syntax works, when it looks as though it shouldn't'. *Environment and Planning B: Planning and Design, 26,* 169–191.
Hillier, B., & Hanson, J. (1984). *The social logic of space*. Cambridge: Cambridge University Press.
Jiang, B., & Claramunt, C. (2002). Integration of space syntax into GIS: New perspectives for urban morphology. *Transactions in GIS, 6*(3), 295–309(http://doi.wiley.com/10.1111/1467-9671.00112).
Jiang, B., Claramunt, C., & Batty, M. (1999) Geometric accessibility and geographic information: extending desktop GIS to space syntax. *Computers Environment and Urban Systems, 23,* 127–146.
Krüger, M. T. J., (1979). An approach to built-form connectivity at an urban scale: System description and its representation. *Environment and Planning B: Planning and Design, 6,* 67–88.
Lang, E., & Schaeffer, P. V. (2001). A comment on the market value of a room with a view. *Landscape and Urban Planning, 55,* 113–120.
Natapov, A., Czamanski, D., & Fisher-Gewirtzman, D. (2013). Can visibility predict location? Visibility graph of food and drink facilities in the city. *Survey Review 45,* 462–471.
Peponis, J., Winerman, J., Rashid, M., Bafna, S., & Hong Kim, S. (1998). Describing plan configuration according to the covisibility of surfaces. *Environment and Planning B: Planning and Design 25,* 693–708.
Plantinga, H., & Dyer, R. (1990). Visibility, occlusion, and aspect graph. *The International Journal of Computer Vision, 5*(2), 137–160.
Shach-Pinsly, D., Fisher-Gewirtzman, D., & Burt, M. (2007). Visual exposure analysis model: A comparative evaluation of three case studies. *Urban Design International, 12,* 155–168.
Stewart, J., & Ghali, S. (1994). Fast Computation of Shadow Boundaries Using Spatial Coherence and Backprojections, Proceedings of SIGGRAPH'94 the 21st. *Annual Conference on Computer Graphics and Interactive Techniques, 231–238.*
Teller, S.J. (1992). Computing the antipenumbra of an area light source. *Computer Graphics, 26*(2), 139–148.
Turner, A. (2003). Analyzing the visual dynamics of spatial morphology. *Environment and Planning B: Planning and Design, 30,* 657–676.
Turner, A., & Penn, A. (1999). Making isovists syntactic: Isovist integration analysis. Proceedings 2nd International Symposium on Space Syntax, Brasilia.
Turner, A., Doxa, M., O'Sullivan, D., & Pen, A. (2001). From isovists to visibility graphs: A methodology for the analysis of architectural space. *Environment and Planning B: Planning and Design, 28*(1),103–121.

Chapter 15
Space Syntax in Theory and Practice

Akkelies van Nes

15.1 Introduction

Most old European cities have a large intact historical centre. In the past many of them had a strong position in terms of politics, trade and cultural activities. These activities from the past mostly have left monuments such as a set of buildings or a street pattern. Intact historical urban centres function as an attractor for cultural activities such as museums, art and crafts, tourism, concerts, and performances. All these activities and their artefacts seem to create the atmosphere and a sense of place (Norberg-Schulz 1971).

As a tool, Space syntax cannot analyse place character, but rather place structure. It can analyse the spatial configuration of past street pattern from old maps and correlate the results with the location of remaining significant old buildings. In this way an application of space syntax provides insight into why these centres were highly vital in the past in comparison with the present situation.

What space syntax measures is the two primary all-to-all (all street segments to all others) relations. On the one hand it measures the to-movement, or accessibility of each street segment with respect to all others. On the other hand it measures the through-movement potential of each street segment with respect to all pairs of others. Each of these two types of relational pattern can be weighted by three different definitions of distance (shortest paths, fewest turns, least angle change paths) which can inform urban design and planning decisions where pedestrian experience is of concern and to anticipate zones of urban activity. Each type of relation can be calculated at different radii from each street segment, defining radius again either in terms of shortest, fewest turns or least angle paths (Hillier and Iida 2005, pp. 557–558).

The space syntax method can be applied on a wide scale level in research on built environments—from the organisation of furniture in a room up to metropolis,

A. van Nes (✉)
Faculty of Architecture, Delft University of Technology, Delft, The Netherlands
e-mail: A.vanNes@tudelft.nl

making possible to compare built environments with one another from a spatial point of view. Similarly, the method is a useful tool for comparison of the spatial changes in a before-and-after situation of structural urban changes in an area. However, while the method is a tool for explaining the physical spatial set up of buildings and cities, the interpretation of the results from the spatial analyses must be done in correlation with an understanding of the societal processes and human behaviour.

Space syntax is under constant development. Its contribution to theories on built environments and methodology develop at the intersection of natural, social and technical sciences. So far, research projects range from anthropology or cognitive sciences to applied mathematics and informatics and touch upon philosophical issues. The evolution of space syntax asks for communication not just between various cultural contexts, but likewise between different scientific domains.

Obviously, there is a relationship between how human beings organize their life and shape spaces for their activities. What matters is the degree of accessibility, visibilities, adjacencies, openness and enclosures in built environments. At present, there exists few precise predictable theories on how cities function. Most writings on urbanism, architecture and society seem to have a lack of careful description of urban form, structure, a concise definition of urban space, and a consistent vocabulary. The success of space syntax seems to depend on at least two things: a concise definition of space and high degree of falsifiability and validation. The space syntax method's independence of context makes it applicable on all types of built environments, independent on types of societies, political structures and cultures. Therefore it is recognised to be a sustainable solid analysis and research method on built environments on various scale levels of a wide range of different cultures. In a geodesign processes, it can provide interesting insight into the relationship between social, cultural, and economic patterns through analyzing various design scenarios.

15.2 The Basic Platform for the Space Syntax Method

Independent on cultures and architecture, all built environments have in common a set of private and public spaces. Public spaces open up for movement from everywhere to everywhere else. Private spaces are spaces inside buildings and gardens, connected to the public ones in different degrees.

Urban public space is mostly linear. Figure 15.1 illustrates how a built environment's street grid is presented as the longest and fewest sets of axes interconnected with one another. The longest sight line in each urban space is represented as one axis. These axes are the units for calculating the inter-relationship between them. The reality is thus simplified as a spatial model consisting of a set of axes in order to calculate how each urban space relates topologically, geometrically as well as metrically to all other urban spaces.

15 Space Syntax in Theory and Practice 239

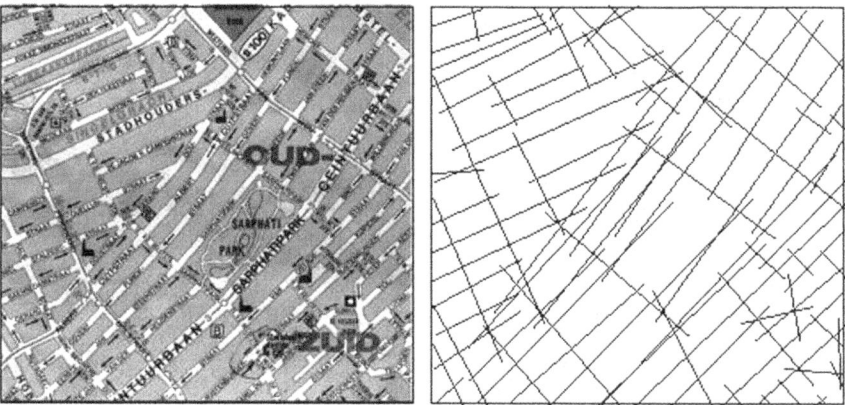

Fig. 15.1 An axial representation of an urban area in Amsterdam

15.3 The Axial Analyses

A global integration analysis implies calculating how spatially integrated a street axisis in terms of the total number of directional changes to all others in a built environment. Figure 15.2 shows a simple settlement, town X, consisting of a main street with side streets and smaller back streets. On the right side the town's street network is represented as a set of fewest axial sight lines. Each direction change is a topological step. The right lower corner shows an integration analysis of the settlement carried out by the computer program Axman, developed for space syntax. Various integration values are represented with grey scales. The black axes are the most integrated, while the light grey axes are the most segregated. As shown in the lower left corner, the justified graph illustrates how the system can be experienced from the most integrated street. In the case of the fewest line axial map, the lines are represented as nodes and the intersection of lines as nodal connections (Turner et al. 1993, p. 425). The grey scales of the nodes are the same as that used in the axial integration analysis.

Figure 15.3 provides an example of how to calculate the interrelation of one street to all other streets in town X. The back street (axis number 3) is represented at the bottom of the justified graph, thus at step zero. First calculate the sum topological depth from the back street to all other streets. If one change of direction occurs away from the back street, one street or space is found. If two changes of direction occur from the back street, three spaces are found. The sum depth of the back street is calculated by multiplying the number of spaces by the number of depths and then taking a sum of all the values. Thus, the more integrated a street is, the shorter topological distance it has to all other streets. Similarly, the more segregated a street is, the greater the topological distance is to all other streets.

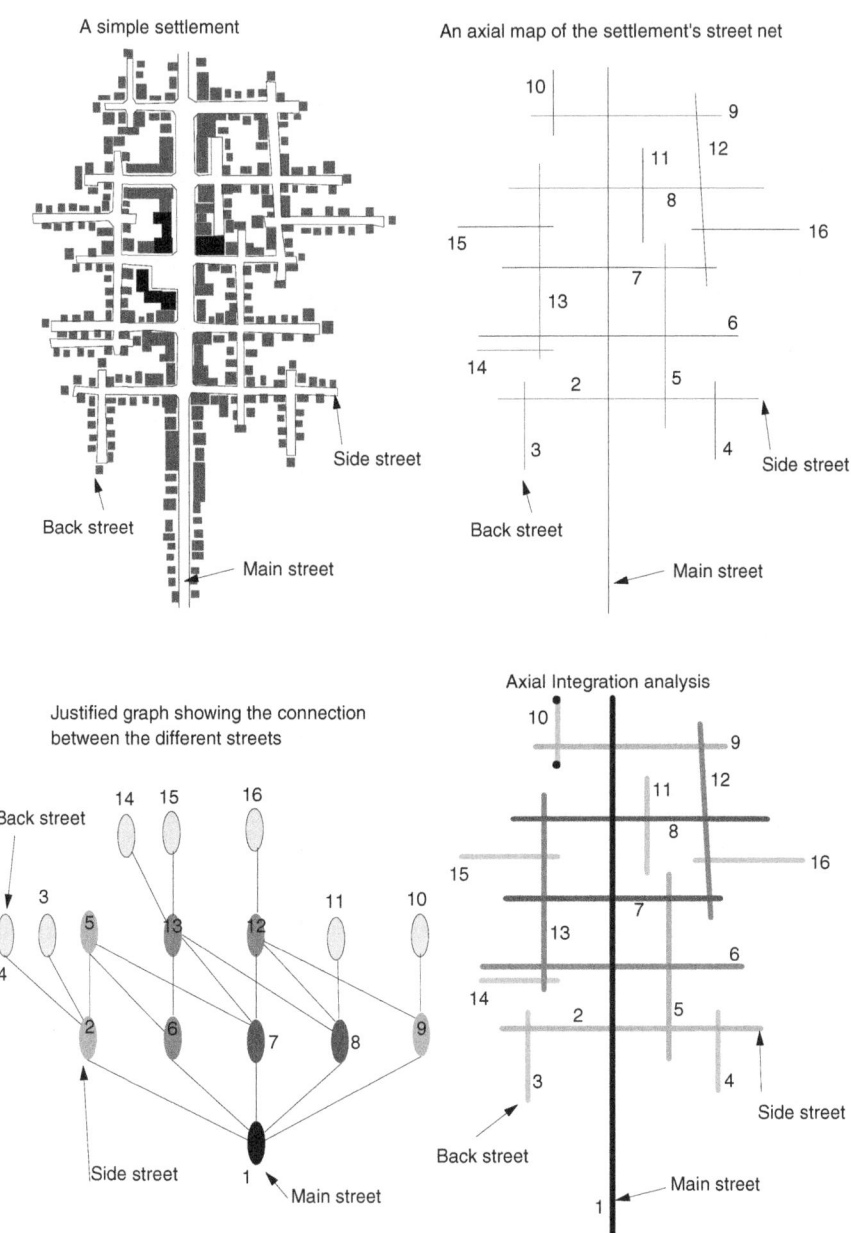

Fig. 15.2 A global integration analysis of town X

Figure 15.4 shows a comparison of the justified graphs of the back and the main street. Topologically shallow graphs such as the main street case have high integration values while topologically deep graphs such as the back street case have low integration values.

15 Space Syntax in Theory and Practice

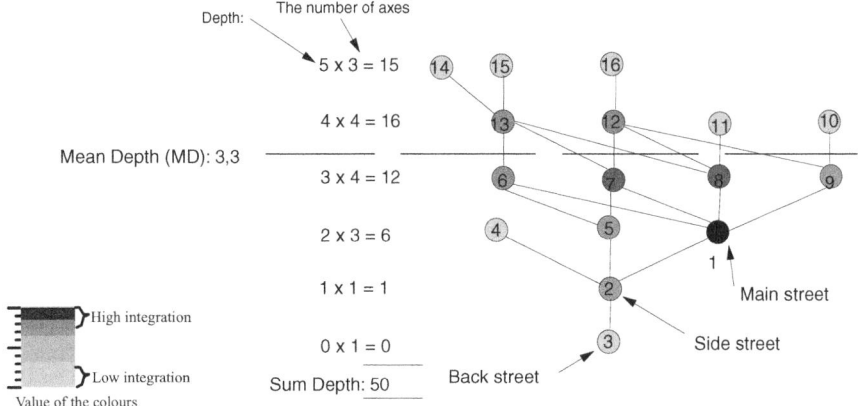

Calculating axial integration:

Mean depth for each axe (MD):
MD = sum depth/k - 1
k = number of axes in a system
sum depth = the topological depth from each axe to all other axes
Dk = diamond value

Calculating the back street axe:
(MD) = sum depth/k -1 = 50/16 - 1 = 3,3

Real asymmetry (RA) = 2(MD - 1)/k - 2 =
2(3,3 - 1)/16 - 2 = 0,3333333333

Real relative asymmetry (RRA) = RA/Dk =
0,3333333333/0,251 = 1,3280212483

Integration value of the back street: 1/RRA =
1/1,3280212483 = **0,753**

Fig. 15.3 How global integration is calculated

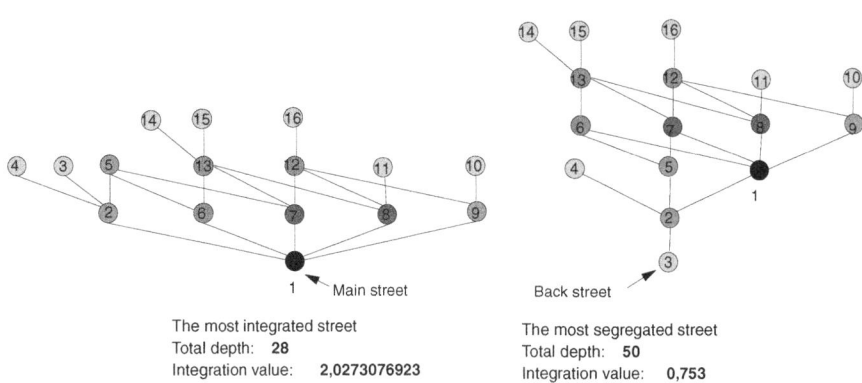

Fig. 15.4 Comparison of the justified graphs of the back and the main street with global integration

Commercial activities seem to take place in the most globally integrated streets (Hillier et al. 1993; van Nes 2002, p. 287). Dwelling areas are mostly located in the segregated areas (Hillier 1996, p. 175). There are two problems with a global integration analysis. One is the way a built environment's outskirts become very segregated, and the other is that many cities consists of several centres. A global

Average Mean Depth for the whole settlement:
Mean Depth from the most segregated street - Mean Depth from the most integrated street: 3,33 - 1,75 =   **2,525**

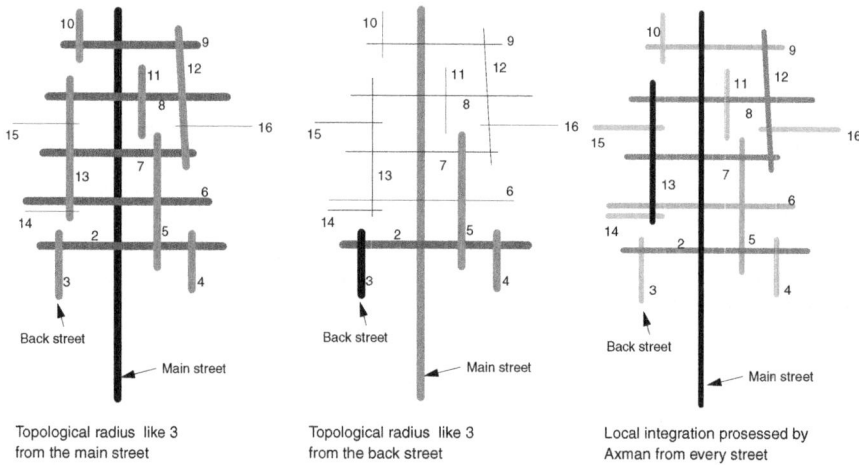

Fig. 15.5 Local integration calculations of town X with radius 3

integration analysis highlights only one centre. Therefore, a local integration analysis is needed using a local integration radius.

Figure 15.5 illustrates what local integration with a radius of 3 looks like in town X. In calculating the value of the main street (Fig. 15.5, left), the axes number 14, 15 and 16 are left out from a local integration analysis with a radius of 3. In the case of the back street (in the middle of Fig. 15.5), only 5 axes are included in the local integration analysis. The system to the right shows a local integration analysis of every street in town X processed by the program Axman. The value of the axis in a topological radius of 3 from each street is calculated.

Studies have shown that flow rates of pedestrians through cities correlate with local integration values while vehicle flow rates correspond with global integration values (e.g. Hillier et al. 1998). Moreover, local integration gives indication of local shopping areas in a city. However, applying local integration analyses on Dutch cities tend to show weak results. For example, in Leiden city center, most long main streets are curved. Axial analyses count each change of direction as one topological step, even though the angle between two axial lines is close to 180°. In this way, Leiden's centre is modelled as a fragmented street network consisting of many short axial lines. As follows, the integration values with various radii will not be the highest in these kinds of areas, which do not always correspond with shop locations. For example, the curved shopping street Haarlemmerstraatis not particularly highlighted as a highly integrated street in either the global or local integration analysis. It is only indicated in the radius-radius integration.

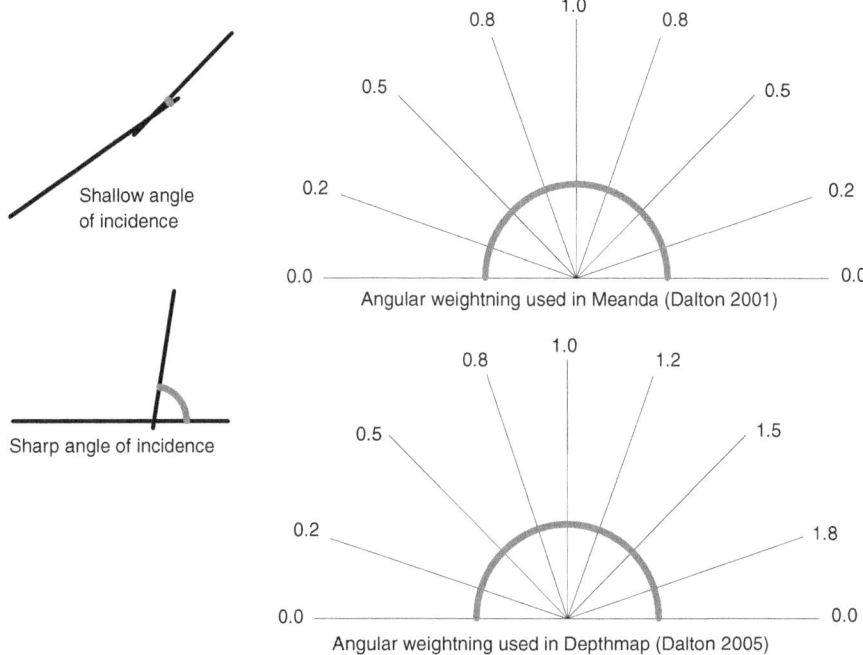

Fig. 15.6 Angular weighting

15.4 The Angular Analysis

The angular analysis is essentially an extension of visibility graph analysis and axial analysis (Turner 2001, p. 30.1). Here, each axial line is weighted by the angle of its connections to other axial lines. As shown in Fig. 15.6, two axes which meet at an angle of almost 180° have a shallow angle of incidence while two axes with almost 90° between them have a sharp angle of incidence (Dalton 2001, p. 26.7).

The angular relationships between streets play a role in the way people orient themselves in built environments. This is empirically supported by the research of Ruth Conroy Dalton who carried out experiments on how street angles influence people's choice of routes at road junctions. She concluded that people tend to conserve linearity through their routes, with minimal angular deviation (Conroy Dalton 2001, p. 47.8). There is a competition between the desire to select the simplest route and the desire to maintain the shortest route from origin to destination. As soon as the difference in angles become too great, the shortest route will win out over the simplest (Conroy Dalton 2001, p. 47.12–47.13). Therefore, geometrical distance, such as the "least angle change" of direction towards one's direction plays a role in way-finding through cities. Hence, it has to be taken into account together with the topological distance (i.e. "the fewest turn").

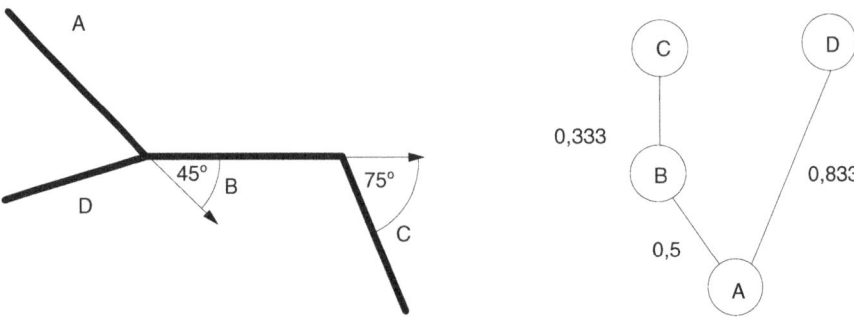

Fig. 15.7 The calculation of angular choice

The axial lines represent the longest and fewest sightlines in the area. For example a curved street consists of several axial lines. A segment line represents a street or road link between two junctions. The degree of curviness does not count in this case. When three or more lines intersect, a junction is defined. In the axial line representation a curved street segment consists of a "junction" every time there is a change of direction.

Consider four street segments connected to one another at different angles, as illustrated in Fig. 15.7. The depth from segment A to B is 0.5, since they constitute a turn of 45°. The depth from segment A to C is 0.833, the sum of the turn of 45° from segment A to segment B and the turn of 30° to segment C. When calculating the angular mean depth, or the local angular integration, the case shown in Fig. 15.7 is used as an example. The angular mean depth from street segment A is calculated as follows:

$$\text{angular mean depth of}(A) = \frac{(B)0.5 + (C)0.833 + (D)0.833}{(3)} = 0.722$$

The least angle analysis seems to be the best predictor of movement, followed closely by the fewest turns (the results from the global and local integration analyses). Metric distance comes far behind the two first ones (Hillier and Park 2007). When correlating angular choice (geometric distance) with topological distance in Leiden, it is possible to identify the local main routes in local areas and through the whole city. Figure 15.8 shows a local angular analysis with a topological radius of 3 of Leiden. The two main shopping streets Haarlemmerstraat and Breestraat are clearly highlighted as being the locally most integrated streets for the whole city.

15.5 Angular Analyses with Metrical Radii

Figure 15.9 shows some examples of how various types of radii affect various types of street grids. The street grid consists of a strict orthogonal street grid and an organic street grid. When applying the two-steps analysis to the thick line, all streets

15 Space Syntax in Theory and Practice 245

Fig. 15.8 The local angular analysis of Leiden

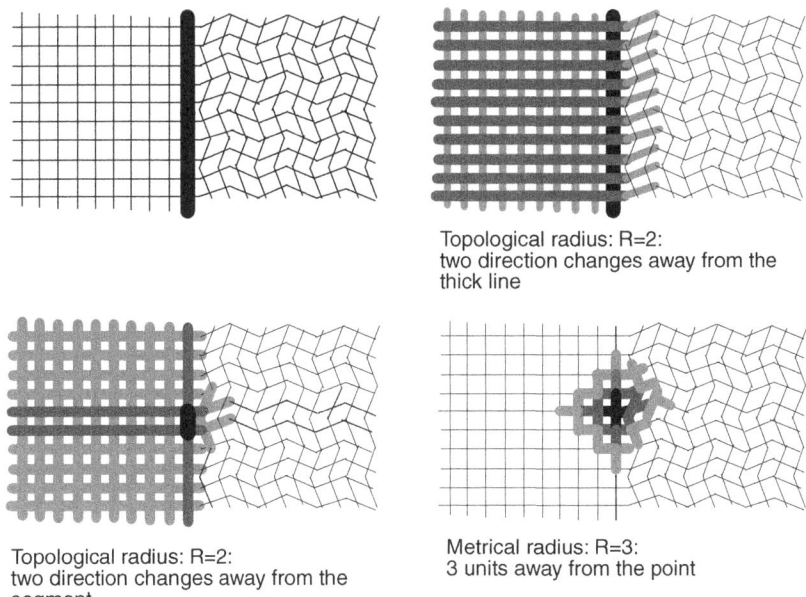

Topological radius: R=2:
two direction changes away from the thick line

Topological radius: R=2:
two direction changes away from the segment

Metrical radius: R=3:
3 units away from the point

Fig. 15.9 The difference between topological and metrical radius

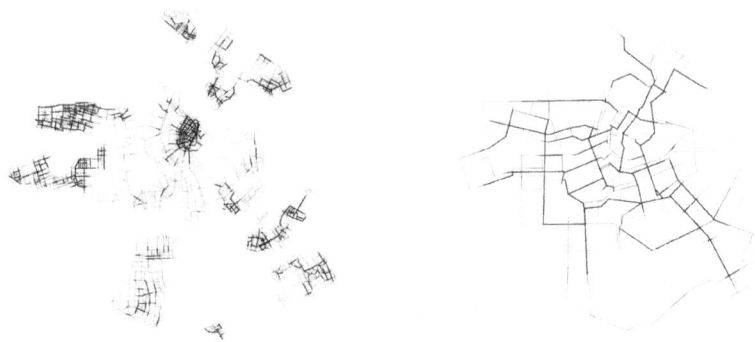

Fig. 15.10 Analyses of Amsterdam with metrical radii

in the strict orthogonal grid can be reached within two direction changes. Likewise, when applying the two-steps analysis to a street segment (below left in the figure), more or less the same local catchment area can be found as in the previous example. The only difference is that the organic street grid is less accessible than the previous example. When applying the metrical radius from a point, the strict orthogonal street grid as well as the organic street grid has more or less the same catchment area. The next step is to show what metrical radiuses add to the analyses of topological and geometrical distances.

Figure 15.10 shows an example of Amsterdam. For a small metrical radius (like 800 m), the old historical city centre is highlighted to be very vital. This area is neither highlighted in the traditional spatial integration analyses, nor in the angular analyses.

Probably this centre was originally developed to have everything accessible within a shorter metrical distance than the today's situation. Today it is the touristic centre of Amsterdam, where everything is easily accessible on foot. When using a larger metrical radius (for example a unit of 8,000), the main routes in the Pijp area in Amsterdam are highlighted as being the most vital streets for the whole city. For Amsterdam's inhabitants, this area is known to be the most vital and lively city centre for the whole of Amsterdam. The area offers a large variation of shop types and has very lively streets at all times of the day. Everything is easily accessible by bicycle, tram, car, and partly by foot.

In order to demonstrate what a metrical radius implies for urban centrality when combining it with topological and geometrical distances, an example of a new and an old town is used. The Dutch new town Lelystad is 40 years old. It has around 74,500 inhabitants. The car traffic routes and pedestrian and bicycle routes are separated. Figures 15.11 and 15.12 show angular integration analyses with a low and a high metrical radius for Lelystad's mobility network. As can be seen in Fig. 15.11, the streets where the small local supermarkets are located are highlighted in black. Conversely, as shown in Fig. 15.12, the main routes between the various local areas

Fig. 15.11 Angular analyses of Lelystad with a low metrical radius

are highlighted in black. These routes are trafficked predominantly by vehicles. Lelystad's main mega shopping mall is located in the middle, where the density of the integrated main routes is the highest.

When applying the same analyses to the old Dutch town Hilversum, founded in 1424, a different structure can be seen. Hilversum has around 84,500 inhabitants. In the analysis with a low radius in Fig. 15.13, the town's local as well as central shopping streets are highlighted in black. When applying a high radius, as shown in Fig. 15.14, the main routes through the urban areas are highlighted in black. These routes cross the local centres highlighted in the small radius analysis. As presumed, an optimal location for shops is along streets accessible to its vicinity as well as along main routes connecting various neighbourhoods with one another. Therefore, the variation of shops in local shopping areas tends to be higher in old towns than in new towns.

Figure 15.15 shows the principles of the main route network in an old and a new town. In the top of the figure, the principles of a traditional urban area are shown. The centres with a low radius are located either on or adjacent to the main routes. These centres are easily accessible by foot as well as by car. The principles of urban centres in a post-war neighbourhood are shown. The main route network in post-war neighbourhoods is separated from the local centres whereas it is integrated

Fig. 15.12 Angular analyses of Lelystad with a high metrical radius

with the local centres in pre-war neighbourhoods. Thus far, the results indicate that the spatial conditions that define vital urban centres are the presence of a topologically integrated street network, with the least deviation of angular direction change on its main routes through the area within a short metrical distance to its potential customers.

15.6 The Micro Scale Analyses

Micro scale spatial relationship in urban studies is about the relationship between buildings and street segments. More precisely it is about demonstrating how dwelling openings are connected to the street network, the way buildings' entrances constitute streets, the degree of topological depth from private space to public space, and inter-visibility of doors and windows across streets.

A registration of the *topological depth between private and public space* is done by counting the number of semi-private or semi-public spaces from the private space to the public street. If an entrance is directly connected to a public street, it has no spaces between private and public space. Then the value or depth is zero. If there

Fig. 15.13 Angular analyses of Hilversum with a low metrical radius

is a small front garden between the entrance and the public street, it gets the value one since there is a space between the closed private space and the street. Moreover, if the entrance is located on the side of the house and it has a front garden or covered behind high hedges or fences it has a value two. Entrances from back paths covered behind a shed have a value like three. It is the topological steps between the street and the private spaces that are counted. In a study on the dispersal of burglaries in Alkmaar and Gouda, the degree of permeability was used as a rule of thumb. In those cases where a flat's front door or main entrance was permanently locked but was provided with a doorbell or calling system, it was registered as a private space. As regards to flats with open main entrances, the number of semi-private spaces was counted up to the apartments (López and van Nes 2007).

Each side of a street segment must be registered separately. There exist many streets where entrances are for example directly connected to the street on the one side, while there is a flat on the other side with an upper walk gallery. In street segments with different depth values between private and public spaces, the average value can be used. The diagram in Fig. 15.16 illustrates various types of relationship between private and public spaces. The black dots illustrate the private spaces, while the white dots illustrate semi-private and semi-public spaces.

Fig. 15.14 Angular analyses of Hilversum with a high metrical radius

A street's *degree of constitutedness* is about the degree of adjacency and permeability from buildings to public space (Hillier and Hanson 1984, p. 92). If, and only if, a building is directly accessible to a street, then it constitutes the street. Conversely, when a building is adjacent to a street, but its entrance is not accessible directly from the street, the street is un-constituted. Naturally, there must be no other buildings with entrances directly connected to this street (Fig. 15.17). Few people tend to sit or stand for a long time in un-constituted streets due to a lowered level of safety (Alford 1996). In his PhD thesis, Shu's research results showed clearly that un-constituted streets were affected more by criminal activities than in the constituted ones. Moreover, entrances covered behind high fences and hedges have little visibility from neighbours (Hillier and Shu 2000).

The density of both entrances and windows in buildings with an active function (dwelling, office, shop, etc.) on ground floor level can to some extent indicate a degree of street liveliness. However, high density of entrances connected to a street does not imply high *inter-visibility*. There is a distinction between *constitutedness* and *inter-visibility*. The way entrances and windows are located opposite of each other in a street gives a high probability in the way people can observe the street (a natural surveillance mechanism). Figure 15.18 shows some diagrammatic principles on the relationship inter-visibility and density of entrances. The micro spatial

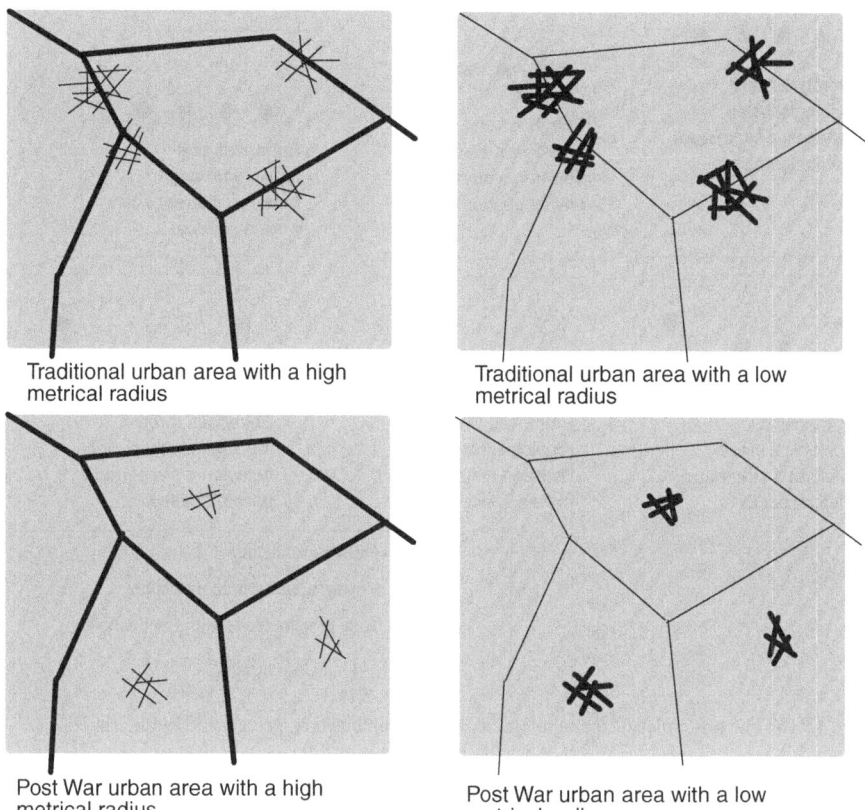

Fig. 15.15 The spatial principles of the location of urban centres in relation to the main route network in a new and an old built environment

conditions influence the quantity and quality of street life and the risks on criminal victimization (van Nes and López 2007).

Micro scale conditions are often neglected in the contemporary planning and design of urban areas. In particular, urban renewal projects, modern housing areas and new large-scale urban development projects often tend to lack adjacency, permeability and inter-visibility between buildings and streets. This has negative effects both on the quality and quantity of the street life and the safety of these urban areas. Urban project developers nowadays tend to build with high density or high floor-space-index and propose large variations of urban functions (dwellings, offices, etc.) in these areas. However, the public-private interface between buildings and streets is often forgotten. Safety and degree of street life depend on how the spatial configuration is on the plinth or built up street sides. Therefore, there is a need to bring micro scale spatial relationships on the research, policy making as well as on the design agenda in the urbanism discipline.

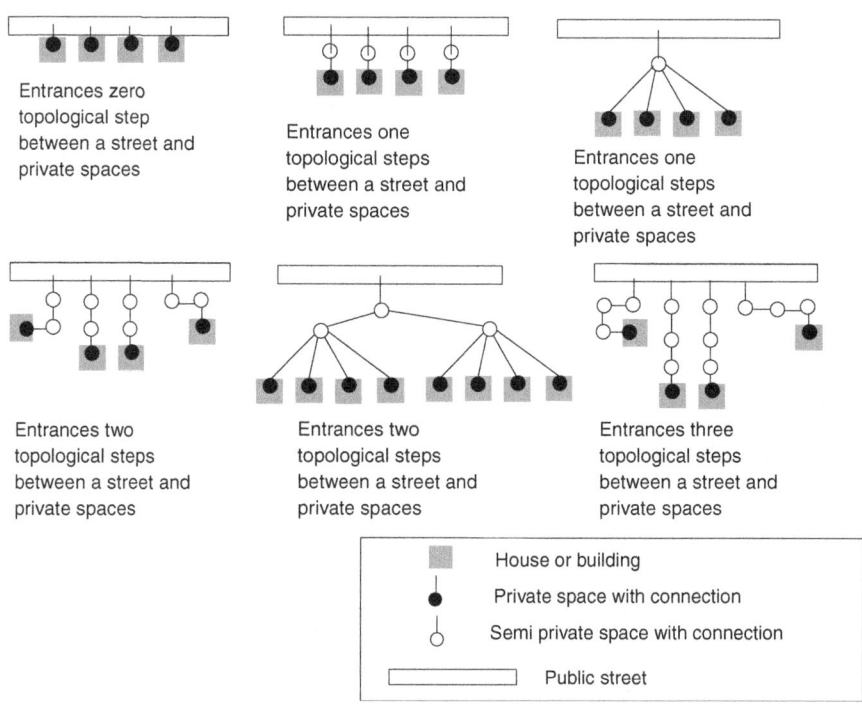

Fig. 15.16 The principles of the topological relationship between private and public space

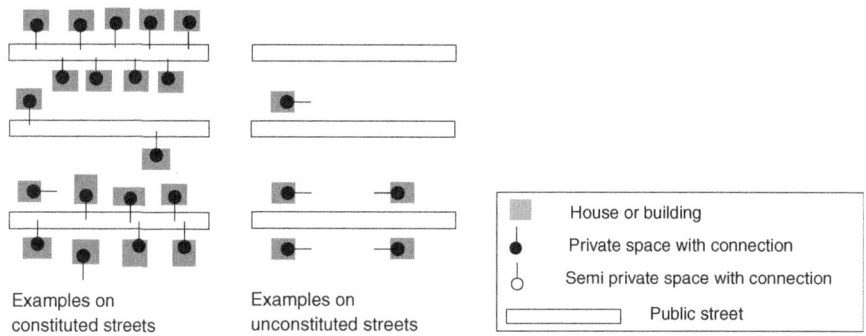

Fig. 15.17 The principles of constituted and un-constituted streets

15.7 Added Value in the Geodesign Process

Space syntax has contributed to an understanding of the spatial structure of the city as an object shaped by a society and conversely how this spatial structure can generate or affect certain socio-economic processes in society. To some extent, it

15 Space Syntax in Theory and Practice

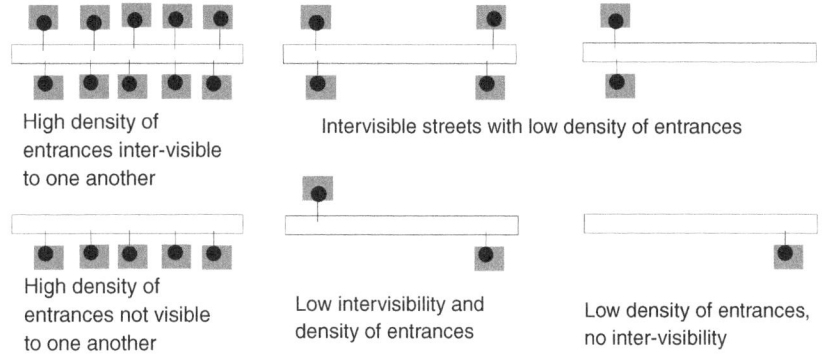

Fig. 15.18 The principles of density and visibility of entrances and windows

is possible to predict some types of economic processes as an effect on urban interventions. Likewise, space syntax provides understandings on the spatial possibilities for certain social activities such as crime, social segregation and anti-social behaviour. It is all about how spatial integration and segregation conditions social integration and segregation.

How a street or an urban area is centrally located can influence its economic attractiveness. *Metric centrality* implies that something is located in the middle of an area with the shortest metric distance to all other points in that area. Sometimes temporal aspects like time use for travel are taken into account. Obstacles like traffic-junctions, bad street quality and a fragmented street network influences the temporal aspects of metric centrality. *Topological centrality* is the most accessible centre in terms of the fewest number of direction change from all streets. The more fragmented street network in a built environment, the weaker the spatial conditions become for a vital economic centre. *Geometrical centrality* occurs along the main route network with the fewest angular deviation from all other streets. The most accessible main route network linking a city's edges towards its centre tends to have the highest flow of through travellers.

For describing social and economic activities in built environments, there is a difference between *economic* and *cultural* centrality. *Economical centrality* is the places where trade, shopping and finances take place. The aim for these kinds of activities is to be both in a metrical and topological central position to all potential customers. Their optimal position depends on the structure of the street net.

As regards to the theory of the *natural movement economic process*, the configuration of the street grid influences the movement rates through an urban street network and where economic activities take place. Attractors, such as shops, retail and large firms tend to locate themselves along the most integrated streets (Hillier et al. 1993, p. 61). Figure 15.19 shows the relationship between configuration, attraction (the location of shops) and movement. It explains how a built environment functions independent on planning processes. Movement and attractors influence

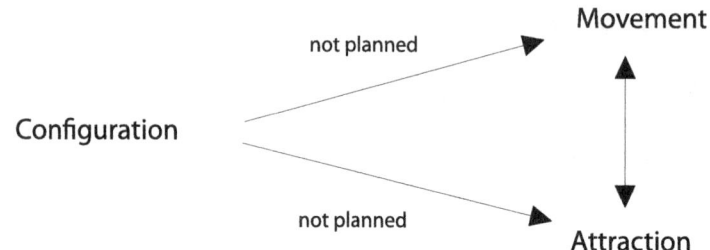

Fig.15.19 The theory of the natural movement economic process

each other. The more people in a street, the more it attracts shops to locate along these streets. The more shops locate along a street, the more they attract people.

If an economic centres' optimal position change through changes on a street network, the location of this centre is likely to change too (Hillier 1999; van Nes 2002, pp. 287–301). Friederichstrasse in Berlin provides an example on this. Before the Wall it was one of Berlin's high streets. During the Wall no shops was located along it, while after the Wall shops returned to Friederichstrasse. At present it is slowly becoming the high street of the new reunited Berlin. Therefore economical centrality is not something static. It is a dynamic process. "At all levels of the hierarchy, centres grow and fade, often in response to changing conditions quite remote from the actual centres" (Hillier 1999, p. 108). Therefore an economic centre is heavily dependent on a street structure, which relates to topological and geometrical centrality. Accessibility to potential customers and catching the through travellers is an issue.

The way human beings act and behave in urban space depends on their motives and intentions. Therefore it can be difficult to predict their behaviour after urban interventions. So far, research has shown that the location of shops, retail, national and international firms and in general the location of urban centres depends on a high degree of integration of the street net on various scales (Hillier 1999; Hillier et al. 1993). Likewise, land values and rent process tend to be influenced by the various integration values of the street net (Desyllas 2000). High degree of accessibility or reach ability is at issue for creating vital urban centres and generating economic activities inside them.

In studies on built form and meaning, dealing with place character and architectural styles, touches upon the limits of space syntax. Therefore *cultural centrality* is a broader issue than economic centrality. Places with a large concentration of historically important buildings and monuments from the past are defined to be cultural centres. The meaning of artefacts and the tradition related to them can be understood from the technical, social, cultural and economic activities that took place in the past (Moudon 1997).

What is happening in extreme segregated streets? In research on urban space and human social interactions, it is difficult to set out general statements on how the spatial layout of a built environment can provoke criminal activities, create social segregation, steer political processes and provoke anti-social behaviour. In contrast with market rationality, where the intentions tend to be unambiguous, social rationality concerns a wide range of intentions. Even though occurred incidents can be understood from a spatial point of view, an understanding on space and crime depends at least on understanding the behaviour of local inhabitants and an insight of the social composition. Sometimes the social composition of the inhabitants can overrun the spatial generative power.

In general segregated streets have more complex routes to all other streets in a city. Areas with segregated spaces, with urban grids visually broken up and with few dwelling entrances constituting streets are often affected by crime and social misuse (Hillier and Shu 2000, p. 232; Hillier and Sahbaz 1984, p. 456). The same investigations prove that spatial organisation can generate movement according to co-presence and co-awareness in the built environment.

A space syntax approach can at least identify the spatial properties to understand why some already established urban areas have a high level of crime and social misuse. These identified spatial properties provides at least precise understandings on how spatial configuration plays a part in broader social processes of perceived and actual decline. Thus space syntax can illustrate that the spaces of a built environment can affect human behaviour (Hillier 1996, p. 184). A configurative approach makes one understand that the means a built environment offers are physical while its ends are functional—not visa versa.

Even though space syntax offers precise concepts to operate with, it cannot analyse the place character, the sphere and the symbolic meaning of the built form. However, space syntax can analyse the configurative structure of their spatial setup as an independent factor of the built form's symbolic meaning. Therefore space syntax deals with place *structure* and not with place *character*. Analyses of place character require a genuine understanding and insight of their society's cultural background and spiritual traditions from the present as well as the past.

While researchers with a phenomenological approach seek to describe the underlying essential qualities of human experience and the world where these experiences happens (Seamon 1994, p. 37), researchers with a space syntax approach identifies the spatial conditions for lively or quiet urban squares, streets, neighbourhoods, etc. Understandings of the spatial conditions of pedestrian flow rates and degree of urban vitality is also an essential component of the sphere of a place. As Hillier and Hanson writes, space syntax is about understanding "the social content of spatial patterning and the spatial content of social patterning" (Hillier and Hanson 1984, p. x–xi). It is a small, but significant contribution towards a comprehensive theory on built environments.

When applying space syntax in urban design or on any kind of physical intervention on the street network or on the building-street interface, it is heavily dependent on evidence from research and theory development on the relationship between space and society. Space syntax makes possible to some extent the evaluation of socio-economic effects of various design proposals for an area, and is a critical piece in the geodesign process, especially when communicating impacts to stakeholders. At present, space syntax has been applied to urban design, urban regeneration and strategic planning projects. Designing with the use of space syntax is slowly entering the design field in regenerating poorly functioning urban areas. Some successful projects are already implemented, such as the regeneration of Trafalgar Square and the South Banks in London. It is a first step towards a scientifically grounded urban design process, rather than working on guesswork or intuition when intervening in built environments.

References

Alford, V. (1996). Crime and space in the inner city. *Urban Design Studies, 2,* 45–76.
Conroy Dalton, R. (2001). The secret is to follow your nose. In J. Peponis, J. Wineman, & S. Bafna (Eds.), Proceedings space syntax 3rd international symposium, Atlanta.
Dalton, N. (2001). Fractional configurational analysis and a solution to the manhattan problem. In J. Peponis, J. Wineman, & S. Bafna (Eds.), Proceedings space syntax 3rd international symposium, Atlanta.
Desyllas, J. (2000). The relationship between urban street configuration and office rent patterns in Berlin. Dissertation, The Bartlett, University College London.
Hillier, B. (1996). *Space is the machine.* Cambridge: Cambridge University Press.
Hillier, B. (1999). Centrality as a process: Accounting for attraction inequalities in deformed grids. *Urban Design International, 4*(3–4), 107–127.
Hillier, B. (2007, October 9). Cities and urban societies: The role of endogenous factors. Power point presentation notes. Lecture given at the Mesoamerical urbanism project conference, Archeological department, Leiden University.
Hillier, B., & Hanson, J. (1984). *The social logic of space.* Cambridge: Cambridge University Press.
Hillier, B., & Iida, S. (2005). Network effects and psychological effects: A theory of urban movement. In: A. van Nes (Ed.), Proceedings space syntax 5th international symposium, Delft.
Hillier, B., & Park, H. T. (2007). Metric and topo-geometric properties of urban street networks. In A. S. Kubat (Ed.), Proceedings space syntax 6th international symposium, Istanbul.
Hillier, B., & Sahbaz, O. (1984). High resolution analysis of crime patterns in urban street networks: An initial statistical sketch from an ongoing study of a London borough. In A. van Nes (Ed.), Proceedings space syntax 5th international symposium, Delft.
Hillier, B., & Shu, C. F. (2000). Crime and urban layout: the need for evidence. In S. Ballintyne, K. Pease, & V. McLaren (Eds.), Secure foundations. Key issues in crime prevention, crime reduction and community safety. London: Institute for Public Policy research.
Hillier, B., Penn, A., Hanson, J., Grajewski, T., & Xu, J. (1993). Natural movement: Or, configuration and attraction in urban pedestrian movement. *Environment and Planning B: Planning and Design, 20,* 29–66.
Hillier, B., Penn, A., Banister, D., & Xu, J. (1998). Configurational modeling of urban movement network. *Environment and Planning B: Planning and Design, 25,* 59–84.
Lopez, M., & van Nes, A. (2007). Space and crime in Dutch built environments. Macro and micro scale spatial conditions for residential burglaries and thefts from cars. In: A. S. Kubat (Ed.), Proceedings space syntax 6th international symposium, Istanbul.
Moudon, A. V. (1997). Urban morphology as an emerging interdisciplinary field. *Urban Morphology, 1,* 3–10.
Norberg-Schulz, C. (1971). *Mellom jord og himmel. En bok om steder og hus.* Oslo: Universitetsforlaget.
Seamon, D. (1994). The life of place. *Nordisk Arkitekturforskning, 1,* 35–48.
Turner, A. (2001). Angular analysis. In: J. Peponis, J. Wineman, & S. Bafna (Eds.), Proceedings space syntax 3rd international symposium, Atlanta.

Turner, A. (2005). Could a road-centre line be an axial line in disguise? In A. van Nes (Ed.), Proceedings space syntax 5th international symposium, Delft.

Turner, A., Penn, A., & Hillier, B. (1993). An algorithmic definition of the axial map. *Environment and Planning B: Planning and Design, 32,* 425–444.

van Nes, A. (2002). Road building and urban change. The effect of ring roads on the dispersal of functions in Western European towns and cities. Dissertation, Agricultural University of Norway.

van Nes, A., & Lopez, M. (2007). Micro scale spatial relationships in urban studies. The relationship between private and public space and its impact on street life. In A. S. Kubat (Ed.), Proceedings space syntax 6th international symposium, Istanbul.

Chapter 16
A Standard-based Framework for Real-time 3D Large-scale Geospatial Data Generation and Visualisation over the Web

Massimo Rumor, Eduard Roccatello and Alessandra Scottà

16.1 Introduction

Geodesign, as a method, even before being conceptualised, has evolved in recent years together with the widespread use of geoinformation and Geographical Information Systems (GIS).

As GIS software and geoinformation have become less expensive and widely available, their use in many activities related to planning and design has increased raising at the same time a strong demand towards technology and data, starting a virtuous cycle. Furthermore, the range of users has increased. The traditional GIS users, coming from planning, geology and geography are now joined by an increasing number of users coming from disciplines such as business, health care, disaster management, education and design activities.

The geo-component brings together planners and geoinformation specialists to a shared picture of reality (also called a "common operational picture"), enabling communication and collaboration between different disciplines. This is of paramount importance when planning and designing, both activities that shape our future world.

This integration is made possible thanks to the evolution of geotechnology, which has become able to support planning and design activities (and this is our understanding of geodesign) at reasonable costs. With respect to the geotechnology evolution, the Internet, multimedia and especially three dimensional representations

---

M. Rumor (✉)
3DGIS, University of Padua, Padua, Italy
e-mail: rumor@dei.unipd.it

E. Roccatello
3DGIS, Padua, Italy
e-mail: eduard.roccatello@3dgis.it

A. Scottà
Geodan, Amsterdam, The Netherlands
e-mail: alessandra.scotta@geodan.nl

of reality are essential features that need to be considered as integral parts of geo-information and GIS. Three dimensional representation can be considered as the "common operational picture" where the communication between planners and geo-information specialists is made possible and communication obstacles are abated.

3DGIS (http://www.3dgis.it/en/) is an Italian GIS software company, born in 2007, built on a long-term experience in geographical information system design and development and in the ICT area. 3DGIS is founded on the combination of different professionalisms: computer engineers, graphic designers, architects. The company is committed to producing state of the art solutions with the best possible quality/ cost ratio.

Based on the assumptions mentioned above and upon large experiences in the geoscience field, 3DGIS has started working on 3D modelling and visualization and developed a 3D visualization application, named Cityvu, in 2008. The project for Public Participation in Planning (that will be illustrated later on) for the city of Follonica has been developed using Cityvu.

Cityvu was a plug-in free CityGML viewer, able to show 3D geospatial data directly in the browser, its integration features were limited to embedding in a web page and to open specific link on feature selection, it was available for both web and desktop but it lacked a simple and fast modelling mechanism (indeed, models for the Senzuno quarter, in city of Follonica, were built by hand with a great amount of work) and was very hard to integrate to other 3DGIS products.

The new 3D framework has been developed to solve these problems, using modern technologies like WebGL and HTML 5. This allowed the company to build a set of libraries which can be easily integrated into other products and that could be used to generate 3D features in real time when possible.

This new framework has been conceived, designed and developed and research work is ongoing. The studies and developments upon the framework consider the following fundamental aspects:

- Generic web-based solution assuring distinct software component independence.
- Rapid application development approach and,
- Integration of three dimensional representations of reality.

The three dimensional integration studies consider the automatic generation of large-scale 3D geodata through a Spatial Data Infrastructure (SDI) and the optimisation of large-scale data streaming over the web. The requirements are the availability of a generic framework, to be component and platform independent, flexible for different solutions where the three dimensional component is an integral part. The aim is to reach a wider and diverse community, from traditional GIS users, to users, such as planners, where GIS is not central but a supportive tool for diverse work processes.

This chapter illustrates practical experiences in the geodesign field, starting from the first geodesign experience, dating back to 1989–1990, to a more recent and technologically updated case.

Afterwards the chapter illustrates the evolution of the 3DGIS framework from 2D to 3D WebGIS for large-scale geodata over the web. The main 3D developments are discussed:

- Web based 3D visualization
- Automatic generation of urban 3D environment via SDI
- Optimisation 3D large-scale geodata streaming via the web
- 3D supported citizen participation

The chapter concludes by describing some future work and points of discussion.

16.2 Geodesign in Practice: Evolution from Twenty-Five Years ago till now

In the early stages of GIS development, planners had been among the main users of this new technology. A great number of them used only automated mapping functions, a few used some analysis functions and very few approached planning in a new way. As an example in the city of Padua, Italy, a new plan for the city centre was designed almost completely by the "computer" (Rumor and Gonzato 1989).

All spatial data had been collected and related to the database where all buildings and other relevant urban structures were stored with a set of attributes. On the basis of the attributes and of some spatial constraints a set of maps had been produced to show the "situation" to the planners.

On the basis of the decisions made after the situational analysis and compared with the strategic aims of the new plan, a set of rules were prepared to be applied automatically to all buildings, thereby producing the new plan.

That work, carried out in 1989–1990, was an experience of geodesign.

The geodesign experience of the city of Padua needed a lot of work from the geoinformation specialists that had to develop very specific ad hoc software with a lot of foresight, and also some bravery, from the planners.

Because too much work, time and consequently high costs were required, only few rather large and rich organisations could afford this approach. With the development of GIS and the increasing availability of low-cost geotechnology and geoinformation this approach can be adopted by the great majority if not all concerned organisations.

Another recent example (2009), explained in more detail later in this chapter, is considered.

The city of Follonica, a medium city in Tuscany, had to compare some alternative solutions for the improvement of an urban district. The possible improvements included the construction of a new bike path, changes to public lighting and to the street furniture. It must be told that the inhabitants of the district were very sensitive about the issue. The City Administration then decided to investigate the issue with the use of GIS technologies and accepted a proposal from 3DGIS, which was already working to develop a 3D framework, to build a 3D model of the district and to include in it the proposed improvements so they could be examined as in reality. The inhabitants could also examine the different proposals navigating in their own district's model. The inhabitants were also asked to vote via the web, and very many did and the Administration decided on the basis of the votes. That was geodesign and public participation combined.

16.3 The 3DGIS Framework: From 2D to 3D WebGIS

The 3DGIS framework is the outcome of years of research, design, experience and development. The framework is a software platform where common code organized in modules with generic functionality can be selectively specialized or overridden by developers or users. The framework is founded upon Geo-ICT standards (ISO and Open Geospatial Consortium standards) and use open source libraries to minimize costs.

One of the main state-of-the-art characteristics of the framework is the fact that it has been designed and developed to achieve a Rapid Application Development (RAD) approach for deeply customized WebGIS solutions. Because the framework is completely standards based, distinct component independence and overall software quality is guaranteed. Standard compliant development, though being more effort consuming, allows software updates to the latest technology evolutions without any trauma or high cost migrations. The framework also easily adapts to needs volatility, preserving high efficiency and reducing cost. The framework is organized in modules, providing a wide range of GIS functionality on the web. The functionality can vary from a simple navigation on the map to sophisticated and complex functionalities (authentication, database schema generation, attachment and document management, wide range of editing functionality, third party integration support, etc.). Upon the 3DGIS framework, different products are already available for different sectors: from analysis of topographic maps to networks analysis, roads and disaster management and many others.

The 3DGIS framework and the derived products handle geospatial data in a two dimensional WebGIS environment. Upon the 3DGIS framework different products are already available for different sectors: from analysis of topographic maps to networks analysis, roads and disaster management and many others. It has been used overin the years to develop a wide range of GIS web applications for many users and has evolved significantly.

In 2008 we decided to start extending our solutions including the use of 3D models and the project named Cityvu, a plug-in free CityGML viewer that was launched in 2008 and concluded in 2009 (Roccatello and Rumor 2009).

Cityvu has been used to implement various applications, the most interesting is represented by the case of Follonica, previously mentioned and will be used to illustrate in detail where concepts of geodesign and public participation has been applied in reality.

That experience convinced 3DGIS of the power of 3D technology in the planning process, and fostered the decision to continue working to make 3D functionalities easily available to a great number of organisations. We then decided to take the challenge to integrate 3D representations into the 3DGIS framework, extending the current 2D framework to a 3D WebGIS generic framework. Though proving to be very challenging, after having completed an extensive research phase, developments have been started and the first positive results are demonstrable.

16.4 The Project

16.4.1 Requirements

We all know that 3D models can be used with great benefits in a variety of applications in geography, urban studies and other fields. On the other hand the use of such models often requires us to render significant amounts of three dimensional geospatial data, and is very demanding in terms of computing capability and memory usage. In order to efficiently manage the volumes needed by large-scale model rendering, and to reach reasonable levels of data quality and performance, some optimizations are needed.

While working in a 2D space, most current WebGIS platforms are leveraging open source technologies and are standards based. Open Geospatial Consortium's (OGC) work on standardization has, in fact, delivered a great opportunity around interoperability and web services like OGC Web Map Service (WMS) and OGC Web Feature Service (WFS) that are very common nowadays. Empowered by increased browser's capabilities, currently developed WebGIS are very promising and able to replace traditional desktop GIS for a wide range of applications. While WebGIS advantages are well known, a lot of work needs to be done about 3D data visualization over the Web.

During the past years, browser technology has been updated to support HTML 5 and CSS 3 in order to allow greater user experience, similar to desktop applications. WebGL extensions supports the development of a complete hardware accelerated 3D engine running in the browser without any third party plugin.

16.4.2 Approach to 3D Modelling for WebGIS

Following our products development, we managed to achieve a good experience with WebGIS and we learned that most organizations are able to deploy or have already deployed their spatial data using OGC Web Services, especially using OGC WMS and WFS. At the same time it is now possible to leverage the latest browser technologies in order to augment traditional WebGIS features with a 3D visualization framework. Therefore we needed a streaming framework for the visualization of three dimensional large-scale geospatial data over the web with real-time generation of features, based on OGC Web Services.

Employing this approach, a 3D scene is incrementally built with a dynamically updated run-time, taking into account the movements of the camera and its field of view. To effectively and efficiently achieve this behavior, proper mechanisms of tiling and caching have been implemented.

The framework implementation focuses on textured terrain and streaming of the buildings. Despite the scope limitation, the defined streaming paradigm has general validity and can be applied to more complex 3D environments. The addition of other functionalities on top of the existing ones is straightforward and does not imply substantial modifications to the framework.

The framework's main aim is to achieve 3D visualization and querying of the attributes of the visualized objects with good performance even in low to mid end computers while keeping ease to develop, portability and wide browser support.

Our development platform is bounded to available browser technology. In order to achieve hardware accelerated 3D support in browsers without any plug-ins, WebGL and JavaScript (components of HTML 5) must be supported.

Portability and ease of development have been enhanced using existing foundation frameworks, which allows faster development and a good level of abstraction for each component.

The framework architecture has been designed in order to allow fast extendibility of geographic features support and it is structured with three main components: Foundation, Scene and Query.

The Foundation is the core component. It manages the ellipsoid model and the geospatial projection, supporting a wide range of extensible projection types. Camera viewport is integrated in the foundation component within Tile Manager, which manages the currently loaded set of map tiles and supports dynamic level of details using a quad tree backend. On the top of the Foundation, the Scene component orchestrates external data loading through the Layer Manager. The framework supports data loading from OGC WMS and OCG WFS services with on-the-fly reprojection support, thanks to the embedded proj4 library.

A layer abstraction is used to allow ease of extension for custom features. The framework currently implements Terrain and Building layers with streaming support. Terrain is streamed using tiled height maps via WMS and buildings are generated using WFS data through on-the-fly extrusion.

Multiple data layers is allowed on the top of the Tile Manager. Finally, the Query component allows external data integration and querying.

16.4.3 Optimisation of 3D Large-scale Geodata Streaming via the Web

GIS applications that have to visualize large-scale 3D geodata need to handle a massive amount of terrain, buildings, textures, imagery and so on. While designing the 3DGIS framework these challenges had to be accounted for while also keeping in mind that LIDAR data is nowadays more accessible. Even if computational power and storage size are increasing, the scan resolution is increasing year by year.

The perfect 3DGIS engine needs to adapt and scale from mid, low to high computer performance (e.g. engines designed for gaming) and to work on smartphones and tablets without complex modifications. Even powerful computers and video boards are bounded to a specific number of polygons and models in order to render the whole scene in time. Due to the large amount of data, these cannot be left unmanaged. The framework includes level-of-detail (LOD) support for terrain, LIDAR and generic models. The approach behind the LOD is to simplify the complexity of a 3D object representation as it moves away from the camera. A complex LOD system should also differentiate between continuous (i.e. terrain) and discrete (i.e. buildings, trees) 3D data.

Our terrain streaming system has to load and render a large-scale digital terrain model (DTM) stored as a geo-referenced grayscale height map and accessible via OGC WMS or Tile Map Service (TMS). The system loads the terrain chunk by chunk following the observer's movements and also dynamically updates the scene such that it contains, at each time, only the terrain chunks lying inside the field of view or just nearby to it (Chunked LOD). Terrain is loaded using multiple detail scale where nearest chunks are more detailed than the ones far away from the camera. This is accomplished using a quad tree data structure for space partitioning.

Also LIDAR data must be treated *per se*, given their complexity and their large file size. In addition to the Chunked LOD approach a lot of pre-processing is required such as multi-scale spatial indexing for LIDAR data streaming. Moreover, the framework leverages No SQL techniques for faster data access, while keeping unprocessed LIDAR data in a traditional GeoDatabase for further inquiries.

The result of the development process is a fully functional 3D framework with basic navigation GIS features (as zooming, panning, and identifying). This viewer implements the designed streaming strategy with excellent results in terms of performances (50–60 frames per second during navigation with occasional drops to 30–40 frames per second while loading) and memory usage.

3DGIS has planned a wide adoption of the developed framework into its GIS product, being able to extend traditional WebGIS with a feasible 3D visualization tool (Roccatello et al. 2013).

16.5 3D Supported Citizen Participation

3D models and 3D applications have a huge role in improving citizen participation, given the possibility to recreate the real world, before and after public works, and to evaluate the proposal in an immediate way, more intuitive than a plain 2D map or an expensive plastic model.

Traditional 3D participation tools like still image rendering and video presentations present serious limitations, hindering the advance in geodesign as they set strong constraints for design evaluation. For these reason we consider 3D real-time

Fig. 16.1 Planning the city of Senzuno with different street variations and styles

rendering as the perfect solution, able to let a person correctly evaluate every aspect of a development proposal or intervention.

In 2009, 3DGIS framework was applied to the Senzuno's requalification project in Follonica, Italy. Working alongside the Municipality GIS' technicians, the entire Senzuno's block was recreated as a detail rich 3D model. Once the reality was modelled, various versions of the urban fabric were designed, in order to achieve nine variations on the existent one. As example, Fig. 16.1 shows as example two different versions (traditional and functional) and different variations (or styles) within each version (classical and modern style for the traditional version and concrete/stone and concrete grit style for the functional version). In addition a new path for the cycling lane has been introduced into the model.

3D real-time rendering has been achieved using Cityvu project, which allowed every citizen to view the new urban fabric proposals in Senzuno, without installing

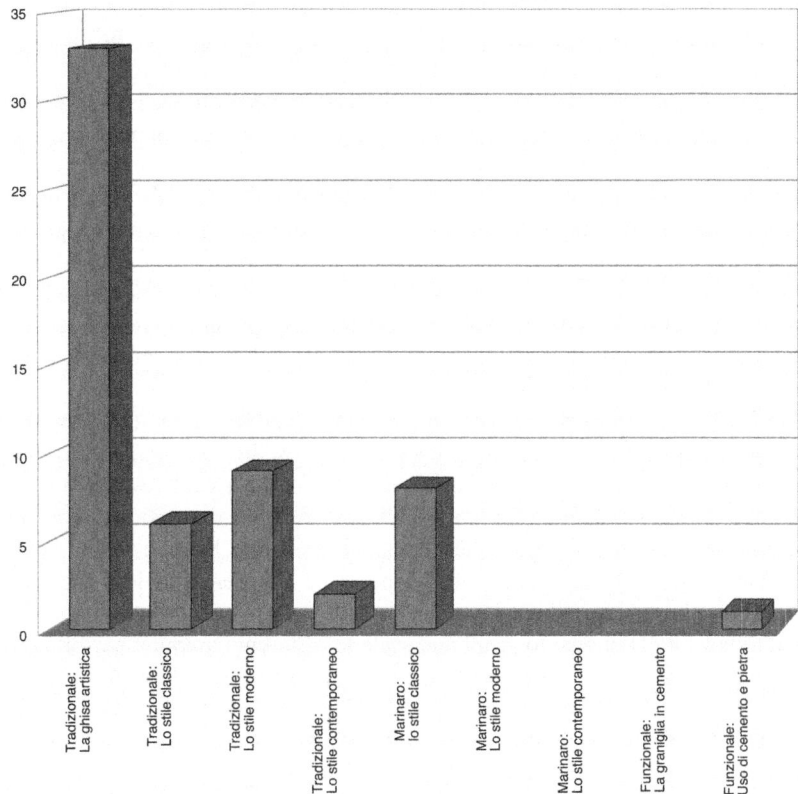

Fig. 16.2 Statistics reporting votes of citizens

any software in their computer. The model has been made walkable, manually or automatically, in order to move the perspective like in a videogame. While navigating the city block, it is possible to change the urban fabric in real-time, using a simple user interface.

Citizen evaluation has been made using a specifically crafted website (still available at the link:http://follonica.3dgis.it/), which required a registration and a code to enable voting for the best design solution. This allowed the Municipality to avoid abuse and restrict votes only to Senzuno's citizens. In order to facilitate the participation of people offline, which usually are the elderly, multiple forum sessions were organized. These forums allowed non-technical people to navigate and vote for their preferred urban fabric solution. Results showing the citizens choices where published on the crafted website, showing the clear preference for a traditional version with artistic style solution (see also Fig. 16.2).

16.6 Conclusions

The 3D framework has been designed to be a solid basis which can be used to develop customized software solutions, built on specific needs. The real-time feature generation of geospatial data via OGC services combined with the powerful results in terms of performances over the web are the state-of-art development factors for the visualisation of the three dimensional data over the web. The next steps will be to confirm the current positive performances results with tests performed in existing scenarios. Afterwards, 3DGIS will integrate the 3D framework into its GIS products (based on the 2D framework), extending traditional WebGIS with a 3D visualisation environment. Besides the 3D WebGIS developments which are already now feasible, support of 3D GIS on mobile devices is also planned. The framework is already designed to adapt and scale on smartphones and tablets, providing already a basis also for mobile devices.

The 2D Web framework is already in production and different 3DGIS products, based on the framework, have been implemented by different Italian customers. The 3D framework has been developed, tested and successful results have been confirmed. The 3D streaming for the visualisation of three dimensional large-scale geospatial data over the web generates features in real-time, based on OGC web services. The framework has been carefully optimised in order to achieve accelerated 3D support in browsers without any plug-ins. Attention to performance issues is essential to have a 3D engine to adapt and scale to different computer performances, and to support mobile devices. The framework includes level of detail (LOD) support for terrain, LIDAR and generic models. The goal is to provide WebGIS with an extended 3D environment, based on OCG services and assuring optimal web performance for 3D visualization. The entire framework is completely standards-based, guaranteeing distinct component independence and overall software quality.

The characteristics of the framework aim to provide an evolution in geotechnology, offering advanced 3D WebGIS support to a wide community (from the traditional GIS to less GIS-oriented users) and at reasonable costs. Only with this approach can geotechnology give a significant contribution to the design and planning activities with the use of geoinformation.

We can conclude that geotechnology for geodesign is already available. What needs to be recalled and further discussed is the fact that data availability is still an open issue: the lack of adequate data sets hinders the application of geodesign in practice. "Adequate" means complete and updated information available at no cost or at a reasonable cost. This implies the availability, among other components, of technologies able to make data collection and maintenance less expensive and time consuming. The data availability issue is an essential one and there is the need to put some effort to increase it. Enabling geotechnology to express its full potential to transform design into geodesign by default, so eventually the 'geo' will disappear as the technology becomes embodied.

References

Open Geospatial Consortium (OGC). http://www.opengeospatial.org.

Roccatello, E., & Rumor, M. (2009). *Design and development of a visualization tool for 3D geospatial data in CityGML format (UDMS Annual)*. Amsterdam: CRC Press.

Roccatello, E., Nozzi, A., & Rumor M. (2013). Design and development of a framework based on OGC web services for the visualization of three dimensional large-scale geospatial data over the web. *ISPRS Archive, XL-4/W1,* 101–104.

Rumor, M., & Gonzato, R. (1989). Computer aided planning: The experience of the city of Padova. International Federation for Housing and Planning IFHP-FIHUAT-IVWSR International Congress Report Chiba-Tokio Japan 1989.

Chapter 17
Crowd Sourced Public Participation of City Building

Ana Sanchis, Laura Díaz, Michael Gould and Joaquín Huerta

17.1 Introduction

As one of the local government responsibilities, urban planning strongly influences how citizens live in their cities and is of civic concern since they are affected by changes in city shape, use and functionality. In the same way, citizens affect city building with their daily performance of actions and decisions on the patronization of city spaces, and use of services and infrastructures.

In this context, many local governments are mandated to invite public reaction to every new city development plan, and citizens are then allowed, and encouraged, to officially state their opinions and concerns. Theories on collaborative planning highlight the benefits of citizen's participation in strengthening public support for policy initiatives. This participation can help identify previously unforeseen concerns and recognize potential conflicts (Healey 1992; Conroy and Evans-Cowley 2006). Local governments must study and pay attention to all public reactions, provide a justifiable response to their concerns and take decisions accordingly.

This concept is not new, we witness many attempts to implement this concept through web pages and on-line documents and forms, but commonly used mechanisms are printed maps and other documents available to be physically consulted in government offices. Citizens must actively seek, ask and wait for these documents—sometimes even consult them *in situ* to be able to physically (hand) write

---

A. Sanchis (✉) · L. Díaz · M. Gould · J. Huerta
Institute of New Imaging Technologies, Universitat Jaume I
Sos Baynat, s/n, Castellón de la Plana, 12071 Castellón, Spain
e-mail: ana.sanchis@uji.es

L. Díaz
e-mail: laura.diaz@uji.es

M. Gould
e-mail: gould@uji.es

J. Huerta
e-mail: huerta@uji.es

and register allegations and concerns. This bothersome and time-consuming manual procedure results in a lack of real public participation in the city's development process.

However, nowadays there are certain conditions that allow pushing the design of urban contexts to another level. A great number of devices and information tools facilitate data manipulation. In this direction cloud computing platforms and the rise of bottom-up activities are leading to the concept of a more participatory design of cities, since they are fast becoming of interest and general concern for every citizen (Batty 2013), who now realize that their opinions can influence government decisions and affect their surroundings directly. Recent initiatives like Brickstarter[1] arose out of this desire to make the process of urban development more accessible and legible. This way, a more diverse public not only has a voice, but can actively contribute in the shaping of the city in a trend that has been coined as *crowdsourced urbanism*.

In this context, one important research topic in Geographical Information Systems (GIS) is related to this concern that, all citizens' voices should be heard in a democracy. This is commonly called Public Participation GIS (PPGIS). This term is used to describe a variety of approaches to making GIS and other spatial decision-making tools available and accessible to all those with a stake in official decisions (Obermeyer 1998). The incorporation of (spatial) information technology into public urban planning processes represents an area of great potential in which better relationships between governments and their citizens can be built. The use of information and communication technologies (ICT) decreases the time and the geography constraints faced by citizens who want to participate, since they are enabled to participate in the location and time of their own choosing (Kwan and Weber 2003; Conroy and Evans-Cowley 2006).

In order to support ICT use, the widely spread European smart phone market, measured by active subscribers of the top 50 networks, was 860 million in 2010 (CMSinfo 2011), provides portable technology advantages such as advanced computing capability and connectivity. Through Global Positioning Systems (GPS), and other embedded sensors, smart phones enable access to location-based information services (LBS) and applications, such as the geo-fence functionalities. A geo-fence is a virtual perimeter for a real-world geographic area. This concept is used to monitor the activity in a certain area, tracking users and raising notifications and alerts when a user is entering or leaving that area providing information about the device location. Among the multiple applications developing this concept, Geoloqi[2], or the ArcGIS Geotrigger Service[3] are good examples of this kind of service.

Moreover, ICT tools can help to develop and build technologically 'smart' cities, since its use brings information closer to citizens (in space, time and content) and encourages and assists them to act in a more participative, environmentally-conscious and even in a more sustainable and 'smart' manner. The provision of

---

[1] http://brickstarter.org/

[2] https://geoloqi.com/

[3] https://developers.arcgis.com/geotrigger-service/

information and knowledge through these mechanisms, enabled by new visualisation and access technologies, support participation and decision making in construction, architecture and planning (Laing et al. 2007).

In this context, the "visual preference studies and virtual interaction in architecture and built environment design" (VISBED)[4] project aimed, among other things, to "explore the potential for the use of multiple-user online visualisation to engage end users in participatory design of public space and public buildings' and has been piloted in Aberdeen, Eindhoven and Zürich.

One of these visualization techniques is procedural modelling, an umbrella term for a number of techniques in computer graphics to create 3D models and textures from sets of rules. Systems like CityEngine are, for instance, capable of modelling a complete city using a manageable set of statistical and geographical data, which are highly controllable by users (Parish and Müller 2001).

Finally, regarding the user's engagement to foster participation in urban development, a new trend is to apply game design techniques to non-game applications. The so-called *gamification* concept is defined as the designing of all kind of applications for business, environmental or other applications following gaming mechanics. Appealing to the emotions induced by problem-solving, exploring-discovering, team-work, recognition, imagination, expression or sharing experiences, gamified applications try to get users (players) playing, and keep them playing. This is solving the engagement gap and setting a participation loop, and thus collaborating attaining a common goal (Werbach 2013).

This work proposes a PPGIS gamified urban planning application. It would allow citizens to be aware of new urban plans and it would encourage them to participate more interactively in the urban development process of their city.

17.2 Approach and Methodology

When a local government studies a new development in the context of its urban area, it receives multiple ideas from its own technicians or from the proposing developers. This results in different proposals that may vary significantly depending on the value given to each parameter affecting the urban design and purpose for the studied area.

Instead of leaving the final decision only to a political—technically advised—board, a mechanism is proposed to open participation to the citizens of the affected study area in these proposals. There is nobody better than the citizens living in the area to identify if a certain proposal meets the area's requirements or presents undesirable features. This information can be very valuable to complement the technical and political vision.

Next we analyse the general scenario and present the data and methodology to design the application to implement this crowdsourced urbanism concept.

---

[4] http://eu-smartcities.eu/content/lack-end-user-engagement-participatory-design-sustainable-public-space-and-public-buildings

Table 17.1 Sample of editable parameters for a city proposal and their impacts

	Element	Parameter	Definition	Relations	Dim	All affecting
N	Network		m (width)	L; R; F	>6	Traffic density
Rd		Roadway	m (width)	=N-Ps-Pv	0; >3	Traffic speed
Ps		Parking	m (width)	=N-Rd-Pv	0; >2.5	CO_2 releases
Pv		Pavement	m (width)	=N-Rd-Ps	>1+1	Parking space needs
Tr		Trees	Y/N	Pv>2		Green space
R	Residential lots		sqm (block division)	N		Per person
Rb		Buildings	n (floors number)		2<n<8	Sun—shadow hours
			m (depth)		10<m<20	...
		Public space	sqm	=R-Rb		
F	Facility lots		Type	N		

17.2.1 Preparing Data

First of all, the area of study will be determined, i.e., a preliminary study of a particular area, such a district or a city. Available data must be studied, organized and integrated into the system model in order to be able to present a rich and coherent scenario, in accordance with all the existing proposals in the system.

One of the most important matters is to define the perimeter of the geo-fence. Geofence is the term describing the boundaries of an area, and can help organize data and events within it. The final goal is to notify users on what is happening in the area and especially, to raise events when they enter or exit the geofenced area.

Next, and in order to build the model for each different proposal, the application administrators must collect and prepare a set of data; for correct urban planning, the road network should be defined. Existing and new buildings should be modelled and their rules (all the elements relations and constraints) defined. Preparation of the necessary data means to model each object (geospatial entities with a role in the application) and the rules that define it, which will take into account relations with other objects and constrains such as limiting values for their parameters.

Depending on the type of proposal, time and financial constraints, the level of detail of the city elements descriptions and rules can vary. However, each rule has to at least allow the modification of the value of a certain parameter in a given range, and render the object and explain the consequences of the modification in terms of its effects on other parameters, according to the chosen value.

A sample of this can be seen in Table 17.1 where three city elements are described. For example, the element 'network' is defined by its total width and classified by its uses: roadway, for wheeled traffic, with a minimum width of 3 m; pavement, for pedestrian traffic, with a minimum width of 1 m for each side of the roadway and, when one of them is wider than 2 m, allows the insertion of trees; and parking space, with a minimum width of 2.5 m, in one or both sides of the roadway. All three should sum up the network's width, and their different values create different configurations which affect, among others, traffic density and speed or related

CO_2 releases. Similar definitions can be made for other elements such as residential or facility lots.

Some of the characteristics of the proposal are not listed in the table because certain parameters, such as square meters of residential building needed, should remain an administrative decision, and cannot be edited or discussed. Is the local government who is in charge of knowing its city needs and sets the main parameters to be accomplished by new developments. Thus, several values and dimensions of the development proposal will be shown, but will not be available for editing. Moreover, the intervals in which the values of the different parameters can be edited will vary from one plan to another, depending on the characteristics and needs of the studied area and its surroundings.

When the local government administrators in charge of urban planning, and therefore of development proposals, have completely defined and implemented the above explained process, geofence boundaries can be assigned. Data is organized and defined according to the system model and the model is ready to be integrated, published and set into the application for the citizens to start using the app.

17.2.2 Using the App

When defining the engagement gap in public participation it is interesting to discuss the definition of the term *crowdsourcing* itself (Howe 2006). It describes the partition of a complex task into smaller sub-tasks, which are typically ones that humans can better handle and capture than computers. The group of people who form a community on a particular task currently connects, acquires and manages feedback through web-based platforms. The challenge of these platforms depends on the coverage and availability of crowdsourced content for a designated area. This means having an active community that provides enough volume of relevant information for a particular geographical (urban) area. The incentives to contribute are various and sometimes hard to determine, but crucial for sufficient data collection. We can find, among others, altruism, professional or personal interest, intellectual stimulation, protection or enhancement of a personal investment, social reward or enhanced personal reputation, creative and independent self-expression or pride of place (Coleman et al. 2009).

A recent trend in application design to engage users is to create potential incentives by adopting gaming techniques (Delacruz and de Souza e Silva 2006; Chatfield 2010). This is being increasingly considered in the context of urban planning. The term *gamification* (Zichermann and Cunningham 2011) is used to describe the set of techniques to entertain and engage users to applications developed in non-entertainment domains applied to particular applications such as capturing data about a certain urban area or the behaviour of the citizens living in it.

As seen in Fig. 17.1, by taking advantage of one of the smart phones' most common sensors, the GPS, a notification will be sent to users when approaching a location that is planned for (re)development. This is a passive action (push notification),

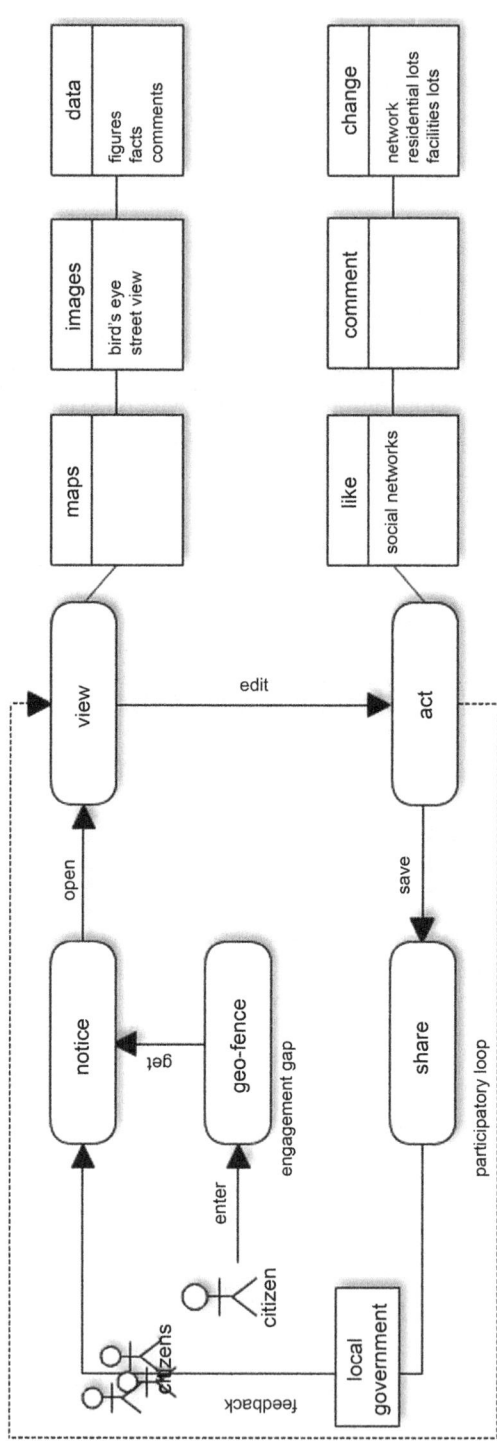

Fig. 17.1 Application workflow

which tries to attract the user's attention to his environment, and thus start using our application and overcome the engagement gap.

If the user decides to open the notification, the officially proposed 3D models, bird's-eye and street view images will be geolocated on his base map, and other information such as explanatory texts and tables with figures or other users comments will be available to be consulted.

Users will then be able to spread the content of the official proposal in social networks, comment and modify some of the proposal characteristics such as road section, number of floors for buildings or type of equipment needed for the area.

While modifying the official proposal, users get feedback about the consequences of their modifications in terms of the affection to other parameters, according to their chosen values for a given parameter in the form of short tips notifications. Thus, citizens can take informed decisions while modifying the original proposal.

This feedback is necessary for better decision making due to the fact that each element and parameter are defined as affecting citizens interests such as traffic density, hours of sun-shade, parking lots or CO_2 releases derived from traffic and are related to each other by dimensional constrains. If a citizen wants to make roads wider, it will affect block size, and therefore, buildings will appear smaller. The app could warn the user that the total residential area is then too small to fit the optimal need for the area. The citizen could then choose to make buildings taller, and the app could warn him that his decision is affecting the hours of sun for gardens and first floors of the buildings. These tips act as internal feedback for users, and aim to encourage more conscientious participation.

When users decide that they like their design, they can save changes and store their project in a personal folder. The 3D model, together with a report showing the graphical and numerical data and the facts derived from them is saved. They can store a number of proposals. If they want to contribute, they can then, from that folder, share their proposal with other users through the application or through social networks.

The objective of sharing users' designs is to motivate citizens through the comparative, competitive, exhibitionist experiences. They will be able to see, comment and vote on other proposals, and thus, proposals are ranked and users receive notifications on the acceptance of their designs together with comments from other users. This feedback is also important for them to stay connected and informed of the project's updates and encourage them to improve their proposals and keep participating and thus providing information.

Finally, they can send their proposals to the local government. If a user wants to officially inform the administration of his opinion on the area's requirements or undesired features he can send it to be stored for the government technicians to study. Citizens will then be notified though the application with an official answer about their concerns.

In any case, government technicians and politicians can consult all of the shared proposals, so they can compare and analyse citizens' opinions and if appropriate, comment and take them into account for the project. This will also update the official proposal, justifying the decisions taken related to citizens' concerns and

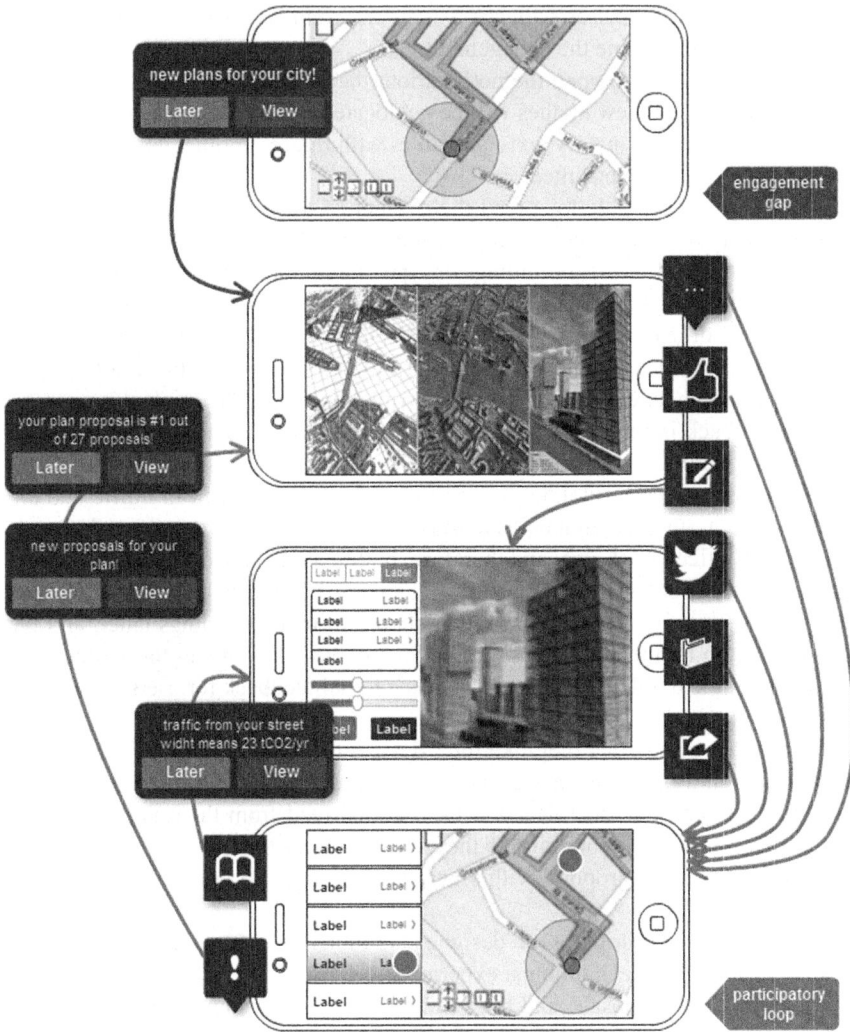

Fig. 17.2 Engagement gap and participatory loop

designs. Citizens will receive, as feedback, a notification with the similarities and differences of their own proposal to the official one, in terms of the parameters values, and the justification of the winning proposal.

All of these types of feedback provided to the users—on the number of participants, the differences from their proposal to the winning project, which design elements are closer and further from the winning ones, the opinions and votes shared with other participants and so on—try to keep users in the participation loop. A more graphical description of this process can be seen in Fig. 17.2.

17.3 Application Design

To facilitate the extensibility and maintenance of the system, allowing its adaptation to different urban environments, types of plans, time and financial constraints, it has to be designed in a modular and interoperable manner.

According to recent ICT trends, this crowdsourced urbanism application architecture would follow the service oriented architecture paradigm. This basically means that the provided data and functionality would be served through web services. The unique entry point to these services should be implemented in the form of a responsive application, for a friendly mobile visualization.

The application must be scalable and extensible, that is, it has to be adapted to new requirements and new data. This means that the application has to be designed in a modular way. Furthermore, these modules should have a well-known functionality and interface so the application is more easily maintained. Most important decisions in this line are to design the architecture and implement the components according to international standards and initiatives. Moreover, underlying technological complexity should be transparent for the final users and they do not need to be aware of the processing of the data consequence of their decisions. For example, when moving a slider for changing a building height, seeing how the building grows is the only thing the users need to get, the underlying processing of the data and rendering mechanisms are not of their interest. Thus, the application provides user with a unique entry point to the system, offering an intuitive manner to handle data,

As seen in Fig. 17.3, project architecture follows the European directive INSPIRE (INSPIRE 2007) which defines a classical service-oriented three-layer architecture. The system includes the required functionality in the form of Network services and the client applications in the form of a mobile application.

This architecture allows us to cope with the different urban environments, types of plans and time and financial constraints, since adding or removing components would add or remove data and/or functionalities.

As an example, Fig. 17.4 describes the functionality and workflow of the first prototype to be developed in the context of the University Jaume I of Castellón Campus, a young medium size compact campus still in development.

From left to right, the unique entry point allows users to access their profiles activity or navigate the map exploring new developments proposed by the University Office for Buildings and Infrastructures Developments (OTOP[5]). Next, options for choosing a proposal through the map from a public list containing all the development proposals, or from the users' list containing only the proposals where the user has performed any action, which can be displayed by project or by other users' activity on those projects. Then, information about the development proposal is displayed: an image, some information about the project and a list of users' activity on that proposal. Next, users' interaction options: navigate the 3D model and capture an image and comment, vote on or share other users' images and comments. Last,

---

[5] https://www.uji.es/serveis/otop/

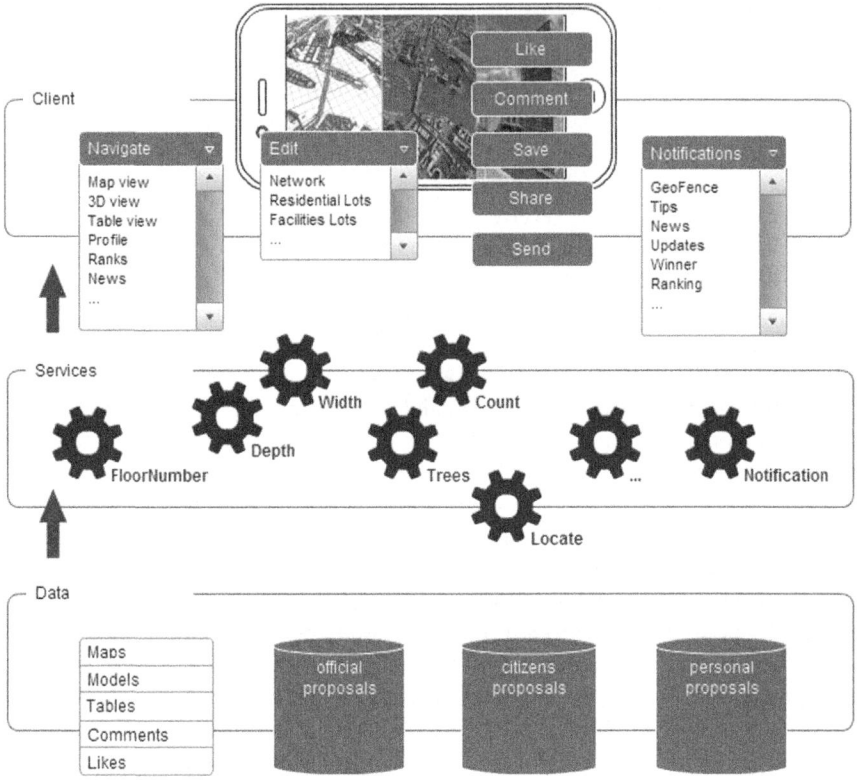

Fig. 17.3 Application conceptual architecture

notifications give users direct access to specific project information. Note how projects and some actions like uploaded images or comments are rated.

Back to Fig. 17.3, this prototype uses some of the concept architecture modules, while leaving others for future development. Data from official proponents (OTOP) and from users is not to be stored separately, but fit within the system model. Services like location and notification will be used to build the client functionalities, which include components from the social media module like share, comment or vote, the map navigation and positioning module and updating and location based components from the notifications module, but it will not use any component from the editing module.

In this context, the modularity of the proposed system will allow module editing in the future, since the data model is fitting the local requirements. If a project type, or time and financial constrains prevents the modelling of the data according to the system or geolocating the data model, geolocation notifications, map navigation or editing components won't be available for use. However, exploring the model, commenting or rating will still be possible. On the other hand, some developers would be able to add new components or modules, both taking advantage of the system data

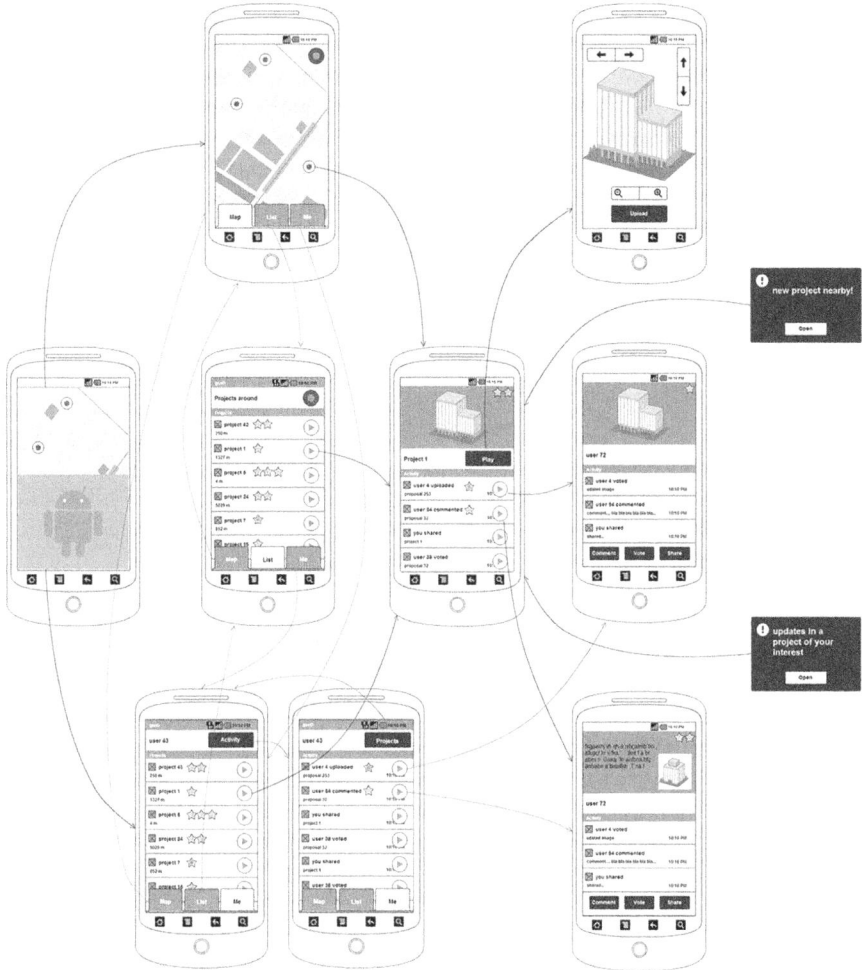

Fig. 17.4 UJI app workflow

model or not, providing functionalities not envisaged by this concept architecture, from as simple as uploading a photo using the smart phone camera to more complex features like displaying information pop-ups by touching a 3D model element.

17.4 Conclusions

We have shown how applying information technologies in general and geospatial technologies in particular act as a facilitator in the participatory design and use of public space and city building. The main goal in doing this is to improve not only the visibility of urban planning but also to improve the citizen participation process to enrich public administration work and achieve a more desirable development scenario of a city. To do so, we described an application that addresses the lack of transparency in the urban planning communication process and enables a mechanism for *crowdsourced* urbanism. The main idea is to increase visibility of the related spatial data and information to facilitate citizen participation.

The proposed application uses location information as a starting point to help users to gather information and provide their opinions, thereby filling-in the engagement gap; using *gamification* techniques it then appeals to user emotions provided by problem-solving experiences, helps citizens to explore-discover an area and its urban development in their surroundings, which enhances team working, recognition, imagination and expression, and allows citizens and neighbours to share these experiences. The ultimate goal of this work is to provide sensitive information about the impact of user suggestions and feedback in the urban planning process and its impact on the final approved plan. All of this contributes towards keeping citizens in the participation loop and moving towards a more desirable community.

17.5 Recommendations

Achieving the main goal of implementing the concept of *crowdsourced* urbanism means working at the intersection of at least these four components: official policy making, location based services, procedural 3D modelling and user-engagement strategies through gaming techniques. The contribution of each one of these should be studied in deeper detail, since they are powerful enough to address important needs. However, they all present their limitations and constrains, and their integration in the exact level of complexity and usefulness is the key to building a successful tool.

References

Batty, M. (2013). Defining geodesign (= GIS + design?). *Environment and Planning B: Planning and Design, 40,* 1–2.
Chatfield, T. (2010, January 17). Fun Inc: Why games are the 21st century's most serious business. The Guardian. http://www.guardian.co.uk.
CMSinfo. (2011). European mobile market. Telecoms Market Research. Accessed 27 June 2013.
Coleman, D. J., Georgiadou, Y., & Labonte, J. (2009). Volunteered geographic information: The nature and motivation of producers. *International Journal of Spatial Data Infrastructures Research, 4,* 332–358.
Conroy, M. M., & Evans-Cowley, J. (2006). E-participation in planning: An analysis of cities adopting on-line citizen participation tools. *Environment and Planning, 24,* 371–384.
Delacruz, G., & De Souza e Silva, A. (2006). Hybrid reality games reframed. *Games Cult, 1*(3), 231–251.
Healey, P. (1992) Planning through debate: The communicative turn in planning theory. *Town Planning Review, 63,* 143–162.

Howe, J. (2006, June 14). The rise of crowdsourcing. *Wired Magazine*. http://www.wired.com.

Kwan, M.-P., & Weber, J. (2003). Individual accessibility revisited: Implications for geographical analysis in the twenty-first century. *Geographical Analysis, 35*(4), 341–353.

Laing, R., Conniff, A., Craig, T., Galan-Diaz, C., & Scott, S. (2007). Design and use of a virtual heritage model to enable a comparison of active navigation of buildings and spaces with passive observation. *Automation in Construction, 16*(6), 830–841.

Obermeyer, N. J. (1998). The evolution of public participation GIS. *Cartography and Geographic Information Systems, 25*(2), 105–112.

Parish, Y. I. H., & Müller, P. (2001). *Procedural modeling of cities*. Proceedings of the 28th Annual Conference on Computer Graphics and Interactive Techniques, pp. 301–308.

Werbach, K. (2013). *Gamification. Coursera course*. Philadelphia: University of Pennsylvania.

Zichermann, G., & Cunningham, C. (2011). *Gamification by design. Implementing game mechanics in web and mobile apps*. Sebastopol: O'Reilly Media.

Part IV
Adopting Geodesign Thinking

Chapter 18
Geodesigning 'From the Inside Out'

Kitty Currier and Helen Couclelis

18.1 Introduction

Geoplanners and geodesigners have been using GIS since its inception, first mainly as a mapping tool, then more generally for data management, analysis and visualization, and more recently also for synthetic tasks such as process modeling and dynamic 3D landscape representation. The emergence a few years ago of *geodesign* as a concept and as a set of advanced techniques has further increased the value of GIS for planning and design. Geodesign has been variously defined as "designing with nature in mind" (Dangermond as quoted in Esri 2010, p. 6), "changing geography by design" (Steinitz as quoted in Esri 2010, p. 7), and "a design and planning method which tightly couples the creation of design proposals with impact simulations informed by geographic contexts" (Flaxman 2010). Wikipedia authors provide a couple of additional definitions along similar lines, but then go on to comment:

> [G]eodesign builds greatly on a long history of work in geographic information science, computer-aided design, landscape architecture, and other environmental design fields—and it's still somewhat unclear whether geodesign differs greatly in substance from existing efforts. (http://en.wikipedia.org/wiki/Geodesign)

The growing interest in geodesign—however defined—is boosting decades-old hopes of landscape architects, land use planners, and spatial planners more generally that GIS will help transform these fields. This has not been the case so far, despite the wide adoption of GIS and related technologies in design-oriented offices and agencies across the world. As recently as 2010, Goodchild noted that "in the four decades that have elapsed since its birth, this notion of GIS as improving the process of design has become less central" (p. 8). Indeed, as a science-based approach, GIS involves aspects of scientific inquiry such as measurement, modeling, simulation, optimization, visualization, and the study of uncertainty. Missing for the most part are notions that are integral to the design perspective, 'soft' aspects

---

K. Currier (✉) · H. Couclelis
Department of Geography, University of California, Santa Barbara, CA, USA
e-mail: currier@geog.ucsb.edu

such as intentionality, purpose, and function, and the recognition of functional and other, not directly spatial, relationships that must hold among disparate physical parts. Purpose and function in particular are rarely explicit themes in GIS, though Couclelis (2010) suggests that they are analytically relevant in two ways: (1) as the motivations for constructing a particular plan, representation, or model; and (2) as fundamental properties of most entities commonly represented in a GIS, i.e. artificial or more generally, human-configured entities. These qualitative dimensions of the design perspective are commonly ignored in GIS applications as being ill-compatible with objective scientific analysis, yet the purposes—often conflicting—that motivate planning and design decisions, and the environmental functions that these decisions may promote or inhibit, are part and parcel of geodesign.

The extraordinary recent expansion of GIS tools, methods and data at the service of geodesigners has rendered moot several earlier critiques of GIS originating in the planning community. Many planners and geographers feared that GIS in planning-related fields would benefit the skilled and the already privileged and further disenfranchise the poor, the uneducated, and those on the margins of society (Pickles 1995). This critique led to the development of the subfields of Public Participation GIS (PPGIS) and—for well-defined groups of stakeholders—Participatory GIS (PGIS), both of which are still going strong (Jankowski and Nyerges 2001). There were also issues of privacy and surveillance, of imposing a technocratic view of the world on other people's perspectives, of affecting societal priorities by focusing on what is easily measurable, and so on. These other concerns too have been to a large extent resolved, to the point that most of those who used to be the critics are now often using GIS themselves (Nyerges et al. 2011).

Planning Support Systems (PSS) emerged in the early 1990s as a response to the increasing complexity of planning in societies that value both the diversity of opinions and the scientific grounding of public decision-making (Brail and Klosterman 2001; Geertman and Stillwell 2009). They were enabled by major improvements in computational resources and geospatial data availability, and relied heavily on the rapid expansion and increasing sophistication of GIS. The main purpose of PSS is to integrate the societal and technical aspects of planning with the computational bonanza of our age, and are thus, at least in concept, one of the best incarnations of the idea of geodesign to date. But the adoption of PSS has been slow, indicating problems yet to be resolved. Also, certain ways of thinking characteristic of design remain elusive. For example, sketch*ing* (the process, not the product) is the designer's way of working out her or his notion of the purpose of the object being designed, its function as an expression of the activities or processes that object is intended to support, and the configuration of spatial parts that will afford that function. Sketching is not about producing a design but a *concept* for a design, mixing map-like parts with diagrammatic parts, with pictorial representations, rough drawings, abstract geometric shapes, and textual annotations. While sketching, the designer will sometimes think aloud, or discuss each meaningful stroke of the pen with a colleague, cross out parts of the emerging gestalt, and start over. This thinking process, which freely draws upon qualitative and quantitative, and spatial and non-spatial elements, does not easily lend itself to tool development. Equally intractable is the

Table 18.1 Contrasting the dominant analytic stance of GIS with the synthetic stance of the Design sciences. (Source: Couclelis 2009)

GIS and traditional sciences	The design sciences
Analysis	Synthesis
From instances to principles	From principles to instances
Causal	Goal-oriented
Descriptive	Prescriptive
Positive	Normative
IS	*OUGHT*

designer's ability to see disparate spatial parts as a *proximal space*—a spatially distributed whole connected through functional, social, ecological, and other relations. Together, the qualitative and non-spatial aspects expressed in sketching and the apprehension of proximal space are essential aspects of design, yet remain very difficult to capture in GIS-based tools (Couclelis 1991).

Missing perhaps from the still young geodesign literature is a view of design deriving from Simon's (1996) famous essay on the "Sciences of the Artificial". Unlike the traditional analytic sciences, the sciences of the artificial concern objects that would not have existed but for an agent's intention to serve particular purposes through a design or artifact that can support desired functions. As Simon notes, even something as simple and physical as a tin can cannot be fully described or understood unless its purpose and function as a fluid container is also taken into account. This connection between intention and product applies to any artifact, from the rough stone implement of a prehistoric society to the most advanced achievement of today's engineering. It applies equally well to non-material things such as plans and formal models, as these, too, are products of human ingenuity designed for a purpose. Table 18.1 summarizes some important contrasts between the traditional analytic sciences and the synthetic 'sciences of the artificial'.

The ongoing research project partly outlined in this chapter hopes to contribute to geodesign by developing a new, systematic way for integrating such 'soft' aspects of design as described above with the 'hard', science-based capabilities of today's GIS. In many ways our proposed approach, which we call 'perspectives mapping', is similar to that of other planners and geodesigners. Best known is probably the framework developed and applied by Steinitz (2012) over several decades, and the six 'questions' at its center. Perspectives mapping differs from this and other well-known efforts in two important respects. First, it takes seriously Simon's (1996) idea that design is a distinct kind of science, requiring a distinct approach. This contrasts with the often ambivalent attitude towards design of even prominent practitioners. For example, Ervin (2008) wonders whether design is an art or a science or just a kind of problem solving. Secondly, perspectives mapping has direct linkages with certain more theoretical aspects of geographic information science and beyond, and could eventually benefit from these associations. These two aspects will be examined in more detail in the discussion section.

The methodology described here is applied to the early phases of the design of a spatial plan for managing activities in a marine area, where several stakeholders hold partially divergent views on critical aspects of the issue. Perspectives mapping

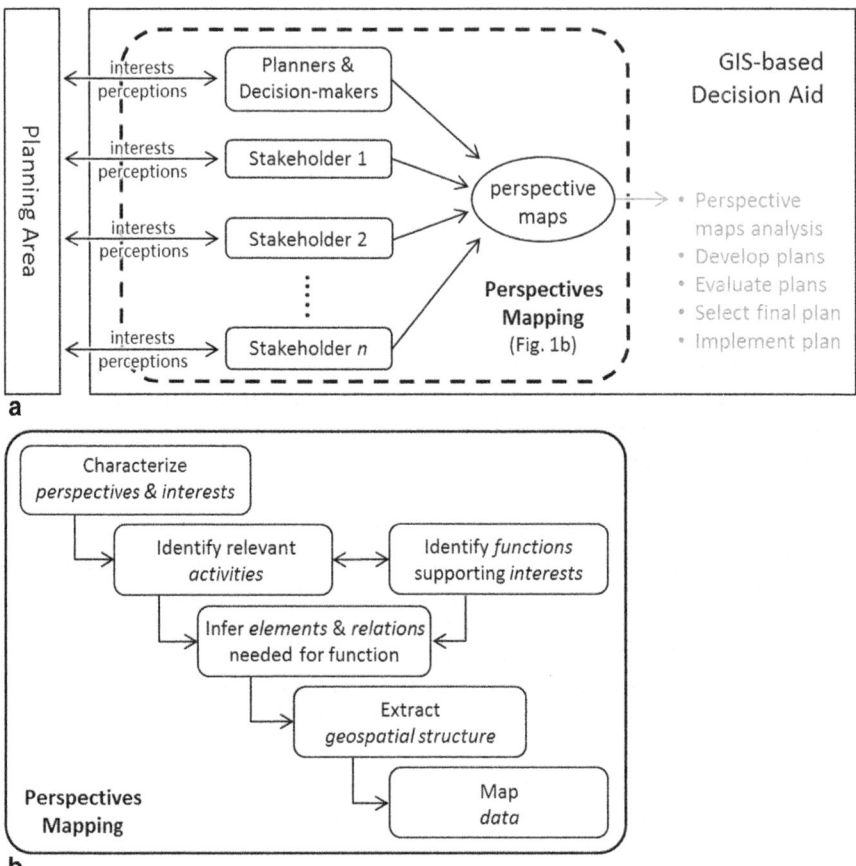

Fig. 18.1 a (Top) Generalized design process. **b** (Bottom) Workflow of the perspectives mapping methodology

concerns the merging of geospatial and non-geospatial information relating to the views of individual stakeholders. Its place in the overall design process is indicated in Fig. 18.1a. The methodology is described in Sect. 3, while Sect. 2 outlines the marine planning context of this effort. Section 4 broadens the discussion and is followed by a brief conclusion.

18.2 'Soft' and 'Hard' Aspects of Coastal and Marine Spatial Planning (CMSP)

Coastal and marine spatial planning (CMSP) is an environmental management approach that limits what human activities may occur in certain areas (Pomeroy and Douvere 2008). CMSP provides the application framework for this research, though the principles would apply equally well to terrestrial environments. In a typical CMSP process information is collected that describes the nature and distribution of biological and physical factors such as benthic habitats, ecologically important species, and oceanic currents. In addition, information is collected that describes the human uses of coastal and marine space including commercial, cultural, military, and other uses. This information is often provided by stakeholders through surveys, participatory mapping, interviews, public meetings, workshops, and other forums. While some of the information may contain direct geospatial references, much stakeholder feedback lacks these references while expressing economic, ecological, and cultural values of the marine resources at stake. Such 'soft' information regarding stakeholders' perspectives is difficult to quantify, georeference, and integrate with biophysical data.

As in terrestrial environmental planning, geospatial tools are often used in marine planning to facilitate the development, evaluation, and selection of policies, regulations, and other interventions that will affect the coast and seascape. Spatial Decision Support Systems (SDSS), Planning Support Systems (PSS), Public Participation GIS (PPGIS) and Participatory GIS (PGIS) are used to address specific tasks in the planning process. We refer to these diverse approaches as GIS-based Decision Aids (GDA). These tools serve two primary functions: to allow participants to explore a problem for better understanding; and to help generate and evaluate alternative solutions taking into account different objectives and trade-offs (Densham 1991).

Planning for the marine environment differs from terrestrial planning in several respects. Ocean systems tend to be less well studied than terrestrial systems (Agardy 2000), resulting in greater uncertainty for decision-making. The space relevant to marine planning—including everything between the sea surface and ocean floor—is volumetric rather than planar, and many of the resources in question are invisible from the surface. The ocean is often viewed as a commons, where the same space may support commercial, recreational, and other uses (Gilliland and Laffoley 2008). Marine boundaries are often invisible and highly permeable to people, animals and pollution, transported under their own power or by oceanic currents. As a result, the spatial extent of areas considered in marine planning is often much larger than in terrestrial planning, with areas covering tens of thousands of square kilometers common (e.g., Gleason et al. 2010).

These differences have led developers to create GDA that are tailored to address specific tasks in marine planning. For example, geoprocessing and analysis of common marine data types is facilitated by the Marine Geospatial Ecology Tools (http://mgel.env.duke.edu/mget). Management, visualization and communication of spatial data are facilitated by online geo-portals and atlases such as the California Ocean Use Atlas (http://marineprotectedareas.noaa.gov/dataanalysis/atlas_ca/viewer/). Optimization modeling that maximizes conservation goals while minimizing cost is performed by Marxan with Zones (http://www.uq.edu.au/marxan/latest-r-d), a tool used in both marine and terrestrial planning. Engaging stakeholders in the design proposal and evaluation process is the main function of SeaSketch (http://www.seasketch.org), a Web-based GDA that is used to conduct the present research.

The research outlined here seeks to expand the scope and type of information that can be considered in a GDA, and to integrate that information within a coherent geodesign framework suitable for environmental (in this case: marine and coastal) planning. More specifically it addresses the question of how to systematically integrate 'hard' geospatial information with 'soft' information on different stakeholder perspectives, along with information concerning associated stakeholder activities and marine environment functions that is not explicitly geospatial. Examples include information about the interactions, relationships, and functions of the environmental entities involved but also—critically—about the values, interests, perceptions, and purposes of the people affected. Indeed, things in the world can be significant to different people (or to the same people under different circumstances) in different ways. A coral reef may provide sustenance to a fisher, an ecotourism destination to a SCUBA dive guide, and a navigation hazard to a container ship captain (who may also be a SCUBA diver when off duty). None of these possibilities are inherent, observable attributes of the coral reef such that would normally be found in a GIS database. A reef's location, extent, biological composition, and morphological type may appear as attributes of a digital object representing a reef, yet none of these indicate how that reef may be significant within the larger social and biophysical world.

We plan to apply and evaluate the methodology in a collaborative CMSP effort currently underway in the Hauraki Gulf, New Zealand (HGF 2011). Managers of the Hauraki Gulf Marine Park, which spans some 13,000 km^2, face challenges of balancing recreational, commercial, and cultural uses of this popular marine area surrounding the busy port of Auckland. The purpose for this CMSP process is to establish areas where different activities may take place in such a way as to reverse environmental decline while taking into consideration the economic, recreational and cultural needs and wishes of different stakeholder groups. The web-based collaborative geodesign tool SeaSketch will be used at different stages of the project to support and enhance stakeholder contributions. The tool accommodates both maps and text-based commentary that references specific information that may be spatial or not. SeaSketch allows stakeholders to visualize geospatial data layers, design their own plans, analyze plans, assemble and share maps with others, annotate their maps and designs, and discuss their maps and designs with other stakeholders in a text-based forum.

18.3 Methodology: From Perspectives to Maps

Like most kinds of spatial planning, the CMSP process consists of a number of standard phases such as collecting and analyzing data, developing and evaluating alternative plans, deciding on a plan, and getting it implemented (Fig. 18.1a). SeaSketch can contribute to any of these steps to the extent that they involve stakeholders, and it emphasizes in particular the early steps of gathering and organizing information relating to the planning area, where stakeholder knowledge and perspectives are most critical. Accordingly, our research also focuses on that initial phase, and in particular on the problem of deriving consistent and information-rich digital 'perspective maps' from stakeholders' inputs that merge qualitative, non-geospatial and geospatial information. The methodology that we describe in this section is novel as far as we know and is inspired by Simon's (1996) vision of artifacts (plans in this case) as primarily the products of user intentions and perspectives. While developed with SeaSketch in mind, perspectives mapping could be conducted using any other GDA with similar collaborative functionality.

The objective of this initial phase is first, to elicit the necessary qualitative, non-geospatial information from each participant, and then to connect this in a systematic way with the appropriate geospatial information, eventually resulting in a digital map. In this section we describe a methodology for establishing a direct path from a stakeholder's perspective to the specific set of geospatial measurements entailed by that perspective. Initially, information is collected regarding stakeholders' perceptions: their understanding of how the relevant systems function, and their analysis of the parts and relations that allow those functions to exist. The planning professionals and other key experts may be treated as 'super-stakeholders' who contribute quantitative and qualitative information of the same kind as other participants. Their views are weighted as appropriate in later phases of the planning process. It is important to document this information in the context of a GDA such as SeaSketch that allows the 'soft' aspects of the stakeholders' perspectives to be represented in the same system where quantitative geospatial data are managed.

For the data collection and analysis phase of the project, the following workflow is planned (Fig. 18.1b). The information may be in narrative form, elicited in interviews or web-based surveys, provided in free-hand digital sketches or diagrams, as rough sketch maps, etc., and will be managed in SeaSketch. Focus groups of three to four individuals from each stakeholder group will be convened, and each will be asked to provide information on the following elements:

a. their *perspective* or *interests* related to the coastal marine environment
   (e.g., commercial fishers depend upon marine environment for livelihood, economic security);
b. the *activities* that they engage in, related to these interests
   (e.g., boating, fishing);
c. the *functions* that the coastal marine environment must provide in order for them to engage in these activities
   (e.g., support viable fishery, allow unrestricted movement between fishing locations and shore facilities);
d. the *spatial* and *non-spatial elements* that enable these activities and functions, including the necessary properties of each and their spatiotemporal relations
   (e.g., adequate transportation, fishing gear, personnel, shore facilities within reasonable distance to fishing locations, regulatory zones that permit boating and fishing activities, suitable marine habitat and ecological relationships for supporting target species).

Based on the information provided by the focus groups, the final two steps will be to:

e. identify the *geospatial structure* and appropriate models that may be inferred from the information in step (d), above;
   (e.g., species' life history models, food web models, habitat connectivity and configuration);
f. map the relevant *geospatial data* in appropriate detail, along with the appropriate spatiotemporal granularity and extent
   (e.g., classified benthic habitat layer, population and migratory data for target species, maximum distance from shore facilities).

The maps obtained for each stakeholder or focus group will be the main inputs of the next stage of the planning process (not described here) in which stakeholders' maps are compared to identify commonalities and conflict hotspots (Fig. 18.1a).

18.4 Discussion

Our methodology builds upon years of work in geodesign in the broad sense and on related areas of geographic information science but also contributes certain novel aspects. The emphasis on participant perspectives is not new. Most GDA allow some consideration of intentionality as expressed in the purposes, perspectives, interests, intentions, viewpoints, goals, objectives, weightings, etc. of participants in a design process. Indeed, as Simon (1996) made clear, without the element of intentionality there is no design. In a planning context, Operations Research formalized this idea in the methodology of multi-objective optimization (Ligmann-Zelinka et al. 2008). The collection edited by Brail and Klosterman (2001) provides several examples of planning support systems that take into account stakeholders' differing perspectives, and so do the methodologies developed in the fields of PGIS and PP-GIS (Kingston 2011; Ramasubramanian 2011). Some planning-oriented work using multi-agent models (e.g. Ligtenberg 2006) represents not only the differing interests of stakeholders but also their varying perceptions and beliefs about the design issue at hand. Finally, several different GDAs along the lines of SeaSketch support different aspects of communication among stakeholders, the expression and clarification of their ideas about a project graphically or in words, conflict resolution, and the negotiation of possible solutions. However, with rare exceptions, participant perspectives are treated as inputs to the geodesign process and not as an integral part of its structure.

What may be new about perspectives mapping is the notion that a direct, systematic path could be traced from a participant's intentions to 'hard' geospatial data, each step being systematically documented using an appropriate GDA. Starting with (1) each stakeholder's characteristic interest in a specific design problem, the methodology (2) elicits information about activities relevant to each, (3) focuses on the abstract notion of *function* and associated activities corresponding to that perspective, (4) infers, in sketch form, the structure of physical and non-physical parts and relations that supports that function, (5) extracts from the above the geospatial structure underlying the function in question, and (6) implements the latter at the appropriate spatiotemporal granularity and level of detail, to the extent allowed by available data. By contrast, the traditional representation of the planning process is one of many possible variations of the following sequence: (1) Identify and analyze problems in the study area; (2) set goals; (3) generate alternative solutions (plans); (4) evaluate plans; (5) choose a solution; (6) implement and monitor. While planning textbooks will immediately add that there should be stakeholder participation in at least some of these steps (as is indeed legally mandated in the US and elsewhere), there is no indication in this impersonal-sounding sequence as to how this should be done. Whose problems are identified and analyzed? Who sets the goals? How, and by whom are the alternative solutions generated? And so on. Steinitz's (2012) version of this model may be the only one that even mentions the investigation of function as one of six necessary steps.

While it is too early in the life of our project to assess the feasibility and usefulness of this procedure, we are encouraged by the fact that it relates to other work in geographic information science and beyond. Insights from that work may be applied to extending the functionality of SeaSketch, the main GDA to be used in this research, to improve the latter's ability to support interaction between all participants in the planning process, at all stages of the 6-step perspectives mapping procedure outlined here. Further, the methodology itself derives from a conceptual framework for geographic information ontologies proposed by Couclelis (2010), currently being formalized. The framework, which integrates analytic and design-oriented thinking, is intended to facilitate the generation of user-oriented models of different kinds that are tailored to the problem-solving context. It consists of a number of levels that may roughly be described as follows:

a. What is the purpose of the model being built?
b. How should it work? (function?)
c. What structure is needed to get it to work in the desired way?
d. What are the necessary parts of that structure?
e. What information (data) do we need to represent or build that structure?
f. What spatiotemporal frame is most appropriate?

This sequence may be thought of as a reusable 'pattern' (or as a sequence of six linked patterns) in the sense of the notion of 'pattern language' as proposed in the design sciences by Alexander et al. (1977). In Alexander's words, "Each pattern describes a problem that occurs over and over again in our environment, and then describes the core of the solution to that problem, in such a way that you can use this solution a million times over, without ever doing it the same way twice" (p. x). This idea born in architecture was adopted by computer science and several other fields to denote a general solution to a recurring kind of problem that may be customized for each specific application. We note that recent work on ontology engineering design patterns (Gangemi and Presutti 2009), which aims at providing tools for more productive access to the wealth of information available of the web, is inspired by the same source. This broader technical interest in an idea originating in design may bode well for GIS as the geodesign support tool *par excellence*. Further, more distant connections that cannot be discussed here go back to AI and the work on scripts, frames and schemas (e.g. Schank and Abelson 1977) that may also be thought of as patterns in the same sense, while automated planners such as Hierarchical Task Networks (HTN) provide templates for connecting activities with functions and eventually spaces (Lekavy and Navrat 2007). We mention these rather remote connections as they may provide potential leads for new kinds of geodesign tools. While our ideas are still tentative, we would like to think that someday the planning methodology presented here may help close the circle from design to ontology engineering to AI and back to design.

18.5 Conclusion

We have outlined a methodology for systematically integrating several of the intangible, non-geospatial or otherwise easily measurable 'soft' aspects of collaborative design and planning on the one hand, with 'hard' geospatial data on the other. While developed with SeaSketch in mind, the perspectives mapping methodology is quite general and should be compatible with any geodesign context that calls for translating participants' intentions, perspectives, and qualitative forms of knowledge into digital representations. Perspectives mapping traces a path from stakeholder interests to functional considerations to functional structure to the geospatial aspects of that structure, such that geospatial data of the appropriate kind, detail, dimensionality and spatiotemporal granularity corresponding to each perspective may be selected for analysis and visualization purposes. While still untested, the methodology is backed by work in geographic information science and beyond that leaves room for hope that it may be implemented as a general-purpose approach for planning and design problems. A possible objective for future work is to develop a suite of geodesign tools specially formulated to support this methodology. More immediately realistic is to extend where possible the functionality of the SeaSketch software along the lines described in this research. Either way, our objective is to promote the notion of geodesigning 'from the inside out', that is, the idea of designing as a distinct science that does justice to the intentional creators of designs.

References

Agardy, T. (2000). Information needs for marine protected areas: Scientific and societal. *Bulletin of Marine Science, 66*(3), 875–888.

Alexander, C., Ishikawa, S., & Silverstein, M. (1977). *A pattern language: Towns-buildings-construction*. Oxford: Oxford University Press.

Brail, R. K., & Klosterman, R. E. (2001). *Planning support systems: Integrating geographic information systems, models, and visualization tools*. Redlands: Esri Press.

Couclelis, H. (1991). Requirements for planning-relevant GIS: A spatial perspective. *Papers in Regional Science, 70,* 9–19.

Couclelis, H. (2009). The abduction of geographic information science: Transporting spatial reasoning to the realm of purpose and design. In K. S. Hornsby, C. Claramunt, M. Denis, & G. Ligozat (Eds.), *Spatial information theory, 9th international conference, COSIT 2009* (pp. 342–356). Berlin: Springer.

Couclelis, H. (2010). Ontologies of geographic information. *International Journal of Geographical Information Science, 24,* 1785–1809.

Densham, P. J. (1991). Spatial decision support systems. In D. J. Maguire, M. F. Goodchild, & D. W. Rhind (Eds.), *Behavioral modeling in geography and planning* (pp. 403–412). London: Longman.

Ervin, S. (2008). To what extent can the fundamental spatial concepts of design be addressed with GIS? Presentation delivered at the NCGIA Specialist Meeting on Spatial Concepts in GIS and Design, 15–16 December 2008, Santa Barbara, California. http://ncgia.ucsb.edu/projects/scdg/docs/present/Ervin-presentation.pdf. Accessed 15 Sept 2013.

Esri. (2010). *Changing geography by design: Selected readings in GeoDesign.* http://www.esri.com/library/ebooks/GeoDesign.pdf. Accessed 31 May 2013.

Flaxman, M. (2010). GeoDesign: fundamental principles. 2010 GeoDesign summit, 6–8 January 2010, Redlands, CA. http://video.esri.com/watch/106/2010-geodesign-summit-michael-flaxman-geodesign-fundamental-principles. Accessed 31 May 2013.

Gangemi, A., & Presutti, V. (2009). Ontology design patterns. In S. Staab & R. Studer (Eds.), *Handbook on ontologies* (pp. 221–243). Berlin: Springer.

Geertman, S., & Stillwell, J. (Eds.) (2009). *Planning support systems: Best practice and new methods.* New York: Springer.

Gilliland, P. M., & Laffoley, D. (2008). Key elements and steps in the process of developing ecosystem-based marine spatial planning. *Marine Policy, 32*(5), 787–796.

Gleason, M., McCreary, S., Miller-Henson, M., Ugoretz, J., Fox, E., Merrifield, M., McClintock, W., Serpa, P., & Hoffman, K. (2010). Science-based and stakeholder-driven marine protected area network planning: A successful case study from north central California. *Ocean & Coastal Management, 53*(2), 52–68.

Goodchild, M. F. (2010). Towards geodesign: Repurposing cartography and GIS? *Cartographic Perspectives, 66,* 7–22.

Hauraki Gulf Forum. (2011). *Spatial planning for the Gulf: An international review of marine spatial planning initiatives and application to the Hauraki Gulf.* Hauraki Gulf Forum. http://www.aucklandcouncil.govt.nz/EN/AboutCouncil/representativesbodies/haurakigulfforum/Documents/Spatialplanforthegulf.pdf. Accessed 15 Sept 2013.

Jankowski, P., & Nyerges, T. (2001). *Geographic information systems for group decision making: Towards a participatory geographic information science.* New York: Taylor & Francis.

Kingston, R. (2011). Online public participation GIS for spatial planning. A. Nyerges, H. Couclelis, & R. McMaster (Eds.), *The SAGE handbook of GIS and society* (pp. 361–380). London: SAGE.

Lekavy, M., & Navrat, P. (2007). Expressivity of STRIPS-like and HTN-like planning. Lecture notes in artificial intelligence, vol. 4496. *Agent and multi-agent systems technologies and applications* (pp. 12–130). Berlin: Springer.

Ligmann-Zelinka A., Church, R. L., & Jankowski, P. (2008). Spatial organization as a generative technique for sustainable multiobjective land-use allocation. *International Journal of Geographical Information Science, 22*(6), 601–622.

Ligtenberg, A. (2006). *Exploring the use of multi-agent systems for interactive multi-actor spatial planning.* Wageningen: University of Wageningen.

Nyerges, A, Couclelis, H, & McMaster, R. (Eds.) (2011). *The SAGE handbook of GIS and society.* London: SAGE.

Pickles, J. (Ed.) (1995). *Ground truth: The social implications of geographic information systems.* New York: The Guilford Press.

Pomeroy, R. & Douvere, F. (2008). The engagement of stakeholders in the marine spatial planning process. *Marine Policy, 32*(5), 816–822.

Ramasubramanian, L. (2011). PPGIS implementation and the transformation of US planning practice. In A. Nyerges, H. Couclelis, & R. McMaster (Eds.), *The SAGE handbook of GIS and society* (pp. 400–422). London: SAGE.

Schank, R., & Abelson, R. (1977). *Scripts, plans, goals, and understanding: An inquiry into human knowledge structure*. Hillsdale : Lawrence Erlbaum Associates.

Simon, H.A. (1996). *The sciences of the artificial* (3rd ed.). Cambridge: MIT Press.

Steinitz, C. (2012). *A framework for geodesign: Changing geography by design*. Redlands: Esri Press.

Chapter 19
People Centered Geodesign: Results of an Exploration

Simeon Nedkov, Eduardo Dias and Marianne Linde

19.1 Introduction

19.1.1 Smart City and Smart Citizens

Smart cities are often defined as urban areas that are made efficient, safe and sustainable by applying sensor networks, algorithms and feedback loops (Neirotti et al. 2014). This research adopts an alternative view where "urban smartness" is a function of its citizen characteristics (Boonstra and Boelens 2011; Boyer and Hill 2013; Greenfield and Kim 2013; Townsend 2013). It recognizes the potential of bottom-up initiatives and it credits the capacity and desire of assertive, engaged and connected citizens to participate in the urban planning process (Van den Berg 2013). In this research we place citizens at the center of the smart city discussion. While sensors measure physical properties, the connected citizens, also known as the networked society, often value and strive for improvement of "softer" urban qualities such as pleasant and beautiful living environments, vibrant urban fabric, serendipity, pleasant chaos, etc, which are crucial for the short- and long-term vitality of the city, but are difficult, if not impossible to quantify with physical measurements.

Unfortunately, attempts by connected citizens in The Netherlands to meaningfully engage with top-down planning institutions are frustrated, as they have few means to engage with the urban planning process offline and even less opportunities to do so online (Van den Berg 2013).

Despite extensive research in information technologies for facilitating the communication between citizens and authorities, such as Public Participatory GIS (PPGIS), there still exists a gap between these groups. Most PPGIS efforts fail to leave permanent marks on newly formed urban plans. In their recent study on challenges

---

S. Nedkov (✉) · E. Dias
VU University Amsterdam, Amsterdam, The Netherlands
e-mail: s.b.nedkov@vu.nl

M. Linde
TNO Built Environment, Delft, The Netherlands

in PPGIS adoption, Brown and Kyttä (2014) acknowledge PPGIS' failure to inform the place-making process by stating that, despite considerable amount of research in this field, "... the future challenge for the development of PPGIS tools and methods will be to provide opportunities to achieve discourse and collaboration, rather than simple collection of spatial data."

In this chapter we argue that past efforts to engage the public in participating in the urban planning process by deploying digital GIS tools focus primarily on applying the workflow of a given (geo) instrument to the place-making process. To be adopted, citizen engagement instruments need to add more to the decision making process than raw data produced by the citizens. Kahila and Kyttä (2009) observe that the adoption of citizen participation instruments "depend on the willingness of the planners and decision makers to use the produced experiential knowledge and the new methods in their work." Efforts to streamline the decision making process should therefore go beyond the mere application of geographical information technology and pay attention to the process' and people's needs. We therefore propose the application and operationalization of the Steinitz (2012) geodesign framework which offers a decision making workflow that includes the citizens, urban planners and designers, and information technology disciplines in equal measures. By broadening the scope to encompass other fields of study—and deemphasizing the role of digital geographical information technology—we seek to move away from mere technology innovation and move towards process innovation.

19.1.2 Context and Previous Work: PPGIS

Public participation in urban planning is a mature field that dates back to Arnstein's (1969) ladder of participation. It has recently come to the forefront through a renewed desire of urban dwellers to influence their environment. Armed with connected technologies, they form a formidable force that contests traditional place making methodologies and actors.

Augmenting citizen's participation with geographical information technologies is equally mature and most commonly referred to as Public Participation GIS (PPGIS). Structural adoption in urban planning practices is, however, lacking. Despite a solid body PPGIS research and a recent increase in the available technical means through which citizens can partake, such as Web 2.0 technologies and mobile computing, PPGIS still fails to make a meaningful dent in future urban plans (Brown and Kyttä 2014). The main critique of PPGIS is that it focusses primarily on the technological aspects and challenges (Craig et al. 2002) and less on the processes they aim to enhance. PPGIS' focus on GI instruments results in a technology "push" that forces the decision process and the people involved in it to adapt to the instruments. As a result, PPGIS are primarily used at the beginning of deliberation process to collect information from citizens instead of enacting an on-going discussion.

Houghton et al. (2014) state that "although the role of ICT in place making is emerging in the communication fields (Gordon and Manosevitch 2011), much remains experimental, in that few occasions of computer simulations and community involvement have led to real results in place." Brown and Kyttä (2014) underline this by stating that "there have been few published studies wherein PPGIS data has been used as a means to engage stakeholders in an iterative process that provides for the review and refinement of the mapped results as part of the larger planning cycle." They remain positive, however, stating that "the potential of PPGIS as a foundation for iterative public participation, rather than a singular by-product of a public participation process, appears large" and advise that future research should "explore the use of PPGIS throughout a complete planning cycle, from scoping to alternative development to decision to monitoring."

Saad-Sulonen and Horelli (2010) argue that achieving successful public participation requires a broadening of scope and the inclusion of other, sometimes non-technical, disciplines. They put forth the term *ICT-mediated citizen participation in urban issues* "as a neutral concept" that "is not tied to any particular field" that comprises the fields of governance, citizen activism, community development, urban planning, geography, information systems and interaction design. Viewed through this frame, it is clear that geographical thinking and technologies play a supportive, instead of leading, role in urban deliberation processes. Putting GI technologies in a broader context explains the slow rate of adoption as it becomes clear that GI technologies are a small part of a larger "ecology of tools" for participatory planning (Wallin et al. 2010).

19.1.3 Research Goals

This research seeks to erect a transparent, traceable and inclusive dialogue between citizens and urban planners. To do so, it infuses the urban planning process with (1) high-quality digital online/offline geographical information services and instruments, and geospatial analyses, and (2) organizational and process innovation that are inspired by the affordances of the networked society: openness, inclusiveness, traceability and intuitive access to information (Castells 2000).

As stated above, technology is not a panacea and needs to be considered in a broader context. We turn to geodesign as it puts information technologies and the various urban planning disciplines and stakeholders on an equal footing. We investigate how geo-information instruments can best be applied to operationalize the various geodesign phases and how they can facilitate information exchange between the involved stakeholders. We focus on the role and needs of citizens in each geodesign phase and ask how technology can best be applied to ease their interaction with the process and the involved stakeholders, and ask the following questions:

1. What geographical information do citizens need;
2. Whether they would understand and value the presented information;
3. Whether it improves their awareness of the situation and place-making context.

These questions are addressed by non-experimental research (observational, case-study and survey) in which citizens are invited to co-design a cycling bridge in a neighborhood in Utrecht, the Netherlands.

19.2 Theoretical Framework

19.2.1 Geodesign

This exploratory study is streamlined under the emerging concept of geodesign. Geodesign is an attempt to use analytical tools to inform and improve the design process interactively (Dias et al. 2013). Geodesign attempts to refocus the application of Geographic Information Systems (GIS) and Planning Support Systems (PSS) into being more design oriented. We build primarily upon Steinitz' (2012) geodesign framework where concrete steps to improve spatial planning design are formulated in the form of models that first aim to understand the context of the situation and then go on to describe an iterative design process. Each model is operationalized by a number of questions:

1. *Representation Models* describe the study area by collecting and presenting relevant geographical information about the problems at hand. Main question: *how should the study area be described?*
2. *Process Models* describe the relationships among the study areas' functional (e.g. how often do citizens visit green areas) and structural elements (e.g. current pollution levels caused by the current infrastructure) by asking *how does the area operate?*
3. *Evaluation Models* judge the area's relationships based on cultural knowledge of the decision-making stakeholders and people of the place (e.g. acceptance levels of noise from the highway or windmill visibility, appreciation of green spaces). Leading question for this model: *is the current study area working well?*
4. *Change Models* describe and investigate how the area can be changed. The outcome of Change Models is a set of scenario's or design alternatives that describe different solutions to the challenges identified in the first steps.
5. *Impact Models* investigate the effects of the proposed changes by asking what difference the changes might cause. Impact Models reuse the Process Models to assess the changed conditions.
6. *Decision Models* evaluate and choose which of the scenarios (designed in step 4) to implement based on its effects on the environment and the stakeholder's evaluation models (based on the cultural knowledge and context of the decision makers). Leading question is *how should the study area be changed?*

These models, and thus questions, are visited three times in consecutive order. In the first pass we try to understand the context within which we operate by answering *why?* questions. The result of the first pass is a description of the situation presented as a common operational picture to the involved stakeholders. During the second pass, which goes from Decision Models back to the Representation Models we seek to answer *how?* questions as an investigation of viable mitigation strategies for the problems and challenges scoped in the first pass. The last pass goes through the models again starting at 1 and ending at 6 to actually carry out the geodesign study and reach a decision. The framework is implemented rarely in a linear fashion and it is common to jump questions, or reverse orders or carry out parts of it, but it helps understand what is needed to purposely change geography by design (Steinitz 2012).

19.2.2 Citizen Oriented Geodesign: An Information Flow in the Context of the Smart City

As stated previously in the introduction, we propose that the "Smart City" only exists when inhabited by "Smart Citizens" (Hemmet and Townsend 2013). These are citizens that willing and able contribute to the discussion and solutions regarding their living space. They are, as primary users of their urban surroundings, best equipped to judge and opine about the meaning of livability and other spatial quality concepts (Kyttä et al. 2013).

As defined by Lee et al. (2014) geodesign is an iterative planning method by which an emerging design is influenced by knowledge derived from the involved stakeholders and geospatial technologies. Different from traditional planning processes, where context-analysis, design, and evaluation are separated into explicit steps, geodesign integrates the exploration of ideas with direct evaluation. The design impact is assessed with geospatial technology (simulations, modeling and visualization) in an iterative manner. Intermediate results and designs are shared with the stakeholders whose input defines the direction and parameters for the next iteration. This iterative approach, backed by geographical information, is a solid basis for the sought dialogue as it allows stakeholders to follow and steer the process as it evolves. This promises a fitter, more context-sensitive design solution that is ultimately more acceptable by the process participants who cooperated in its maturing. And when local citizens can part take in the planning process, acceptability is an imperative criterion.

We operationalize the geodesign phases described above by decomposing them into an information flow that feeds off citizen produced data where experiential information plays a center role. This information flow is schematized in Fig. 19.1

The following sections elaborate on how we propose specific questions or specific information products to be explored in the geodesign process in order to fully involve the citizen perspective in envisioning and defining a new living environment.

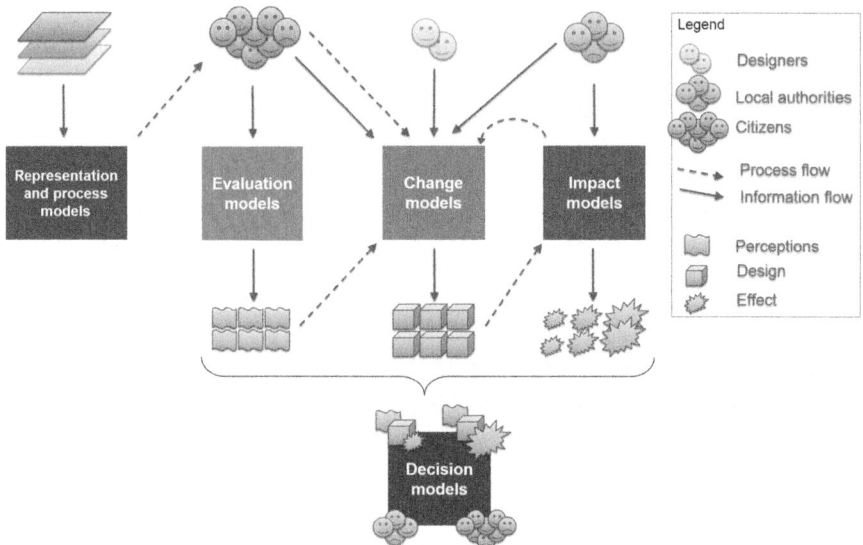

Fig. 19.1 Information flow illustrating the implementation of the geodesign concept to the participative process

Representation and Process Models

The information flow starts with a representation of the area and how it currently operates. This is a collection of digital maps representing the features (e.g. street network, green spaces, elevation) and processes (e.g. noise levels, air pollution, traffic intensity). The map themes, resolution and extent depend on the specific issues of the smart city study at hand. This is the current situation and free from judgment or value.

Evaluation Models

The evaluation model is represented by the citizens perspectives, the citizens apply judgment to the area on different functions, such as green/recreation, transport, safety (systems thinking) and process their evaluation into a map directly, indicating where it performs well (e.g. attractive green areas) or areas where it performs poorly (e.g. feel unsafe).

We propose additional analysis to uncover trends (e.g. heatmaps), relationships between the different stakeholders input data (conflict/consensus) and/or between stakeholder opinions and the physical characteristics of the area (paradoxes). The outcome of these analysis are maps that are based on the knowledge of the people of the place, their values and compared (if appropriate) against the physical characteristics. We define this step as: create insightful case-based digital maps.

Table 19.1 Questions to explore when defining a common picture of the situation and setting up the citizen oriented geodesign smart city study

Citizens	Local authorities
How do they perceive the space?	What are the available resources?
How do they use the space?	Who are the stakeholders?
What are their needs and priorities?	What are the constraints and limitations?
	How much room is there for participation?
What are the constraints, what is the context,	
Who are the players?	
What is at stake? (*define roles, expectations, expertise and competencies*)	

In our information flow (Fig. 19.1), the purple crowd represents the people of the place, e.g. citizens, who have different expectations, perceptions and opinions about their living environment and its future. Different technologies can be used to collect and inventorize the body of knowledge that constitutes the variety of opinions and perceptions. Saad-Sulonen and Horelli (2010) defined the individual perceptions of things that citizens know from their experiences as experiential knowledge and it is comparable to the concept of "local ecological knowledge" in environmental planning (Davis and Wagner 2003; Berkes et al 2000). It is crucial to have an understanding of the collection or global trends.

We define this collection of different stakes as the Common Operational Picture (COP) and the access and sharing of the COP improves situational awareness (SA), the awareness and understanding of opinions different from one's and the capacity to understand the relationships between ideas and the physical geographical aspects. To develop the COP it is important to query all actors involved, from the citizens to the local authorities (e.g. municipality). The questions necessary to explore at this setup stage of the "People oriented geodesign: smart city" are summarized in Table 19.1.

From this information we are able to understand who are the intervenient in the smart city dialogue, and specially what are their roles (e.g. project manager, information architect, mediator, [domain] expert), expectations, expertise and competencies. These can also be local citizens (mandated by a local organization or not), who are able and willing to take care of a part of the process. The roles and decision steps should be matched to concrete tools (and steps within the design and planning process).

After this collection, it is important to set up online and offline visualisation and information provisioning system(s) to be populated with gathered spatial information (COP), that includes input from citizens and stakeholders, and also relevant base datasets derived from the stakeholders queries. These are additional map representations of the physical characteristics that provide insight into constraints and opportunities (e.g. land-use maps, pollution levels).

Change Models

At this moment there is insight into the case study, how it operates, how it is perceived by its inhabitants and the actors involved. The next step is the start of the collaborative and iterative process to determine possible alternative (smart) futures for the area. Information needs at this step may include:

1. Prioritize actions (areas of change) according to people's needs, using the evaluation models results from above;
2. Enable participants (including citizens) to propose changes in process systems (traffic, housing, etc), informed by legal and regulatory spatial constraints (e.g. master plan);
3. Synthesize or fuse one or more proposed changes into a collection of changes (proposed design or change model), this can include also proposals by professional designers or local authorities as they are also part of the participative process.

The change models are proposed by citizens who are motivated and skilled to propose the changes, by designers (professional planners or architects, usually hired by the local authorities) or by the local authorities.

Impact Models

An advantage of a digital workflow and using digital tools is the ability to import the proposed change models (landscape designs) directly to impact modelling procedures. Examples of impact models include noise estimations for a new road, air quality changes, or even aesthetics (the impact on horizon/landscape from placing a new windmill can be assessed using realistic 3D landscape models). In the citizen oriented geodesign, citizens are informed of impacts their ideas may have which develops awareness of consequences and therefore understanding and acceptability of the decision in the following step.

Decision Models

In citizen oriented geodesign, we propose that the decision step revisits the evaluations models (where the citizens evaluated where and how the area is functioning well or needs improvements), the change models and the impact models for a constructive dialogue. It is not expected that a decision is done in one iteration, so an important aspect in the people centered geodesign is that the different alternatives and discussions are logged (since subsequent meetings with more or different citizens may occur) therefore a consistent versioning system should be established that records, among others, the proposed design alternatives and the discussions around them. In this way, the process, discussion and rationale leading to any decision is available and transparent.

19.3 From Theory to Practice: Implementing and Testing the Information Flow

19.3.1 Case Study: Information Provision

The information flow and the information exchange strategies outlined above were implemented in a landscape intervention project in the Dutch neighborhood Lunetten near Utrecht, the Netherlands, in which citizens are invited to decide on the type and location of a cycling connection. The current connection is up for refurbishment due to upcoming widening of the highway it traverses. The governmental authority responsible for the widening and refurbishing of the bridge is seeking consultation and discussion with the public living in the adjacent neighborhood Lunetten. Citizens were invited to evaluate the current connection and voice their opinions on the location and type of the new connection and how it can be improved in terms of facilities i.e. wind shields, lightning, decoration, width, presence of a green lane, or others.

19.3.2 Information Exchange—Consultation Evening

The consultation process was designed in collaboration with the national authority responsible for the construction works (Rijkswaterstaat in dutch) and representatives of the citizens in what turned out to be a conflation of the four geodesign groups: people of the place (citizens), designers (local authority) and ourselves as information technologies and geographical sciences. The information exchange process aimed to inform citizens about the developments in order to raise their situational awareness, solicit input from citizens, craft the initial common operational picture and set first steps towards iterative co-design process. The participative process consisted of three phases:

a. Online citizen consultation for profiling the citizens, asking them to rate the current connection (evaluation model), voice their opinion on the locations of the new connection (impact models), voice their opinion about the type of future connection: tunnel or bridge (change models), voice their opinion about the facilities they desire to see and indicate their digital and geographical literacy. The information gathered through this pre-evening questionnaire was used to set the context of the information exchange.
b. Information evening in two parts: a short plenary introduction in which the local authority would update the citizens on the progress of the road widening, and group discussion sessions during which people have the opportunity to consume information about the area (representation and process models) and the new situation (change model) through different interactive maps and, combined with a display of the information from the online questionnaire, discuss the situation with their fellow citizens. Around 80 citizens visited the information evening (Fig. 19.3).

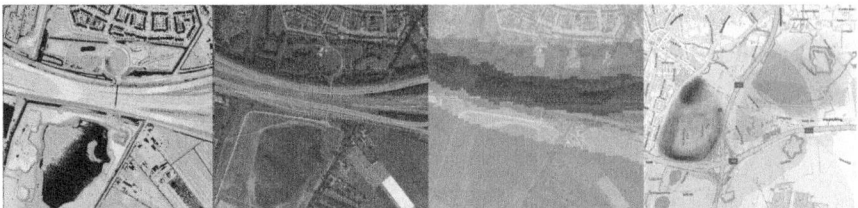

Fig. 19.2 A selection of maps explored by the participants during the information evening: **a** shows the relief of the area, **b** denotes the extent of the highway widening (*red*) and the locations of individual lanes (*orange*), **c** current air pollution map, and **d** heatmap with evaluation of current green spaces by citizens

Fig. 19.3 Impression of the information evening. **a** and **c** Group discussion with participants where mediators enabled citizens to explore geographical information. **b** Citizen exploring a current noise pollution map using a natural user interface device

c. A post-event evaluation aimed at measuring citizens' appreciation of the presented information and the chosen setup.

In designing the information evening, a preliminary and informal inquiry of the citizens' situational awareness and the information they had access revealed it to be limited. It was therefore decided to prepare basic geographical information about the new situation and the area surrounding the bridge. The maps presented to the participants included, among others:

- A height and relief map (Fig. 19.2a), aimed at increasing participants' understanding of the terrain;
- Two situation maps depicting the highway extensions: one with detailed extension plan (all lanes and adaptations) and a simplified version that shows only the extremes of the extended road Fig. 19.2b);
- Air and noise pollution maps to inform citizens about the current values of air and noise pollution (Fig. 19.2c);
- Perception maps that indicate the preferences of fellow citizens collected through a PPGIS spatial questionnaire (Fig. 19.2d).

The information provision was facilitated by geographical information (GI) experts during the second part of the information evening (Fig. 19.3a, 19.3c). Each GI expert was equipped with a laptop and a basic GIS to display the information listed above. Participants were invited to split in smaller groups and consume the information, suggest additions (deliver input) and discuss the current and future situation, including possible bridge placement options, with fellow citizens. In addition to consuming information, citizens also had the option to request simple spatial analyses such as distances, area and height calculations and add their own ideas to the map.

The experts were instructed to start the discussion by showing a map of the context (the planned widening of the highway), assuming that with a better understanding of the future situation, the citizens are better equipped to discuss the options for the accompanying cycle bridge. Since we are interested in supporting the decision process with geographical information and tools instead of adapting the process to their presence we instructed the experts to show only information that supports the ongoing discussion.

It is important to note that no concrete plans were presented about actual designs. We decided to abstain from showing these in order to give citizens the space to brainstorm and talk about the more general topics such as location and desired facilities instead of design-specifics such as colors and visual representations.

19.4 Results and Discussion

Information about people's perception and valuation of the evening was collected through participative observation of each citizen group as it discussed the situation and through informal conversations with participants after the group sessions, semi-structured interviews of the GI experts (mediators) and through a post-evening questionnaire Table 19.2)

It is important to note that the post-evening questionnaire was answered by a sample of the people who were present during the evening (the total population of participants was around 80 and the sample who completed the questionnaire of 11). Results from the questionnaire should be observed with care due to the small sample size and likelihood of self-selection issues. Respondents who fell discontent with the evening might be reluctant to contribute further (and fill in the questionnaire), and the collected answers may suffer from positive bias. Nevertheless, these results can be considered as anecdotal support of the chosen mode of inquiry and a reason to continue future research in this direction.

In general, the questionnaire results support other observations (from participant observation and interviews) and the combination of results yielded the following observations:

Interactive Maps are Valuable When asked about the interactive maps, participants state that they appreciate them as they give them a better view of the case study's physical environment. In particular, the height map seems to contribute considerable towards this end. Respondents furthermore find the maps a valuable addition to the discussion and the decision making process as a whole. The presented maps are deemed easy to understand and contribute positively to spatially oriented discussions.

Table 19.2 A selection of the results of the information evening evaluation questionnaire. Statements rated in a five point scale (agree to disagree, including neutral and not applicable). N=11 (respondents); Total population=80 (participants in the evening)

Statements/rating	++	+	0	−	n.a.
The group discussion created new insights	3	4	2	1	1
The discussion gave me insight into others' ideas	4	4	2	0	1
I have a good feeling about the discussion	1	6	2	0	1
I now understand how others see the situation	2	5	2	1	1
Others understand my position better	0	3	4	1	3
I am satisfied with the discussion	0	7	2	0	1
The discussion results were useful	0	3	5	1	1
I have a better understanding of the opinions of others	2	2	4	1	1
I feel other participants are happy with the discussion	0	2	7	0	2
My knowledge of the situation increased	1	5	3	0	1
Interactive maps are valuable in the decision making	6	3	1	1	0
Interactive maps are difficult to understand	0	3	1	6	0
Interactive maps facilitate discussion	3	5	3	0	0
By using interactive maps I understand the area better	4	4	2	0	1
Interactive maps improve the discussion	4	1	5	1	0
Interactive maps improved the "common operational picture"	4	4	2	0	1
My knowledge of the situation is improved	3	4	1	2	1
Access to analyses (measuring distances, areas and heights) advance the discussion	2	1	3	0	4
Drawing on the map gives the participants better insight into each others' ideas and remarks	2	2	3	0	4

Situational Awareness Improved Although participants indicate that they feel an improvement in situational awareness (stated preference) we did not measure whether this is really the case (revealed behavior). Nonetheless, we believe that increased citizen satisfaction in the process is a positive achievement.

Valuable Group Discussion Supported by Geo-Info When asked to value the group discussions, participants rated them moderately high as they gave them ample opportunity to gain insights in other people's ideas and convictions about the situation. Citizens report, however, that their understanding of other people's reasoning has not improved and indicated that they found it difficult to judge whether fellow citizens understand them better.

Other Observations On the topic of access to information respondents indicate that they wish to look into all the information that was presented and generated in the decision process thus far: results of initial questionnaire, results of plenary session, geographical information, perception maps of fellow citizens, results of discussion sessions. One citizen thought it outrageous that geographical information about the highway expansion is presented this late in the decision making process. The people who filled-in the questionnaire indicate that they would like to consume this information in a digital manner through electronic documents and interactive web maps. Some also indicated they want to use a high-quality website through which to access the data. These results should be treated with extreme care, however, as people who may prefer paper documents may have not have the opportunity to fill in the digital questionnaire (the self-selection issue reported above).

In addition to answering the closed questions, respondents were generous in answering open questions. A selection of the relevant remarks is summarized below:

a. Citizens indicate that, although they enjoyed and valued the evening, they are skeptical of the effectiveness of the deployed process in case the issue under consideration is controversial or based on a conflict. They furthermore indicate that professional mediation is of paramount importance in making these evenings meaningful.
b. Some show little faith in the structureless approach practiced by the citizen group that interacts with the local authority and indicate that they should start over, arrange for a problem owner and a project group that controls a budget and has clear tasks, obligations and privileges.
c. In addition to geographical information, people want access to financial information and a rough financial picture. They indicate that a lack of budget constraints renders the deliberation process meaningless as it implies "anything is possible" while in reality that is seldom the case.
d. While some citizens showed great agility and readiness to participate and voice their opinions, others indicated that they need more guidance and content/material to work with. Some called for mediators to lead these discussion while others requested ready-made designs by experts as they are unable to come up with own solutions to the stated problems.

19.5 Conclusions and Further Work

This chapter presents an information flow aimed at improving the communication between citizens actively seeking opportunities to participate in urban place making processes and policy makers leading and implementing changes to the urban environment. These interactions are placed in the context of geodesign. The introduced tools and methodologies were evaluated through observations, questionnaires and interviews. The main findings indicate that citizens appreciate an increase in information exchange by way of access to geographical information and interactive maps.

Based on the performed experiments, conducted questionnaires and on-the-ground inquiries with citizens and stakeholders we observed that geographical information presented on interactive maps is a valuable communication channel between stakeholders. The citizens of Lunetten were eager to participate in the decision making process and actively seek insight into its progress by attending information evenings, requesting information and proactively participating in discussions. They should therefore be treated as capable and able stakeholders; while they may lack expert knowledge, they have valuable contextual and experiential information that may benefit smarter urban plans.

Ultimately, the success of citizen participation efforts depend on the willingness of urban planners to act on the input. Despite an increased interest from citizens to participate and recent advancements in geographical information technologies, barriers to adopting geographical information instruments and input received from the citizens still exist. We observe that there exist copious amounts of platforms that collect data from citizens. Data, however, does not equal information. Despite advancements in recent geographical information dissemination and visualization technologies, managing and analysing geographical information remains a specialist task. Currently, local authorities seeking input from citizens lack specialists who are able to translate the collected data into actionable information. Urban planners are unable to use the collected data which in turn results in a deliberation process that is void of citizen input.

Future research should therefore investigate how to enable the permanent presence of a dedicated information specialist in the citizen participation process who, in addition to preparing each discourse session are also tasked with managing the information flow after and in-between sessions.

This chapter discusses the new civic reality created by the networked society and proposes to integrate it with geodesign and decades old urban planning practices. Still there are limits and challenges to citizen participation. A major limit relates to the typology of problems. Citizens should participate in issues concerning visible or direct livability (e.g. contributing to aesthetic discussion, functions and amenities provision and traffic solutions as car free days or cycle network) but their participation in expert optimization issues will be limited (such as sewage systems or waste disposal optimization) and should not be expected.

Acknowledgements This research is funded by the research programme Urban Regions in the Delta (URD), part of the VerDuS-programme ('Verbinding Duurzame Steden') of the Netherlands Organization for Scientific Research (NWO). The authors would like to thank Geodan for facilitating the use of the table PC hardware and software, Phoenix. We would also like the Bewoners Organisatie Lunetten and A Living Wall for allowing us to participate in their events. Last but not least, we would like to thank all the anonymous citizens who participated in the information exchange and whose feedback allow us to improve.

References

Arnstein, S. R. (1969). A ladder of citizen participation. *Journal of the American Institute of Planners, 35*(4), 216–224.

Berkes, F., Colding, J., & Folke, C. (2000). Rediscovery of traditional ecological knowledge as adaptive management. *Ecological Applications, 10*(5), 1251–1262.

Boonstra, B., & Boelens, L. (2011). Self-organization in urban development: Towards a new perspective on spatial planning. *Urban Research & Practice, 4*(2), 99–122.

Boyer, B., & Hill, D. (2013). Brickstarter. Sitra. http://www.brickstarter.org/Brickstarter.pdf. Accessed 12 Dec 2013

Brown, G., & Kyttä, M. (2014). Key issues and research priorities for public participation GIS (PPGIS): A synthesis based on empirical research. *Applied Geography, 46*, 122–136. doi:10.1016/j.apgeog.2013.11.004.

Castells, M. (2000). *The information age: Economy, society and culture*. Malden: Blackwell.

Craig, W. J., Harris, T. M., & Weiner, D. (2002). *Community participation and geographic information systems*. London: Taylor & Francis.

Davis, A., & Wagner, J. R. (2003). Who knows? On the importance of identifying "experts" when researching local ecological knowledge. *Human ecology, 31*(3), 463–489.

Dias, E.S., Linde, M., Rafiee, A., Koomen, E. and Scholten, H. J. (2013) Beauty and Brains: integrating easy spatial design and advanced urban sustainability models. Chapter 27. In S. Geertman et al. (Eds.), *Planning support systems for sustainable urban development* (pp. 469–484). Berlin: Springer.

Gordon, E., & Manosevitch, E. (2011). Augmented deliberation: Merging physical and virtual interaction to engage communities in urban planning. *New Media & Society, 13*(1), 75–95. doi:10.1177/1461444810365315.

Greenfield, A., & Kim, N. (2013). *Against the smart city (The city is here for you to use)* (1.3 ed.). Do projects. Helsinki: Finland.

Hemment, D., & Townsend, A. (2013). *Smart citizens* (Vol. 4). Manchester: Future Everything Publications.

Kahila, M., & Kyttä, M. (2009). SoftGIS as a bridge-builder in collaborative urban planning. In D. S. Geertman & P. J. Stillwell (Eds.), *Planning support systems best practice and new methods* (pp. 389–411). Netherlands: Springer.

Houghton, K., Miller, E., & Foth, M. (2014). Integrating ICT into the planning process: impacts, opportunities and challenges. *Australian Planner, 51*(1), 24–33. doi:10.1080/07293682.2013.770771.

Kyttä, M., Broberg, A., Tzoulas, T., & Snabb, K. (2013). Towards contextually sensitive urban densification: Location-based softGIS knowledge revealing perceived residential environmental quality. *Landscape and Urban Planning, 113*, 30–46.

Neirotti, P., De Marco, A., Cagliano, A. C., Mangano, G., & Scorrano, F. (2014). Current trends in smart city initiatives: Some stylised facts. *Cities, 38*, 25–36.

Steinitz, C. (2012). *A framework for Geodesign: Changing geography by design*. Redlands: ESRI press.

Saad-Sulonen, J. C., & Horelli, L. (2010). The value of Community Informatics to participatory urban planning and design: A case-study in Helsinki. *The Journal of Community Informatics,* 6(2) http://ci-journal.net/index.php/ciej/article/view/579/0.

Townsend, A. M. (2013). *Smart cities: big data, civic hackers, and the quest for a new utopia.* New York: W.W. Norton.

Van den Berg, M. (2013). *Stedelingen veranderen de stad.* Amstelveen: TrancityValiz.

Wallin, S., Horelli, L., & Saad-Sulonen, J. (2010). *Digital tools in participatory planning.* Finland: School of Science and Technology, Aalto University.

Chapter 20
Enhancing Stakeholder Engagement: Understanding Organizational Change Principles for Geodesign Professionals

Lisa A. McElvaney and Kelleann Foster

20.1 Introduction

Geodesign is gaining visibility as a valuable way to assist communities with complex land-based planning and design issues to bring creative change to a place. Given the emphasis on the use of geospatial technology in the geodesign process, one might think that technology is the key factor in the success of geodesign. While technological advancements are undoubtedly important, the thoughtful involvement of people is of equal importance, as evidenced in this current definition:

> Geodesign is an iterative design method that uses stakeholder input, geospatial modeling, impact simulations, and real-time feedback to facilitate holistic designs and smart decisions. (McElvaney 2013)

Harvard professor Carl Steinitz states that geodesign must be decision-driven, not data-driven, which places people in a key position in the geodesign process. Since less has been documented about the human component of geodesign, and Dr. Steinitz recently called for more research on this topic (2012), this chapter seeks to deepen the understanding of human behavior within the context of geodesign and community change. Dr. Steinitz uses "people-of-the-place" as the term for a community's stakeholders in his book on geodesign (2012). This chapter further defines the various groups of people who need to be involved, as well as their roles in the geodesign effort.

A major innovation presented in this chapter is how the geodesign process, in particular when utilized in community planning and design, can be informed by the insights and expertise of a related profession whose focus is assisting people with change initiatives. The field of organization development and change (OD&C)

L. A. McElvaney (✉)
Business Transformation Consulting, Redlands, CA, USA
e-mail: dr.lisa.mcelvaney@gmail.com

K. Foster
College of Arts and Architecture, Pennsylvania State University, University Park, PA, USA
e-mail: kxf15@psu.edu

provides a body of knowledge and practices, and an understanding of working with people that can be incorporated into the geodesign effort. It is both a broad and deep profession; therefore only the most salient issues related to enhancing stakeholder engagement, including why people resist change, and how to increase participation and build support for new ideas, are discussed.

Although some professionals involved in geodesign have significant training and experience in stakeholder engagement (e.g., landscape architects, planners), many others may have little formal training or experience in how to help a community go through a change effort (e.g., geographic scientists, information technologists). Furthermore, even those with training may not realize the intricacies of why people embrace or resist change. Therefore the guidance outlined here should be valuable to both categories of professionals, whether they are new to the field or are seasoned professionals.

20.2 Methodology

Two bodies of knowledge provided the foundation for the content of this study: (a) geodesign and participatory planning, and (b) organization development and change.

The first step in this study involved a qualitative comparison between the processes used by these fields for engagement with their population or audience, which was conducted over a series of meetings between the authors. Although the engagement processes necessarily address different contexts and audiences, the authors discovered a remarkable similarity in overall objectives and desired outcomes, which suggested that the expertise of the OD&C field could provide valuable insights for use in the geodesign process.

The second step involved soliciting feedback from seven respected geodesign pioneers who are actively using the geodesign process in professional practice. Six[1] of the pioneers responded to this short survey:

1. What are the top three typical challenges you've encountered when engaging the community (local or key stakeholders) during the geodesign process?
2. Please provide a couple of examples where the engagement with the community (local or key stakeholders) went well, and why.

Insights from the OD&C field were then applied to the geodesign challenges to develop suggested approaches that can help professionals attain better community engagement outcomes, including higher levels of participation, more willingness to change, and higher levels of adoption of both the geodesign process itself, and the recommendations it produces.

---

[1] Geodesign pioneer survey respondents: Gustavo Arciniegas (MAPSUP), Jaap de Kroes (MAPSUP), Michael Flaxman (Geodesign Technologies, Inc.), Carl Steinitz (Harvard University), Doug Walker (Placeways, LLC), Paul Zwick (University of Florida).

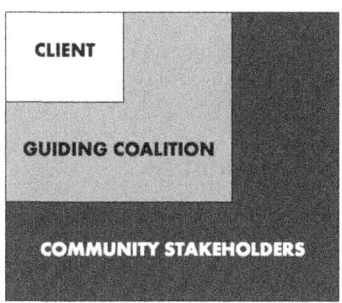

Fig. 20.1 Three interrelated groups comprise the people-of-the place, which is the first of the four Steinitz groups referenced in Sect. 20.2.1, first paragraph

20.2.1 Defining the People and Groups Involved in the Geodesign Process

Steinitz (2012, p. 4) considers four groups of people to be essential for effective collaboration throughout the geodesign process: (a) the people-of-the-place, (b) design professionals, (c) geographic scientists, and (d) information technologists. Together these groups comprise the "geodesign team."

The authors recognize that people from these groups may play different roles in the geodesign effort, depending on their expertise and whether or not they are members of the community. To capture the distinct differences between consultant and client, for example, and to elaborate on the different roles held by the people-of-the-place, the authors have defined four groups for this discussion: (a) the external consulting team, (b) the client, (c) the guiding coalition, and (d) the community stakeholders. The last three clarify Steinitz's "people-of-the-place" component, which is not a singular entity.

The *external consulting team* is comprised of three of Steinitz's groups, design professions, geographic sciences, and information technology professionals, who have been hired or invited to work with the community. Their backgrounds might include planning, architecture, landscape architecture, geospatial sciences, engineering, hydrology, and any number of scientific fields. The specific professional composition of the team will depend on the challenges the community is facing.

The *client* is comprised of those who have summoned or hired the external consulting team, such as municipalities, planning agencies, developers or private landowners. The client may include professionals with sufficient domain expertise to evaluate the work done by the external consulting team.

The *guiding coalition* is comprised of leaders that, ideally, represent all of the potentially affected factions and interests in the geodesign study area (Orton Family Foundation [OFF] 2011). Many of these leaders will be local residents of the community, however some may have a stake in the outcome but not reside there, such as leaders from regulatory or other special interest groups. Additionally, leaders from the client group are typically part of the guiding coalition (see Fig. 20.1).

The *community stakeholders* group includes all area residents, and others who may not live in the area but who have an interest in the change initiative, such as people from regional government agencies, non-profits, and developers. The client and the guiding coalition are also part of this larger group (see Fig. 20.1).

20.3 Findings

Geodesign pioneers reported that utilizing the geodesign process and its associated geospatial technologies can increase participant understanding of the issues, reduce conflict, enhance change visualization, expand awareness of options and change impacts, and shorten the time required for decision-making. However, these experienced design and planning professionals still reported having a number of stakeholder engagement challenges, many of which are common in conventional planning processes.

The authors consolidated feedback from the geodesign pioneers survey, reviewed community participation literature, and reflected on their own professional experiences to identity four common challenges associated with the human component of geodesign.

20.3.1 Challenges Associated with the Human Component of Geodesign

The first challenge is to secure adequate stakeholder representation in the geodesign process. If a geodesign project is meant to involve the community, questions of community standing and who will have a voice are key. This challenge entails securing effective interactions with appropriate representatives of the community. Who is considered "appropriate" varies depending on the project type, the issues and concerns central to the geodesign study, and how polarized the community is at the onset. When key leaders or subgroups do not participate or are not adequately represented in the geodesign process, complications can arise, particularly in the form of resistance to findings and recommended changes.

The second challenge is in building stakeholder confidence in the geodesign process itself. Due to the complexity of community development, it is often a negotiated and complicated process. Adding geospatial technology to the mix is typically new to community members. If the geodesign process is perceived as too complex, many people in the community may not understand it. As a result, stakeholders may not trust the process or its findings. Additionally, stakeholders may distrust the external consultants facilitating the process, since they are often "outsiders." They may also distrust the process due to real or perceived disparities in the power structure within the community. These power disparities may be related to who is informed about the process and who is invited to be involved in the geodesign effort. There can also be issues related to how well the process considers the alignment of resources and community expectations.

The third challenge is rooted in the difficulties associated with recognizing and integrating community values into the geodesign process. While there is a general understanding that design solutions should align with community values (OFF 2011), the need to capture this subjective information and turn it into indicators that can be used in geodesign analyses is a current challenge for geodesign.

And the fourth challenge relates to project duration and participant time commitments. Due to the complexity of community change efforts, an appropriate amount of time is needed at the onset and then for each step in the process. Compressing the process can be problematic given the cognitive and practical limits on how much people can be asked to assimilate at each meeting. There are also challenges related to individuals' schedules. The overarching challenge is keeping participants stimulated and willing to be engaged for the project's duration.

The remainder of this chapter provides guidance to help professionals better understand why some of these challenges happen, how to address them, and more importantly, how to avoid them in the first place.

20.4 Discussion

Organizational change professionals recognize that to enhance stakeholder engagement one has to know a great deal about the stakeholders: who they are, what they care about, what motivates them to act, and how change actually happens, both in general and within specific communities.

This guidance is for professionals who intend to utilize the geodesign process for planning and design efforts at various scales. Stakeholder engagement can be difficult if these efforts are in one's own town or region, and coming in from the "outside" invites a whole additional set of challenges.

20.4.1 Change Agents and the Adoption of Innovations

When professionals from a variety of specialties form a geodesign team to assist a community, many will be external consultants. These professionals will be in the role of "change agents," which in the world of innovation diffusion research (Rogers 1995) means they are external to the system they are seeking to influence regarding an innovation-decision. Diffusion researcher Everett Rogers (1995) defines an innovation as, "an idea, practice, or object that is perceived as new by an individual or other unit of adoption" (p. 11).

In a geodesign effort, community stakeholders are being asked to adopt the geodesign process as a means to examine and potentially change aspects of their place. Two of the challenges revealed in the geodesign pioneers survey are associated with introducing the geodesign process and winning the trust of the people. The geodesign process will often be new to community stakeholders, and they will have to decide whether or not to go along with it. During the geodesign process new design ideas for the community will emerge, and these potential innovations will have to be adopted or rejected by the community. Together this amounts to a great number of innovation-decisions to be made by a community and its leaders.

As both researchers and professionals can attest, whenever an innovation is introduced to a social system, some percentage of the population resists acceptance (Rogers 1995). Adopting new innovations—such as inviting new ways of thinking and living that require changes to the world as it is—can be very hard for people. Sometimes just the idea of change potentially happening is enough to upset segments of the population who then respond by refusing to engage in the geodesign process. It is critical to have an understanding of why people can have a difficult time with change.

20.4.2 Seven Core Principles of Human Change

To assist geodesign professionals in understanding the seemingly mysterious behavior of humans in the face of change, the authors have developed seven core principles of human change (Table 20.1).

Principle #1: Most people resist change if they don't believe it is necessary or worth upsetting the status quo. Throughout history, humans have strived to build organizations and communities that function effectively over time (Trompenaars and Hampden-Turner 1998). This entails bringing people together to decide on structures and processes that enable them to achieve certain desired outcomes. Whether the desired outcome is to build a successful company or a thriving community, people put a lot of time and effort into creating processes and systems that function with a certain level of stability and predictability. Most people don't like change that upsets existing processes and relationships that, *as far as they know,* are still working (Dannemiller Tyson Associates [DTA] 2011; Kotter 1996).

Principle #2: Since change is inevitable, the most successful human systems are those built to adapt to change. In contrast to our desire for stability in our human endeavors, science and experience tells us that change is actually the truest element of our existence. Every aspect of our world is evolving to some degree at every moment (Diamond 2005). Historically, environmental disasters, depletion of resources, migration of people into new areas, and other external forces brought about change in communities. People either adapted or they didn't survive (Banathy 1996; Diamond 2005).

Principle #3: Most cultures do not change easily, especially those that have been in place for a long time. Groups that adapted to change typically gathered their knowledge and 'lessons learned' and passed that on to others. The accumulation of these solutions and lessons learned eventually constituted a *culture* for the people in that group, be it a tribe, a region, or a nation. Culture develops as a means of solving problems and reconciling dilemmas within a particular human

Table 20.1 Seven core principles of human change

#1	Most people resist change if they don't believe it is necessary or worth upsetting the status quo
#2	Since change is inevitable, the most successful human systems are those built to adapt to change
#3	Most cultures do not change easily, especially those that have been in place for a long time
#4	Since change potentially threatens the stability of the whole system, change is often perceived of as dangerous
#5	Potential changes can be perceived as a threat to a person's sense of self
#6	People tend to resist change when they don't understand the process used to reach the conclusions and proposed changes
#7	People resist change that is imposed upon them but will support change they have helped to design, or that they believe sufficiently takes their needs into consideration

group (Trompenaars and Hampden-Turner 1998). Culture provides solutions to the limitless number of problems that people face every day, so culture creates a level of stability and predictability that allows members of that culture to survive, and hopefully thrive, in the face of inevitable change. As a result, most cultures do not change easily.

Principle #4: Since change potentially threatens the stability of the whole system, change is often perceived of as dangerous. Another primary principle of human reaction to change comes from the field of systems theory. It is recognized that one aspect of a system can't be changed, for example, without it having some effect on the rest of the system (Banathy 1996; Nadler 1998). Change in one part of a community could be viewed as dangerous to the stability of the whole community. Typically this perception is deeply embedded in the culture of a community or organization, and can run counter to stated goals. For example, some groups that encourage innovation can simultaneously be highly-risk averse (McElvaney 2006). In reality, change often *does* threaten the interests of some members of a social system, so resistance from those members is to be expected.

Principle #5: Potential changes can be perceived as a threat to a person's sense of self. Another reason for resistance to change is that people tend to become attached to ideas or places that have personal value (OFF 2011). If a person associates his or her identity with a particular idea or place, and potential changes appear to threaten that idea or place, the person may feel personally threatened and may exhibit strong resistance to the changes.

Principle #6: People tend to resist change when they don't understand the process used to reach the conclusions and proposed changes. Decisions that are seemingly made in a mysterious fashion are resisted because people affected by the changes typically want to decide for themselves whether the processes and conclusions seem to be legitimate, in light of their own concerns, values, and rationale (Nadler 1998).

Principle #7: People resist change that is imposed upon them but will support change they have helped to design, or that they believe sufficiently takes their needs into consideration (DTA 2000). This is particularly true in Western cultures where people expect to have more control over decisions that impact them (Trompenaars and Hampden-Turner 1998).

20.4.3 Overcoming Stakeholder Resistance to Community Change

Given the variety of reasons why people resist change, how can professionals engaged in geodesign create the conditions for successful community change? Many in community planning and design have provided approaches to address this (e.g., Sanoff 2000; OFF 2012) however, in their book *Whole-Scale Change*, Dannemiller Tyson Associates (2000) provide a model that delves deeper into human behavior to outline "the conditions necessary to get a real paradigm shift" (p. 16) in an individual, organization, or as applied here, in a community. They refer to it as the DVF Model: $D \times V \times F > R$ (Dissatisfaction × Vision × First Steps > Resistance).

Dissatisfaction: The first step in lasting change is to ensure that people understand why change is needed, and become dissatisfied with the current situation. What are the problems? What are the key drivers for change? Why is it urgent to take action now and not later? For example, in the assessment phase of the geodesign process, trend data illuminates what the status quo will look like in the future. If traffic congestion has been identified as a problem, projected population increases will only exacerbate that problem. Showing that in ten years, travel time will double, may increase willingness to consider alternative futures.

Helping stakeholders reach a certain level of dissatisfaction with the status quo means that they understand clearly why certain things need to change. Creating a high enough level of dissatisfaction with the current state is essential to moving to a future state, because the dissatisfaction helps overcome human resistance to upsetting the existing system (Kotter 1996).

Vision: The second step in lasting change is for community members to develop and share a common desired vision of what the community can become in the future. If the status quo is no longer acceptable and change is needed, what would the ideal future state look like? The focus here is on addressing stakeholder dissatisfaction by creating an aspirational vision and goals that when met, will reduce the community's problems and create a better quality of life (McElvaney 2012). Imagining how much better their community could be helps energize stakeholder action and builds the positive momentum necessary for change to progress (Kotter 1996; Nadler 1998).

First steps: Next, the community needs to agree on "significant, system-wide first steps… to take to begin to move toward the vision" (DTA 2000, p. 17). This might entail deciding which project(s) should be funded first, for example, within a larger planning effort. This sends a clear signal to all stakeholders that concrete actions have been identified which demonstrate clear progress towards the vision.

Resistance: And finally, "*if any of the first 3 elements is zero*, the drive for change cannot overcome the natural forces of resistance" (DTA 2000, p. 17) that exist within any individual, organization, or community.

This DVF Model is an effective tool to use when planning for and engaging in the geodesign process. If resistance to change is anticipated or becomes evident during the process, external consultants can work with their clients to analyze which areas (Dissatisfaction, Vision, or First steps) need to be strengthened to reduce it.

20.4.4 Communication and Innovation Complexity

One of the keys to being a successful change consultant lies in the ability to use a defined process (like Steinitz's (2012) geodesign framework) to engage stakeholders in examining existing conditions and imagining ways to improve those conditions. As discussed above, if the external consultant/change agents are not members of the community, and are significantly different from members of the community, they can have difficulty effectively communicating the value of an innovation to the community stakeholders (Rogers 1995). Consultants typically have specialized knowledge and vocabularies that can confuse and potentially alienate community members; therefore consultants should carefully select language to increase understanding (Kotter 1996), but never "dumb down" their message. Sincere respect for others is very important and if consultants don't truly feel it, it shows. Unintentionally alienating the client, community leaders, or stakeholders can result in resistance to the geodesign effort and its recommendations.

Consultants can also find their own patience tested at times, especially if they can't seem to make themselves understood. Developing the capacity to *listen* and to *understand*—as well as to be an effective speaker—is essential for change professionals (Kotter 1996) leading geodesign sessions and working on geodesign teams.

As discussed in the second challenge outlined by the geodesign pioneers, if community members find the geodesign process and associated geospatial technologies to be too complex, it will likely reduce the speed at which they adopt and support the innovation (Rogers 1995). Given this, and the fact that people tend to resist change when they don't understand the method used to reach conclusions, it is essential that professionals working with the geodesign process slow down if necessary, and emphasize transparency as much as possible (McElvaney 2012).

Steinitz (2012) encourages transparency early in the geodesign process by requiring clarification of how decisions will be made, by whom, and on what basis. Consultants should also plan to review the geodesign process with participants at multiple points, identifying where the group is, and where they are going (Nadler 1998). This will reduce participant overload and help them recognize that progress is being made, even if the process is very complex and takes longer than expected.

20.4.5 The Role of Community Leaders in Stakeholder Engagement and Communicating the Case for Change

Power and politics, and the influence of special interests, often play a significant role in community change efforts. For a variety of reasons, consultants can find themselves the victims of political dynamics within the community. They may be labeled as "outsiders" who do not understand the community, or have their own agenda, and are not to be trusted. An unfortunate potential consequence of this can be rejection of the geodesign process and its findings. The fact that external consultants actually *are* outsiders means that they have to respectfully acknowledge that fact, and then take action to reduce that as a barrier.

One important strategy change consultants employ is to identify, reach out to, and begin building relationships with influential community leaders as quickly as possible (OFF 2012). Strong relationships with community leaders can start to shift the perception of consultants from being "outsiders," to highly trusted "partner-advisors" (Kotter 1996; Nadler 1998). Community leaders are typically responsible for making decisions that affect their followers and the communities they live in. Change consultants understand that one of the most effective means of overcoming community resistance to change is to engage influential community leaders to lead (or co-lead) and strongly endorse the change effort. The authors have named this group of influential leaders the Guiding Coalition (see Fig. 20.1).

Experienced consultants understand the need to meet with the client group early in the process to learn as much as they can about the issues that are prompting the community's need for change, which stakeholder groups could be affected by any proposed changes, and who the influential leaders are. This is typically part of the early stage of needs assessment and requirements gathering.

During these initial meetings, consultants must emphasize the importance of identifying and reaching out to all affected stakeholders in the geodesign process, even if the client would prefer to keep some groups out, or if some groups have previously been disinterested in participating in community planning efforts. Recall that the first common challenge reported by geodesign pioneers is the need to secure adequate stakeholder representation in the geodesign process. Not only is representation critical to discovering local knowledge needed to create useful and desirable design solutions, but human change principles indicate that people are more likely to support change solutions they have helped to design. It needs to be clear to leaders that by excluding certain stakeholder groups from the geodesign process, they put the entire community change effort at serious risk.

While there are now web-based citizen engagement applications available, it is typically not possible to engage every individual citizen throughout the entire geodesign effort. Therefore it is imperative to find representative leaders from all affected stakeholder groups to participate on their groups' behalf. Which leaders should comprise this guiding coalition to secure true representation and community buy-in will best be determined by discussions with the client and with the initial leaders involved in the geodesign effort.

Early meetings with the client are also critical for identifying "opinion leaders," people who may not hold positions of *formal* leadership in the community, but who nevertheless are powerful influencers amongst stakeholders. An opinion leader is "at the center of interpersonal communication networks ... which ... allow him or her to serve as a social model whose innovative behavior is imitated by many other members of the system" (Rogers 1995, p. 27). Some communities have leaders whose opinions are sought on any and all matters, and they need to be identified and engaged early on. Opinion leaders can also be specific to certain sub-cultures and geographic areas within a community, or they can be subject-specific. Consultants who can engage all relevant opinion leaders in the geodesign effort and win their support can anticipate increased stakeholder engagement and increased support for design solutions emerging from the geodesign process.

Identifying stakeholder groups, formal leaders, opinion leaders, and key communication networks; understanding history, power and political dynamics; and uncovering cultural values, norms and preferences are all critical information for working with the human component of geodesign. Gathering all of this can take a great deal of time and effort upfront. Some of the geodesign pioneers reported enlisting the services of local community engagement professionals in their geodesign efforts, partly to reduce their own preparation time and partly because the locals are "insiders" who know a lot about how things really work in their communities. No matter how these key pieces are obtained, external consultants will need all of this information to better understand the communities they are working with, and to develop positive relationships with leaders. It is through positive relationships with influential leaders (both formal and informal) that external consultants have their best means of communicating the value of the geodesign process to community stakeholders, and increasing levels of stakeholder engagement.

20.4.6 The Significance of Culture

To most effectively collaborate with the people-of-the-place, the geodesign team will need to learn more about the culture(s) of the people who reside there. Culture defines how people see and interpret their world. It shapes and reflects their values, assumptions, thoughts, perceptions, feelings, behavioral norms, and external creations (including their technologies). According to Edward Hall in *Beyond Culture* (1981),

> Culture is man's medium; there is not one aspect of human life that is not touched and altered by culture. This means personality, how people express themselves, the way they think, how they move, how problems are solved, how their cities are planned and laid out, how transportation systems function and are organized, as well as how economic and government systems are put together and function. (p. 16)

Among many important factors, culture defines how decisions should be made and how change should happen. For example, in many cultures, the democratic concept of inviting the public to give input and vote on different geodesign options would be inappropriate. Members of these cultures would say, "That's the leaders' job." Deciding what is best for their people is part of the leadership role in certain cultures, and people trust their leaders to make the best choices.

The geodesign team can reconcile this dilemma by first building the guiding coalition of representative leaders who will participate throughout the entire geodesign process to arrive at a recommended community plan. Then the ideas identified via the geodesign process can either be shared with citizens for their approval or not, depending on the role of leaders, and norms for participation and decision-making within the culture. Geodesign will not succeed as a global practice if it is approached as a "one size fits all" process. It must be adapted on a case-by-case basis that takes cultural differences into consideration.

20.4.7 Early Identification of a Shared Community Identity and Core Values

Organizational change professionals have guided organizations through exercises to identify mission, vision, and values for decades. The intention behind these exercises is to examine an organization's current strategy and operations (mission), compare that to a desired future state (vision), and then determine how to change the organization from the current to the future state (new strategy and operations). This approach is very much aligned with typical community planning processes, and specifically, Steinitz's (2012) geodesign framework.

Values are concepts that help to capture what is "good" or "bad" in the subjective view of stakeholders; something which an individual or group considers desirable (Trompenaars and Hampden-Turner 1998). Core values describe the qualities of experience that most stakeholders believe people interacting within a community should have in that environment. Values capture both current conditions—what stakeholders say *is best* about living in a particular community—and help identify what qualities or experiences they feel *should exist*. Once a value is well defined, it becomes "a criterion to determine a choice from existing alternatives" (p. 22). In the community planning and design field, the criteria are often referred to as "objectives," however sometimes these are not rooted in community values. Identifying objectives at the onset of a planning process is considered good practice (Sanoff 2000).

One reason for defining core values early in the assessment phase of geodesign is so they can be utilized as an important source of input to the design process. Currently some design firms do not engage with stakeholders until late in the design process (e.g., at 85 % completion) when designs are nearly complete. There are also design professionals who don't want to involve citizens because they feel those community members lack expertise and/or will slow down the process (Sanoff 2000). Designers should be aware that design ideas that are *compatible* with deeply held community values, beliefs, and past experiences are more likely to be adopted at a faster rate than designs that are perceived as less compatible (Rogers 1995). Speer and Hughey (1995) echo this, saying that "Relationships based on shared values ... produce more meaningful/sustainable bonds" (p. 733).

Community relationships and core values help to form a shared community identity. Early meetings where stakeholders identify who they are as a community and what they value help to reduce stakeholder resistance to change. Stanford University researchers Collins and Porras (1994) discovered something relevant to this discussion in their research on "visionary companies," those companies in their study who have existed for over 50 years, with performance far exceeding their competitors. In the visionary companies, key business factors (such as type of business, processes, and products) were subject to change over time. However, these companies all had well-defined core values and a sense of purpose that were not subject to change, and which provided stability and continuity in the face of the risk and uncertainty that change entails. This approach enabled visionary companies to simultaneously "preserve the core *and* stimulate progress" [italics added] (p. 85).

This concept can hold true for communities too: a shared collective identity, a sense of knowing who you are as a community regardless of a multitude of changes, can act as a foundation that enables stakeholders to better accept and adapt to change. Manzo and Perkins (2006) reinforce this, stating that the identification of common interests means a community is more likely to feel empowered and be mobilized to action (p. 340). Community planning expert Henry Sanoff (2000) also stresses that participants need to begin the process with a shared sense of purpose.

Once community stakeholders have a clear sense of shared identity and core values, design options can more readily be evaluated for alignment. If their collective community identity and core values are upheld in design options, and stakeholders are made aware of that throughout the process, stakeholders should be more willing to accept and make changes to existing practices (such as designs, policies, and zoning or building codes) in order to move the new vision forward.

Translating subjective value statements into criteria for use in geodesign analyses is a current challenge for geodesign. A few geodesign pioneers are experimenting with ways to do this. During the 2013 American Planning Association conference, Doug Walker of *Placeways* identified important outcomes from conducting a Values Mapping Workshop (Walker 2013):

> When you ... do a values workshop, you ask people what do you value, you let them talk, and ... (they) come up with a lot of what planners typically talk about: valuing open space, ... cost of living. But they're also coming up with softer characteristics... "I love the small town feel" or "I love the recreational opportunities", "I love the views." I think this is an exciting trend because if we can start building that into our work, we're going to have more powerful plans, ... (this is) what we should be doing as planners which is building wonderful places .... The technology is now getting to the point where you can start at least attempting to score different scenarios, different proposed plans, different visions, on those kinds of soft characteristics.

Once core community values are well defined, they can become an integral part of the geodesign process, and eventually be encoded in new community plans and land use policies.

More research needs to be conducted, however, into how to effectively translate narrative value statements, which are primarily qualitative in nature, into quantitative models that can be used in geodesign analyses. These models also need to be adaptable so that they can be utilized in different cultural contexts. This is not a simple prospect, however, as Shannon McElvaney explains in his 2012 book on geodesign:

> Creating design with respect to hard science in many ways is the easy part. The soft science of social values is actually harder to quantify and evaluate because it is often qualitative or based on personal views that arise from differences in culture, religion, class, education, politics, or age. (p. 8)

20.5 Conclusion

From the first contact with a potential client and community, professionals working with the geodesign process need to be acutely aware of the importance of the human component of geodesign. People are in the position to either accept or reject the geodesign process and its recommendations. This chapter reviewed a number of deep-seated reasons why people resist change, as well as many actions professionals can take to help communities not only accept, but embrace change.

Utilizing the recommendations outlined in this chapter, professionals can begin to increase the level and quality of stakeholder engagement in geodesign and community planning efforts. Professionals will also have a deeper understanding of (a) why communicating complex concepts from an external change agent role can be difficult, (b) how to increase stakeholder acceptance of the geodesign process, (c) why securing the participation and endorsement of influential community leaders is key to stakeholder representation and building support for change, (d) the significance of culture in all change efforts, and (e) why early identification of shared community identity and core values improves designs and increases stakeholder acceptance of community change.

References

Banathy, B. H. (1996). *Designing social systems in a changing world.* New York: Plenum Press.
Collins, J. C., & Porras, J. I. (1994). *Built to last: Successful habits of visionary companies.* New York: Harper Collins.
Dannemiller Tyson Associates (2000). *Whole-scale change: Unleashing the magic in organizations.* San Francisco: Berrett-Koehler.
Diamond, J. (2005). *Collapse: How cities choose to fail or succeed.* New York: Viking Penguin.
Hall, E. (1981). *Beyond culture.* New York: Anchor Books.
Kotter, J. P. (1996). *Leading change.* Boston: Harvard Business School Press.
Manzo, L. C., & Perkins, D. D. (2006). Finding common ground: The importance of place attachment to community participation and planning. *Journal of Planning Literature, 20*(4), 335–350.
McElvaney, L. (2006). *The relationship between functional supervisor behavior and employee creativity in a project matrix organization* (Doctoral dissertation). Retrieved from ProQuest (Order No. AAT 3218891).
McElvaney, S. (2012). *Geodesign: Case studies in regional and urban planning.* Redlands: Esri Press.

McElvaney, S. (Speaker). (2013). *Geodesign: Strategies for urban planning*. Chicago: American Planning Association.

Nadler, D. A. (1998). *Champions of change*. San Francisco: Jossey-Bass.

Orton Family Foundation. (2011). *Heart & Soul handbook: Building partnerships.* http://www.orton.org/sites/default/files/resource/1670/Handbook%20QG_PARTNERSHIPS_041411_FINAL.pdf. Accessed 2 May 2013.

Orton Family Foundation. (2012). *Heart & Soul handbook: Community planning resources.* https://www.orton.org/sites/default/files/resource/2235/H&S_INTRO_7_27_12_FINAL.pdf. Accessed 3 May 2013.

Rogers, E. M. (1995). *Diffusion of innovations* (4th Ed). London: The Free Press.

Sanoff, H. (2000). *Community participation methods in design and planning*. New York: Wiley.

Speer, P. W., & Hughey, J. (1995). Community organizing: An ecological route to empowerment and power. *American Journal of Community Psychology, 23*(5), 729–748.

Steinitz, C. (2012). *A framework for geodesign: Changing geography by design*. Redlands: Esri Press.

Trompenaars, F., & Hampden-Turner, C. (1998). *Riding the waves of culture: Understanding diversity in global business*. New York: McGraw-Hill.

Walker, D. (Speaker). (2013). *Geodesign: Strategies for urban planning* (Recording by Shannon McElvaney). Chicago: American Planning Association.

Chapter 21
Geodesign in Practice: What About the Urban Designers?

Peter Pelzer, Marco te Brömmelstroet and Stan Geertman

21.1 Introduction

> He [the architect-urbanist] is a participant just like the others (...) More important and better than what you are proposing is to stimulate the imagination of the participants in the team. (Cornelis van Eesteren in a letter to Jaap Bakema 1957, translated from Van Rossem 1993, p. 9).
> Design is a discipline and process where people deliberately create. Design is about purpose and intentions; it's about seeing in our mind's eye what could be, then creating it. (Dangermond 2010, p. 507).

It is often remarked that the burgeoning concept of Geodesign has great potential for all disciplines engaged in changing the future of places (e.g. Flaxman 2010; McElvaney 2012). Through better visualizations, simulations and impact analysis, the planning of cities and regions could be improved, both in terms of process (e.g. increased participation) and outcomes (e.g. more sustainable solutions). What makes the concept of Geodesign unique is that it explicitly attempts to combine design and analytic disciplines. As Zwick (2010, p. 20) remarks about Geodesign: 'it must integrate the design professions with other disciplines—ecology, geography and other earth sciences, real estate and the social sciences'. As a working definition, we conceive Geodesign in this paper as: an approach to visioning, planning or policymaking in which insights and ways of working from a design and analytical perspective are integrated, often supported by dedicated geo-information tool.

This integration is far from an easy task, since some fundamental differences exist between these disciplines. Typically, design disciplines—in this regard urban design and landscape architecture in particular—focus on 'intent or purpose, the

---

P. Pelzer (✉) · S. Geertman
Urban and Regional research centre Utrecht (URU), Utrecht University,
Heidelberglaan 2, 3508 TC Utrecht, The Netherlands
e-mail: p.pelzer@uu.nl

M. te Brömmelstroet
Amsterdam Institute for Social Science Research (AISSR), University of Amsterdam,
Plantage Muidergracht 14, 1018 TV Amsterdam, The Netherlands

creation of something better, beautiful, or both' (McElvaney 2012). For analytical disciplines, like environmental science, geography, real estate, transport planning, etc., the core aim is to understand and describe how a certain object (in this case spatial phenomena) works. This could refer to the situation at a certain moment in time, the future situation, or the situation after an intervention—the impact (Steinitz 1990; cf. Rydin 2007).

Hence, it is not surprising that differences exist between the professional conduct of the design and analytical disciplines. Moreover, the usage of tools is quite different; whereas for a transport analyst a traffic model is critical in assessing obstacles and identifying solutions, urban designers rely much more on paper, pens and visualization software like AutoCad. Geodesign holds the potential to integrate these two worlds by offering a common professional language, since it allows the iterative combination of sketching and drawing with, often quantitative, analysis.

However, similar to the implementation of Decision Support Systems (DSS) and Planning Support Systems (PSS), usage of new tools in practice tends to be problematic and several barriers and bottlenecks have to be overcome (Vonk et al. 2005). In this contribution we try to improve our understanding of the barriers for urban designers to apply Geodesign in practice. Since 'understanding a problem is only halfway to solving it' (Van Aken 2004, p. 20), we propose a set of solutions to overcome the identified barriers. The paper is structured as follows. First, we give some more insight into the difference between design and analytical disciplines. Second, we describe the results our empirical material to study this issue: an experiment with students and interviews with practitioners. We identify four main barriers for urban designers. Subsequently, we provide tentative solutions for these barriers. The paper closes with some conclusions, reflections and suggestions about the direction of future research.

21.2 Barriers for Urban Designers

21.2.1 *Analytical and Design Frames*

There is no straightforward definition of urban design. As Rowley (1994, p. 179) notes: 'having a name for something does not necessarily mean that we understand what it is! (...) Urban design is, surely, a case in point'. Nonetheless, some relevant differences between urban designers and more analytically inclined stakeholders can be observed. Carton (2007) studied the framing of maps by different stakeholders in planning processes. Frames are 'schemata of interpretation' (Goffman 1974) steering the way in which planning actors perceive problems and solutions, and fulfill their tasks. Carton finds that, typically, there are actors with a design frame for whom a map is a design tool, perhaps comparable to the brush and canvas of a painter. On the other hand, there are actors with an analytical frame, for whom a map is a research device. Geodesign aims to support actors with both a design and an analytical frame through GIS-based tools.

Whereas the development of tools to support Geodesign has taken off and its conceptual basis is quite sound, empirical research into its application in practice is still limited. This paper aims to fill this lacuna by making use of two types of research: a structured experiment and semi-structured interviews. The experiment is *control rich*, allowing us to make relatively robust claims about the usage of a GIS-based tool in Geodesign applications, because we can control for variation not related to the tool usage (e.g. a different set-up of the workshop). Since these experiments are conducted with students, they provide little insight in the planning context. Therefore we also conducted complementary semi-structured interviews. These are *context rich*, which make it difficult to measure mechanisms and claims systematically, but provides a way to get in-depth understanding of the praxis of urban design in relation to GIS-based tools and to allow us to understand the barriers and potentials for Geodesign in practice. Whereas the experiment and the semi-structured interviews deal with different tools, what they have in common is that a GIS-based tool is applied in situations where urban design plays a critical role.

21.2.2 An Experiment

We set up an experiment with 55 students (more detailed description in te Brömmelstroet 2013a). The students were selected from three different undergraduate studies; spatial planning, environmental engineering (both second year, Saxion University of Applied Science) and transport engineering (first year, Windesheim University of Applied Science). They were invited to take part in a design competition, organized by TNO, Utrecht University and the University of Amsterdam. Beforehand, we randomly divided them into six groups. In each group there were three to four transport planners, two to three environmental engineers and six spatial planners. To avoid the risk that the six spatial planners would dominate the groups, they were (randomly) divided into urban designers and project economists.

Two weeks before the design competition, each student received a set of information that contained:

- A predefined plan for an urban infill area in the Rotterdam harbor (Fig. 21.1). A map showed the initial planned locations of housing, offices and leisure activities in the area and the accompanying text explaining what ambitions Rotterdam has with this area to become attractive for creative companies and urban families.
- Depending on their role, a text and maps explaining the problems of the existing plan. The environmental engineers received maps of air pollution, external safety and noise nuisance; the transport engineers received maps of capacity problems on the internal and surrounding car network; the urban designers became aware of the notion that the current plan was not attractive and the project economists that the plan was not financially viable.
- The set-up and rules of the competition: 1 hour in which a new plan needs to be developed on paper and with accompanying text.
- Depending on their role, a specific target and possible interventions that can support this.

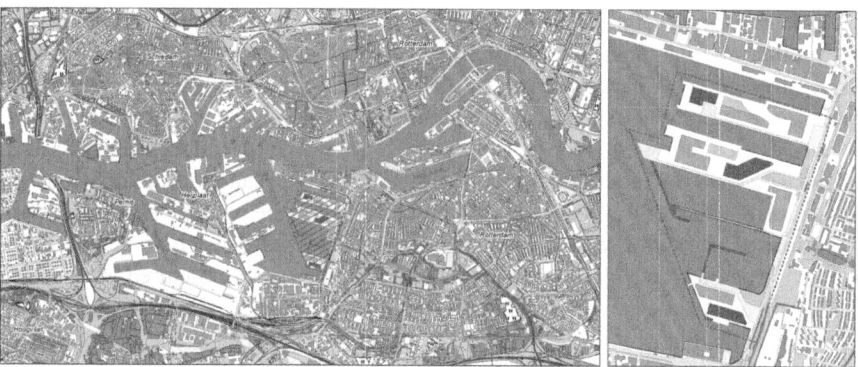

Fig. 21.1 Infill area in the old harbors of Rotterdam (*left*) and original design (*right*); *blue* = offices, *orange* = housing

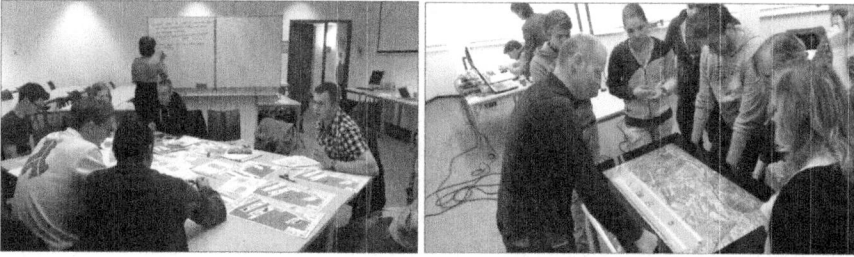

Fig. 21.2 Physical set-up for groups: Table with maps and whiteboard to develop ideas (*left*) and interaction with the calculated effects through a surface table with a chauffeur (*right*)

The resulting plans would be reviewed by experts and for each distinct role a price would be rewarded for the best plan.

On May twenty-first 2013, the groups developed their plans. The six groups worked in turn, enabling us to observe every group. Each group received a short opening statement on the goal and set-up of the competition. Then, one of the authors served as mediator to guide the group through a number of design and analysis loops, which were meant to help them to understand the problem and develop solutions. The groups were first encouraged to develop concrete planning interventions. Hereby, they were supported by a large paper map of the plan area and a whiteboard (see Fig. 21.2). The solutions they developed were translated into the Urban Strategy tool (www.tno.nl/urbanstrategy). This state-of-the-art *Planning Support Tool* links fast computer modeling on a variety of urban dimensions to intuitive visualization of effects of a planning intervention on for instance traffic flows and noise pollution (Dias et al. 2013). It uses a surface table to communicate between the users and the models (see Fig. 21.2 right). By seeing the effects of their interventions, the groups could get a feeling about what works and what does not. Subsequently, they could discuss further improvements. Ten minutes before the end, the groups were urged to draft their final plan and write their accompanying text.

After the session the students had to fill out an extensive survey. Among other things, they were asked to give their opinion on statements that addressed multiple characteristics of Urban Strategy (based on te Brömmelstroet 2010). The participants were asked to respond to each statement on a seven point Likert scale (1 = strongly disagree; 7 = strongly agree). Table 21.1 presents the average outcomes for each of the four roles and their standard deviations.

Looking at the outcomes in Table 21.1, four statements show significant differences between urban designers and the other roles (see also Fig. 21.3). For all other statements, no significant differences were recorded. The project economists score relatively low on all four significant statements. The urban designers seem content with the clarity of the output presented but seem much less satisfied with the fit of the tool with their role and the idea that the tool separates sense from nonsense. The transport engineers are positive about both this fit and the function of Urban Strategy to separate sense from non-sense.

It is an interesting finding that on all other statements, no significant differences could be found. This indicates that there seems to be an agreement across the roles about the general positive perception of usability of Urban Strategy. It also points to the limitations of our experimental setup. Increasing the sample by repeating the experiment could help us to assess if this agreement is indeed present or if it is an artifact of this sample. Also, it shows that although the control-rich environment and possible repetition of an experiment is powerful in terms of internal validity, it is severely limited in understanding the rich context of real world characteristics. The students have not yet been specializing in their roles and have not made a career in their own silo (with their own instruments). Moreover, the students with the role urban designer did not have a background in urban design, but spatial planning. Therefore, we cannot expect to replicate such important real world conditions in the experiment. To increase this external (or ecological) validity, it is necessary to triangulate an experiment with context rich methods. This is what we will discuss in the next section.

21.2.3 Interviews with Practitioners

In addition to the survey, 15 semi-structured interviews were conducted with people involved in spatial planning in the Netherlands. The purpose of the interviews was to further explore the issues designers face while using GIS-based tools. The interviewees had experience with GIS-based tools in a planning or design process and/or were active in practice as an urban designer. The quotes below are all translated from Dutch, whereby inevitably some of the nuances are lost[1]. Most of the interviewees emphasize that usage of (interactive) GIS by urban designers is problematic. As one interviewee noted:

> I had expected that urban designers would endorse it because they are designers and visually focused, but this has disappointed me. (…) whereas they could use geo-information really well to make designs more realistic and better. (Project Manager and GIS Specialist)

---

[1] A full list of quotes from the interviewees is available upon request.

Table 21.1 Results on the evaluation survey of urban strategy

Statements	Environmental engineers (N=11)		Transport engineers (N=20)		Urban designers (N=12)		Project economist (N=12)	
	Average	St. dev	Average	St. dev	Average	St. dev	Average	St. dev
Urban strategy is transparent	5.18	0.751	4.65	0.813	5.09	0.831	4.42	1.24
The communicative value of the output is high*	5.73	0.467	5.00	0.918	5.08	0.900	4.58	1.084
The output is clearly presented*	5.82	0.603	4.95	1.191	5.92	0.793	4.67	0.888
Urban strategy is user friendly	5.64	0.924	4.80	1.005	5.18	0.603	4.92	0.900
The output is credible	5.73	0.905	5.20	0.834	5.33	1.303	5.00	0.739
Urban strategy is comprehensive enough	4.36	0.809	4.85	1.461	4.75	1.545	4.92	0.793
The focus of urban strategy is good	4.90	1.101	5.35	0.933	5.25	1.215	4.67	0.651
The level of detail in the maps is sufficient	5.09	1.578	5.25	1.682	5.92	0.996	4.58	1.881
Urban strategy is easy to understand	5.64	1.362	5.65	0.745	5.83	0.835	4.92	1.084
The tool facilitated evaluating alternatives	5.64	0.809	5.35	0.933	5.33	0.888	4.83	1.403
The tool facilitated creating ideas	5.36	1.027	5.25	0.967	5.17	1.115	5.42	0.515
The tool facilitated sketching ideas	5.73	0.786	5.4	0.754	5.08	1.311	5.25	0.866
Urban strategy fits well in my role*	5.18	0.751	5.80	0.894	4.92	1.443	4.00	1.595
Urban strategy separated sense from nonsense*	4.36	0.924	5.37	0.895	4.75	0.965	4.58	0.996
Urban strategy limited our creativity	2.64	1.433	3.25	1.773	4.25	1.603	3.75	1.485
I understand what is (not) represented by the indicators	4.00	1.414	4.65	1.089	4.50	0.972	4.27	1.104
Urban strategy allowed us to do more in less time	5.91	0.831	5.45	0.826	5.58	0.515	5.08	1.165

*Dimensions marked with (*) reveal significant differences at a 0.05 level between the average from the role urban designer and one of the analytical roles

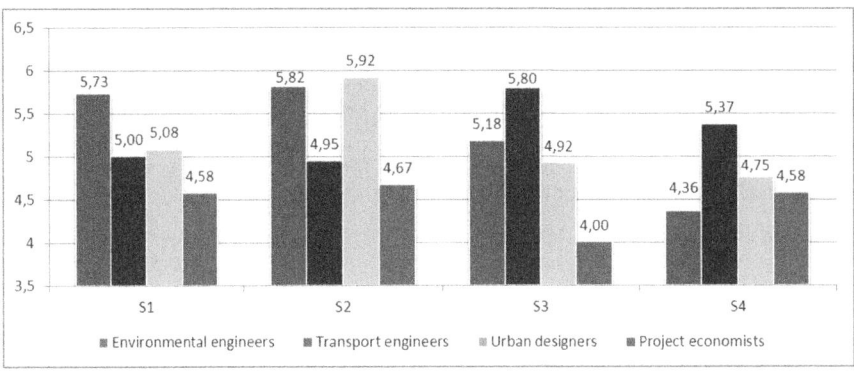

Fig. 21.3 Response to the four significant statements by student by role. The scores are the means per role (for the respective N's see Table 21.1). *Statement 1 (S1):* The communicative value of the output is high. *Statement 2 (S2):* The output is clearly presented. *Statement 3 (S3):* Urban Strategy fits well in my role. Statement 4 (S4): Urban Strategy separated sense from nonsense

The next section discusses four main barriers that came out in the interviews.

Not Everything that Counts can be Measured, and Not Everything that can be Measured Counts

An issue frequently mentioned is that geo-information tools have a bias towards specific themes. Environmental aspects—noise pollution, water quality, etc.—are much easier to model than an abstract and subjective concept like 'spatial quality'. The latter could involve the 'look and feel' of an area. As one urban designer remarks: 'For most urban designers feeling is critical—the atmosphere. How do I live somewhere? Images play a crucial role during the process.' These are aspects that are currently insufficiently covered when GIS is applied in design sessions. Hence, the first barrier:

→*Barrier 1: Unquantifiable aspects that are important for urban designers are currently insufficiently included in the application of GIS-based tools.*

Creativity Versus Analysis

A second barrier is that GIS-based tools are perceived to restrict creativity and freedom for urban designers[2]. This was indicated by the experiment results and supported in the interviews. An urban designer remarks:

---

[2] We are not providing a judgment as to whether or not this notion is true, since the famous Thomas-theorem states: 'If men define situations as real, they are real in their consequences'.

> From a design side, we're on a very abstract level, making sketches in which a meter does not matter so much. I do it roughly and find out later what the exact contours will be. But then [when using an interactive geo-information tool] there is a number that is very precise, with three decimal places. And that does not fit the idea that I have in my head. (...) I would prefer a rough sketch on the table. (Urban Designer with experience of using an interactive GIS-based tool).

He continues to remark about the dynamics of the design process:

> The role of the urban designer in the design process is to keep the flow. It works really well to fix specific points and leave others open. Later on, you can find a solution for these fixed points. You should not solve everything the moment you do something. It stops, it stands still. One wants to work from large to small, from principles to detail. (GIS Specialist)

Another interviewee, who had a fair bit of experience with applying map-based touch tables in planning situations, also emphasized the point of restricting creativity: 'I notice they [urban designers] find it much too realistic, I think they're afraid of it. They think it will restrict their creativity or it is too systematic' (Project Manager and GIS-specialist). The critical question is, of course, to what extent does a GIS-based tool *actually* restrict creativity, or is this only the perception? It is clear, however, that we need to better understand this notion.

→*Barrier 2: A GIS-based tool is perceived to restrict creativity.*

Fuzziness Versus Explicitness

Applying a GIS-based tool leads to an explicit representation of spatial phenomena. For instance, an area either has a leisure or a residential function; within models and tools, boundaries between functions tend to be hard–think of lines and polygons. Fuzziness, such as the blending of land uses, is more difficult to handle in a GIS. In the interviews, however, fuzziness was mentioned as an important aspect of urban design. Integral and explicit evaluation of the problems at hand is sometimes difficult to handle for an urban designer, particularly in the early stages:

> That's the risk of such a table [map-based touch table combined with an interactive GIS] – that one continuously sees everything. It becomes too integrative; it is no longer possible to arrive at one abstract system. One is forced to solve everything at once. (Urban Designer with experience with an interactive GIS-based tool).

A GIS specialist observes something similar:

> Urban designers are used to drawing and sketching rough lines on paper. And that is something that is difficult in a MapTable. One can draw a polygon but in the end you have to click with a mouse and indicate vertices. And then, after a short period of time, they [urban designers] tend to say: "I don't like this". (GIS Specialist)

In summary, the third barrier can be formulated as follows.

→*Barrier 3: Fuzziness is a virtue for urban design that is often not supported by GIS-based tools.*

Innovation Versus Habits

The fourth and final barrier has already been observed earlier in the GIS and PSS debate. New technologies are not easily accepted, since existing habits and perceptions are difficult to change (e.g. Vonk et al. 2005, 2007). This barrier is particularly relevant for the adoption of GIS-based tools by urban designers because it involves quite a radical shift in the way of working. A GIS specialist describes the struggle he has to get urban designers on board:

> We have a touch table, which we attempt to get urban designers to use, because one can use it with the fingers. But that is too much for them. They want to draw with a pen on the table. Although it is easier to draw with their fingers, they prefer a pen because that is what they are used to. I'm always struggling with the question: should I facilitate what they are used to? Or is it better to say: it's a whole new world, we're going to try something different now? Often I start with what they are used to and then say: "we're doing it like this now, but let's try something different". It is easier to get them on board when you give them a handhold. And for them [urban designers] that handhold is, literally, a pen. (GIS-specialist)

Hence, the fourth barrier can be formulated as:

→*Barrier 4: It is difficult to break with old habits.*

21.3 Solutions to Overcome the Barriers

Based on insights from existing literature, our experiment, the interviews and hands-on involvement with applications of GIS-based tools in practice, we have developed some tentative directions for finding a solution. There are two distinct directions for finding improvement: one is by adapting the *tool* and one is by adapting the *process*.

21.3.1 →*Barrier 1: Not All Dimensions that are Important for Urban Designers are Currently Included in GIS-based Tools*

In terms of improving tools, this is arguably one of the most challenging issues. Some spatial phenomena (e.g. 'spatial quality') are hard to capture in a GIS or a quantitative model. It is likely that the solution lies in clever combinations with other types of visualizations, such as images and 3D visualizations or quotes from inhabitants. Some interesting examples of this can be found in Geertman and Stillwell (2009) (in particular: the mapping of landscape values by Carver et al. 2009; gathering localized knowledge through 'Soft GIS' by Kahila and Kyttä 2009; and making use of a 'Virtual Landscape Theater' by Miller et al. 2009).

The process approach to overcome this barrier lies in ensuring that there is sufficient time, resources and attention for aspects that cannot be included in a GIS-based tool. What should be prevented is that GIS becomes 'performative' and that stakeholders start to be dependent on the tool instead of vice versa (Smith et al. 2013). Just as a model specialist or a GIS specialist should be trusted with preparing the tool, urban designers should be trusted as being intuitive and creative experts. In sum, applying a GIS-based tool successfully in Geodesign requires the very careful organization of workshops, dealing with different types of knowledge forms, for instance experiential and systematized knowledge (Healey 2007). The importance of careful knowledge management has also been emphasized by earlier work in the field of PSS (e.g. te Brömmelstroet 2010).

21.3.2 →Barrier 2: A Geo-information Tool Restricts Creativity

The rising use of map-based touch tables in combination with GIS makes it easier to support sessions in which creativity and analysis are combined. However, even when using user-friendly software like SketchUp and Sketchbook Pro, the freedom of a paper and pens is unrivaled. Two solutions could work. First: combining paper maps with a GIS-based tool—something we did in the experiment. Hereby, the intuitive process of drawing with pens and paper is still possible. The problem however is combining the paper drawings quickly and efficiently with a GIS. This is almost impossible in a workshop and could lead to a loss of momentum and a loss of detail in the translation process. The second solution is something to be done much later in time: in the future software and hardware will more than likely become even more user friendly and intuitive.

One of the strengths of urban designers is that they are able to ignite a creative process in which a plan is developed. This virtue should be captured by Geodesign. Key to this is to continuously emphasize that tools are *supporting* the process; they are a part of it, but they do not *steer* it. It's the people that analyze, develop ideas and sketch, not the tool!

21.3.3 →Barrier 3: Fuzziness is a Virtue for Urban Design

Related to overcoming the barrier of creativity is the barrier of too much precision and systematized information. An urban designer has an idea for a tool that fits their demands:

> What if we were to have a table [a map-based touch table] with lines that are thick and flexible? Lines that one could dent and stretch. More like rubber than the 'hard' GIS. Because it takes so much time and consultation to draw lines. Because there is this little blue line in the soil and then it has to be next to that line otherwise it is in the water, etc. Whereas, actually, you would just need to 'go-go-go-go' to have a rough sketch and later on you put it in an AutoCad. I would not know how this should work exactly, but this is more from our [urban designers'] perspective. (Urban Designer with experience with an interactive GIS-based tool)

This idea relates to the more intuitive geo-design tools identified for the previous barrier. The idea of 'rubber' lines is interesting and innovative and deserves further attention in the future.

As important as a tool that appreciates fuzziness, is a process which is open to it. One of the key elements is not to start too early with complicated and integral calculations. Whereas these are necessary in developing a plan—and inevitably have to be conducted somewhere in the process—it should be very carefully decided *when* in the process these are conducted. To an important extent this depends on the urban designer (To what extent is he/she able to deal with complicated calculations?) and the planning issue (Are there strict financial or environmental restrictions for the urban designer?). Also, when detailed calculations are needed later on in the planning process, this should be acknowledged from the start (Mouter and Pelzer 2013).

21.3.4 →*Barrier 4: It is Difficult to Break with Old Habits*

The fourth and final barrier is arguably the most persistent but it has one big ally: time. The future generation of urban designers is now using tablets, smartphones and Google maps. The step towards GIS-based tools will in the future arguably be much smaller. However, to enhance the integration of Geodesign and urban designers, it would be relevant to pay attention to the framing of a tool. GIS-based tools are often perceived to belong to GIS specialists or environmental analysts. A concrete intervention to overcome this is to use map-based touch tables to discuss sketches and images (for instance scans of drawings on paper or 3D visualizations). In the next stage, this table could be used for more advanced applications of Geodesign (see Pelzer et al. 2013, for an example) (Fig. 21.4).

From a process perspective, it is important to see a Geodesign application as a collective learning experience. The technology, models, indicators, topics, and other disciplines are being interactively explored and by actively linking them, new and shared knowledge is created. This takes time and some of the benefits are likely to be more visible in the long run. It is very important for all actors, including urban designers, to acknowledge this. A spatial planner looks back on the sessions with a GIS-based tool in which he was involved.

> For me it was a very instructive experience because I was forced into seeing a very different approach. What I found very funny about the sessions we had around the table (map-based touch table) with urban designers was that we had to get used to each other.

Truly fundamental shifts in the ways of working and using tools could only occur, however, if urban designers endorse *and* communicate the potential of Geodesign.

21.4 Conclusion

This paper addressed a critical issue for the future of Geodesign: the persistent disconnection between the potential of Geodesign and the usage of GIS-based tools in practice. Although this will in part be solved by time and technological developments, we also observed some fundamental dilemmas. Our tentative solutions

Fig. 21.4 Application of a map-based touch table in the Province of Utrecht, the Netherlands

for the identified barriers are not written in stone. We identified them to ignite the discussion about the implementation of Geodesign. As in the strongly related field of PSS, it is critical for the future of Geodesign to pay attention to user perspectives (for relevant discussions on PSS see Geertman 2008; te Brömmelstroet 2013b). 'Users' should be defined broadly, including GIS specialists and model developers, but also urban designers and policy advisors. More research into this aspect would also enhance our understanding of the relation between urban designers and Geodesign. The findings from this paper should be tested in more detail and in other contexts. Particularly the quantitative findings from the experiments are indicative, and require more empirical research. Hereby, it would be very relevant to test the extent to which recently developed and more intuitive GIS-based tools overcome the barriers outlined above (for a relevant early attempt see Dias et al. 2013; cf. te Brömmelstroet 2014).

Moreover, another factor that deserves further scrutiny is a holistic perspective of the users involved in Geodesign. In this paper we primarily focused on disciplinary background. The psychological background of the users is, however, also very relevant. Several interviewees indicated that to get people to be willing to use GIS-based tools, character (personality, emotions, etc.) is pivotal. People that are open, assertive and able and willing to think out-of-the-box are more likely to adopt new technology. Moreover, age also seems to be a factor. Younger people tend to have more experience with digital tools and are less used to old ways of working. Further research could provide more in-depth insight into this issue.

We will end this paper with the two persons from the introduction. Cornelis van Eesteren is arguably the most famous urbanist in Dutch history. His 'General Extension Plan for Amsterdam' from 1934 is still being used in 2013 by the Municipality of Amsterdam. Jack Dangermond founded ESRI and is now one of the strongest proponents of Geodesign. What do these two have in common? They both realize that developing a plan for the future requires creativity and collaborative imagination *combined with* analysis.

Acknowledgements We would like to thank the students and interviewees for participating in our research. This research has been made possible by the 'Connecting Sustainable Cities' (VerDuS) knowledge initiative of the Netherlands Organization for Scientific Research (NWO).

References

Carton, L. (2007). Map making and map use in a multi-actor context: Spatial visualizations and frame conflicts in regional policymaking in the Netherlands. Dissertation, TU Delft.

Carver, S., Watson, A., Waters, T., Matt, R., Gunderson, K., & Davis, B. (2009). Developing computer-based participatory approaches to mapping landscape values for landscape and resource management. In S. Geertman & J. Stillwell (Eds.), *Planning support systems: Best practices and new methods* (pp. 47–60). Berlin: Springer.

Dangermond, J. (2010). Geodesign and GIS—Designing our futures. In E. Buhmann, M. Pietsch, & E. Kretzler (Eds.), *Digital landscape architecture* (pp. 502–514). Offenbach/Berlin: Wichmann/VDE.

Dias, E., Kuipers, M., Rafiee, A., Koomen, E., & Scholten, H. J. (2013). Beauty and brains: Integrating easy spatial design and advanced urban sustainability models. In S. Geertman, J. Stillwell, & F. Toppen (Eds.), *Planning support for sustainable urban development* (pp. 469–484). Heidelberg: Springer.

Flaxman, M. (2010). Fundamentals in Geodesign. In E. Buhmann, M. Pietsch, & E. Kretzler (Eds.). *Peer reviewed proceedings digital landscape architecture 2010, Anhalt University of Applied Sciences* (pp. 28–41). Berlin: Wichmann.

Geertman, S. (2008). Planning support systems: A planner's perspective. In R. Brail (Ed.), *Planning support systems for cities and region* (pp. 213–230). Cambridge: Lincoln Institute for Land Policy.

Geertman, S. & Stillwell, J. (Eds.). (2009). *Planning support systems: Best practices and new methods*. Berlin: Springer.

Goffman, E. (1974). *Frame analysis: An essay on the organization of experience*. New York: Harper Colophon.

Healey, P. (2007). *Urban complexity and spatial strategies. Towards a relational planning for out times*. Oxon: Routledge.

Kahila M., & Kyttä, M. (2009). SoftGIS as a bridge-builder in collaborative urban planning. In S. Geertman & J. Stillwell (Eds.), *Planning support systems: Best practices and new methods* (pp. 389–412). Berlin: Springer.

McElvaney, S. (2012). *Geodesign: Case studies in regional and urban planning*. Redlands: ESRI Press.

Miller, D., Vogt, N., Nijnik, M., Brondizio, E., & Fiorini, S. (2009). Integrating analytical and participatory techniques for planning the sustainable use of land resources and landscapes. In S. Geertman & J. Stillwell (Eds.), *Planning support systems: Best practices and new methods* (pp. 317–345), Berlin: Springer.

Mouter, N., & Pelzer, P. (2013). Zwemles voor Planners. In F. Filius, E. Vanempten, C. Uittenbroek, G. Bouma, & S. Reniers (Eds.), *Planning is niet waarde-n-loos*, Papers for the Plandag 2013 in Antwerp, Belgium.

Pelzer, P., Arciniegas, G., Geertman, S., & de Kroes, J. (2013). Using map table® to learn about sustainable urban development. In S. Geertman, J. Stillwell, & F. Toppen (Eds.), *Planning support for sustainable urban development* (pp. 167–186). Heidelberg: Springer.

Rowley, A. (1994). Definitions of urban design: The nature and concerns of urban design. *Planning Practice and Research, 9,* 179–198.

Rydin, Y. (2007). Re-examining the role of knowledge within planning theory. *Planning Theory,* 6(1), 52–68.

Smith, H., Wall, G., & Blackstock, K. (2013). The role of map-based environmental information in supporting integration between river basin planning and spatial planning. *Environmental Science & Policy, 30,* 81–89.

Steinitz, C. (1990). A framework for theory applicable to the education of landscape architects and other environmental design professionals. *Landscape Journal, 9,* 136–143.

te Brömmelstroet, M. (2010). Making planning support systems matter: Improving the use of planning support systems for integrated land use and transport strategy-making. Dissertation, University of Amsterdam.

te Brömmelstroet, M. (2013a). Different process strategies for PSS: Influence of process strategies on added value of Urban Strategy (trial No5), CESAR working document series, no. 7.

te Brömmelstroet, M. (2013b). Performance of planning support systems: What is it, and how do we report on it? *Computers Environment and Urban Systems, 41,* 299–308.

te Brömmelstroet, M. (2014). Drawing versus calculating. The difference in added value between a drawing PSS (Phoenix) and a calculating PSS (Urban Strategy) (trial No. 6), CESAR working document series, no. 9.

Van Aken, J. (2004). Management research based on the paradigm of the design sciences: The quest for field-tested and grounded technological rules. *Journal of Management Studies, 41*(2), 219–246.

Van Rossem, V. (1993). *Het algemeen uitbreidingsplan van amsterdam.* Rotterdam/Den Haag: NAI Uitgevers/EFL-stichting.

Vonk, G., Geertman, S., & Schot, P. (2005). Bottlenecks blocking the widespread usage of planning support systems. *Environment and Planing A, 37,* 909–924.

Vonk, G., Geertman, S., & Schot, P. (2007). New technologies stuck in old hierarchies; an analysis of diffusion of geo-information technologies in Dutch public organizations. *Public Adminstration Review, 67*(4), 745–756.

Zwick, P. (2010). The world beyond GIS, planning, July 2010, pp. 20–23.

Chapter 22
Open Geospending: Bridging the Gap Between Policy and the Real World

Egbert Jongsma

> *Egbert Jongsma is working as an audit manager for the Netherlands Court of Audit (NCA), the Supreme Audit Institution of the Netherlands. He also is head of the Knowledge Center GIS & Audit of the NCA. Egbert Jongsma can be contacted via email: e.jongsma@rekenkamer.nl.*

22.1 About the Netherlands Court of Audit (NCA)

The Netherlands Court of Audit (NCA) is the Supreme Audit Institution of the Netherlands. The NCA strives to assess and further improve the regularity, efficiency, effectiveness and integrity of the State of the Netherlands and the institutions associated with it. It wants to promote and contribute to good governance. Good governance—according to the United Nations—has 8 major characteristics. It is participatory, consensus oriented, accountable, transparent, responsive, effective and efficient, equitable and inclusive and follows the rule of law. It assures that corruption is minimized, the views of minorities are taken into account and that the voices of the most vulnerable in society are heard in decision-making. It is also responsive to the present and future needs of society (United Nations Economic and Social Commission for Asia and the Pacific [UNESCAP] 2009).

The NCA was formally founded in 1814 and celebrates its 200th anniversary in 2014. The NCA is a member of the International Organization of Supreme Audit Institutions, INTOSAI (www.intosai.org). It has been active in various working groups of INTOSAI in the last number of years, like the Working Group on Environmental Auditing, the IT Working Group and the Working Group on the Accountability for and Audit of Disaster-related Aid. In these working groups the NCA has tried to bridge the world of auditors with that of other disciplines like IT and

E. Jongsma (✉)
Netherlands Court of Audit, The Hague, Netherlands
e-mail: e.jongsma@rekenkamer.nl

geospatial information. This comes from the strong belief that auditors should innovate to maintain and strengthen their relevance and added value.

Innovation is not only key to effective auditing, but also to an effective design of government policy.

22.2 Gap Between Policy and Real World

Government policy should be directed at improving the living conditions of its citizens. Therefore, a good understanding of those conditions is essential for assessing where the government should intervene to improve the life of its citizens. Furthermore, when government intervenes it should monitor the results of its interventions. Not only to be accountable for the results realized with public expenditures, but also to be able to learn and to alter interventions when the desired results are not realized.

The NCA reached the conclusion in many audits that government policies are mostly based on assumptions that the proposed measures will work, but not on any true evidence that it will work. There is a lack of understanding the real world. It seems that most policies are designed on a trial-and-error basis: let's see if it works and if not we have to think about something else. When government needs to intervene in situations it has never experienced before, then trial-and-error could be a smart approach. But it is only smart if government is able to monitor the effects of its policy measures in order to redesign its policies when needed. This means government should have access to data, information and knowledge about the living conditions of its citizens. The term used for this information is policy information.

But in many audits the NCA reaches the conclusion that policy information is lacking. There is a lack of information on where public funds have been spent and which results have been realized with these funds[1].

For example, on December, 26th 2004 a tsunami in the Indian Ocean devastated houses, ports and other infrastructure in the coastal areas of 16 countries. More than 200,000 people died.

A group of supreme audit institutions both from countries hit by the natural disaster as from countries involved in the emergency relief formed a taskforce to follow the money. They tried to build an audit trail in order to make sure the money was wisely spent and information about the results became available (Fig. 22.1).

The Task Force on the Accountability for and Audit of Disaster-related Aid didn't succeed in building an audit trail due to the lack of available data and information on national and international level. In the case of humanitarian aid this is a direct result of the complexity of the humanitarian aid sector. Funds for the affected countries are made available by many public, private, multilateral and intergovernmental organizations (Fig. 22.2). Furthermore, they flow via multiple channels and intermediate organizations from one part of the world to the other during which

---

[1] For an overview of various audit reports reference is made to the website of the NCA: http://www.courtofaudit.nl/english.

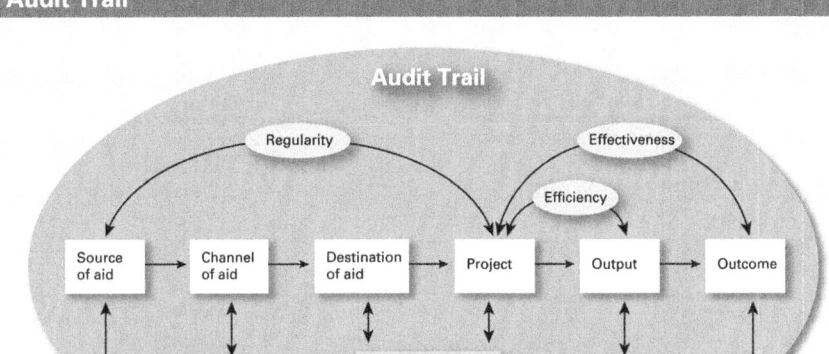

Fig. 22.1 Audit trail humanitarian aid flows, ideal situation

they lose their identity. Meaning that public and private funds are mixed and then split up again, which makes it difficult to follow to the final destination. Data and information regarding the "landing" of aid funds on the final destination was also incomplete and not reliable enough. The Task Force therefore reached the conclusion that if an audit trail is not included from the beginning, reconstructing an audit trail afterwards is not possible or fairly difficult (INTOSAI 2008).

The lack of policy information already is evident from the annual budget plans that ministers present to the Dutch parliament. In those plans ministers are often not able to present a clear relation between policy objectives, policy instruments and the amount of public funds needed to implement the policy instruments. Meaning that ministers are not able to be fully accountable for the spending of public funds nor the results that have been realized with these funds. This also means ministers cannot learn and improve their policies. So the NCA has to repeat its recommendation to improve policy information in order to enhance accountability and learning over and over again.

The NCA decided to become more involved in the design of policy information to be able to influence and strengthen it. This with the aim to enhance transparency and accountability and to enable audits that can focus on assessing the results of government interventions. The NCA in 2010, for instance, contributed to the design of the accountability and audit framework of the aid funds for Haiti collected by a large group of Dutch humanitarian aid organizations, the Samenwerkende Hulporganisaties (SHO) (NCA 2010). In Sect. 22.5 other examples of the NCA efforts in contributing to the design of policy information will be explained.

So how can the gap between policy and the real world been bridged in a structural way?

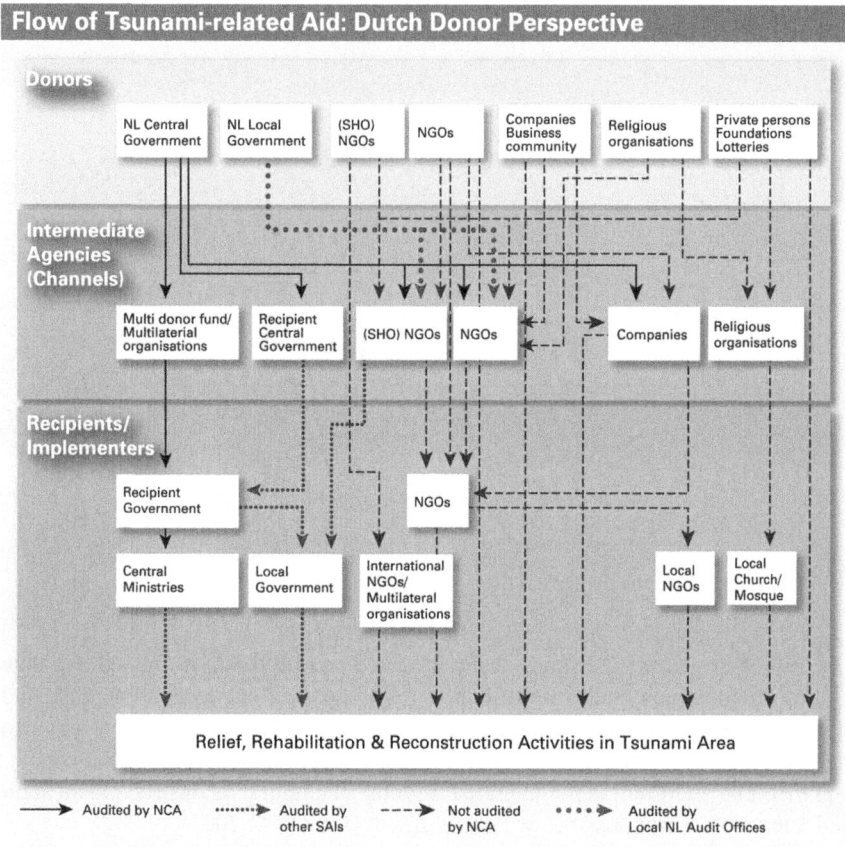

Fig. 22.2 Simplified overview flow of Dutch aid funds for Tsunami-affected countries

22.3 Geodesign Framework Stimulates Understanding of the Real World

The geodesign framework of Carl Steinitz, as visualized below, is a systematic process of decision making directed at (re)designing the physical world we live in (Steinitz 2012). It provides a step-by-step approach around two main questions: is the landscape working well and if not should the landscape be changed? The strength of the framework lies in its broad applicability for decision making by public entities. It stimulates decision makers to first obtain an understanding of the real world as it is. Secondly it stimulates decision makers to assess the impact of possible interventions before deciding whether the real world should be changed.

The experience of the NCA shows that in general policy makers don't have a good understanding of the real world as it as. They lack data, information and knowledge about the living conditions of citizens in the real world. Furthermore, the NCA sees a lack of values and goals based on which assessments can be made if the real world works well. The lack of data, information, knowledge, values and goals also hampers the design of interventions, because interventions should be evaluated on whether they have a desired impact on citizens' lives.

The importance of goals and values for any design was stressed by Carl Steinitz in his keynote speech at the European Geodesign Summit held in 2013 in the Netherlands. He stated that the geodesign framework is value driven and not data driven: values and goals should be starting point of any design. Just as values and goals should be the starting point of any government intervention. As stated before in this chapter, reality is different. The NCA sees that many policy objectives are vague and non-coherent with the proposed policy instruments and the amount of public funds needed to implement these instruments. When objectives and goals are not specific enough it is difficult to know what policy information (data, information and knowledge) is needed for assessing whether policy objectives have been realized.

The NCA believes that designing government policy should include policy information. Assurance on the quality of any design can only be provided when it is possible to assess its impact in the real world. And in many audits the NCA had to conclude that it didn't know whether policy interventions had the desired impact (effectiveness) due to the lack of sound policy information.

22.4 Geodesigning Policy Information: Open Geospending

As stated above, the NCA tries to promote and contribute to good governance. And the key to good governance is sound information. Sound information for understanding the real world and for assessing the impact of policy interventions in the real world.

It is important to realize that a number of developments—like internet, smartphones, social media—have changed our society into a digital one. This also has a huge impact on how policy information is created. Only a number of years ago information was created in a linear step-by-step process under the control of policymakers (Fig. 22.3). The speed of this process depended on the travel speed of paper.

Nowadays information is created from data of multiple sources in a continuous flow. It can be gathered, adjusted and distributed by everyone with an internet connection (Fig. 22.4).

Data, information and knowledge no longer is limited to a selective group of people, due to the fact that so many people have an internet connection and more and more data have become available as open data. The broad and easy access to the internet has led to more transparency, democracy and empowerment of citizens (armchair auditors). The observations and feedback of these empowered citizens can be an enormous relevant source of information for policy makers. It can provide a better understanding of the real world (is it working well?) and moreover it can provide direct feedback on the impact of policy interventions in the real world (is it

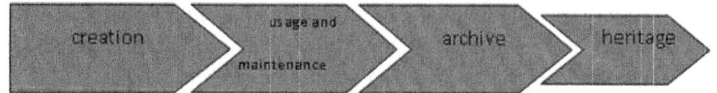

Fig. 22.3 Creation of policy information (paper era)

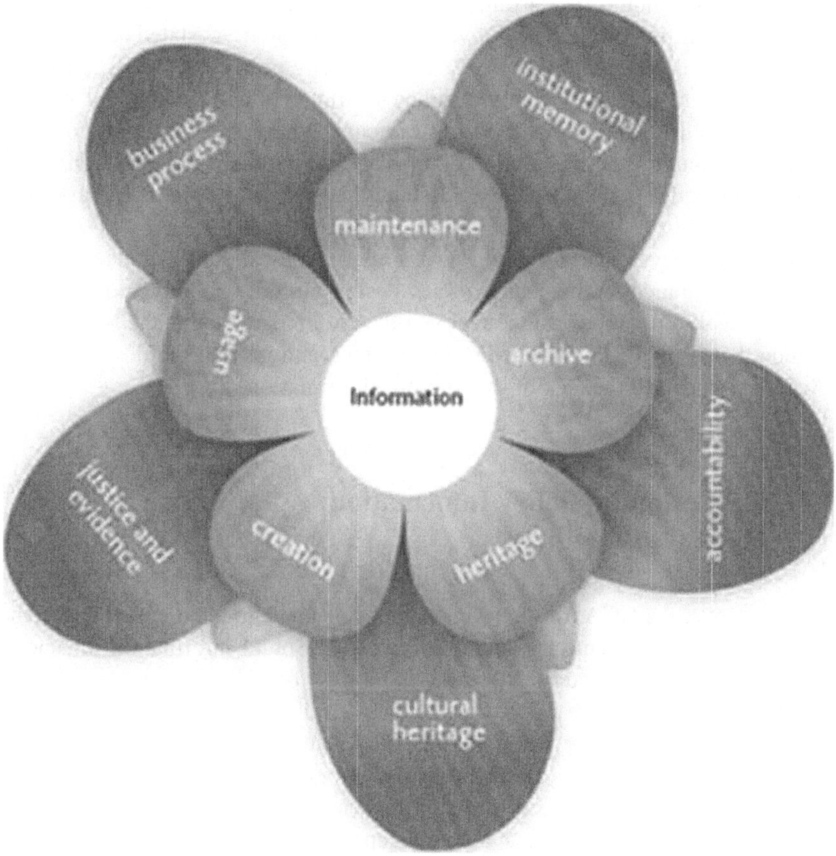

Fig. 22.4 Creation of policy information (digital era)

working better?). According to British prime minister Cameron opening up data to the wider public can lead to an army of armchair auditors that can scrutinize policy interventions and thus preventing fraud, misuse and waste (Cameron 2010).

The digital era also poses challenges. Solely opening up data and making them available on the internet doesn't automatically lead to a better understanding of the real world or to activating armchair auditors. The availability of so much data and information can lead to confusion and distrust of the data and information presented.

The NCA believes that data and information should be made available in a way that supports a better understanding of the real world. It believes that location is the key to understanding the real world: everything happens somewhere. Therefore, the NCA supports the use of new technologies that can integrate data from various sources—including that of citizens—and visualizing them on a map. Geography is, in a way, the missing link between the policy world and the real world. Making it possible to create more coherence between policy goals and objectives, policy interventions and the public funds needed for implementing policy interventions on one hand and the impact of the policy interventions and the expenditures on the other.

An interesting development in this regard is "open spending".[2] Open spending started as a grassroots movement dedicated to opening up financial data of governments to be able to track and analyze public spending. Many governments around the world have embraced ideas on open spending and have started to provide financial data on public spending as open data. Some governments have even decided to provide financial data on transaction level making it possible to track and analyze individual financial transactions. The NCA has decided to join the growing community of public organizations that provide their financial data as open data. As a first step the NCA has published its financial data on transaction level over 2013 on March, 27th 2014 as part of its annual report and accounts.

Combining financial data with location or geospatial data provides an even better insight into public spending and lays the foundation for looking into efficiency and effectiveness of public spending. The NCA uses the term "Open geospending" for this.

22.4.1 Example: Open Geospending, www.recovery.gov

The Federal government of the United States proposed the American Recovery and Reinvestment Act (ARRA) of 2009 to fight the consequences of the financial crisis. The website www.recovery.gov was created under the Recovery Act to show the American public how ARRA funds are being spent by recipients of contracts, grants, and loans, and the distribution of ARRA entitlements and tax benefits. The website displays information about the Recovery Accountability and Transparency Board's activities, as well as data related to the $840 billion stimulus bill (Fig. 22.5).

The website makes it possible to search via zip code, city or federal state. It also enables searching on the map. The data provided provides an insight in the total dollar amount for all the ARRA contract, grant, and loan awards in a specific area. Furthermore, it provides insight in job numbers, the cities receiving the bulk of the money, and the top recipients. The website also stimulates citizens to scrutinize the activities in their neighbourhood and to provide feedback on misuse, fraud and waste via a Complaint form. It also provides insight in whether ARRA money went to the places that had the greatest need:did areas of high crime get COPS funding for instance? The website also provides its data as open data thus making it possible to use it to create charts and graphs or to build own applications (Recovery Accountability and Transparency Board 2009).

---

[2] For more information regarding open spending reference is made to the website https://openspending.org/ accessed on 31 March 2013.

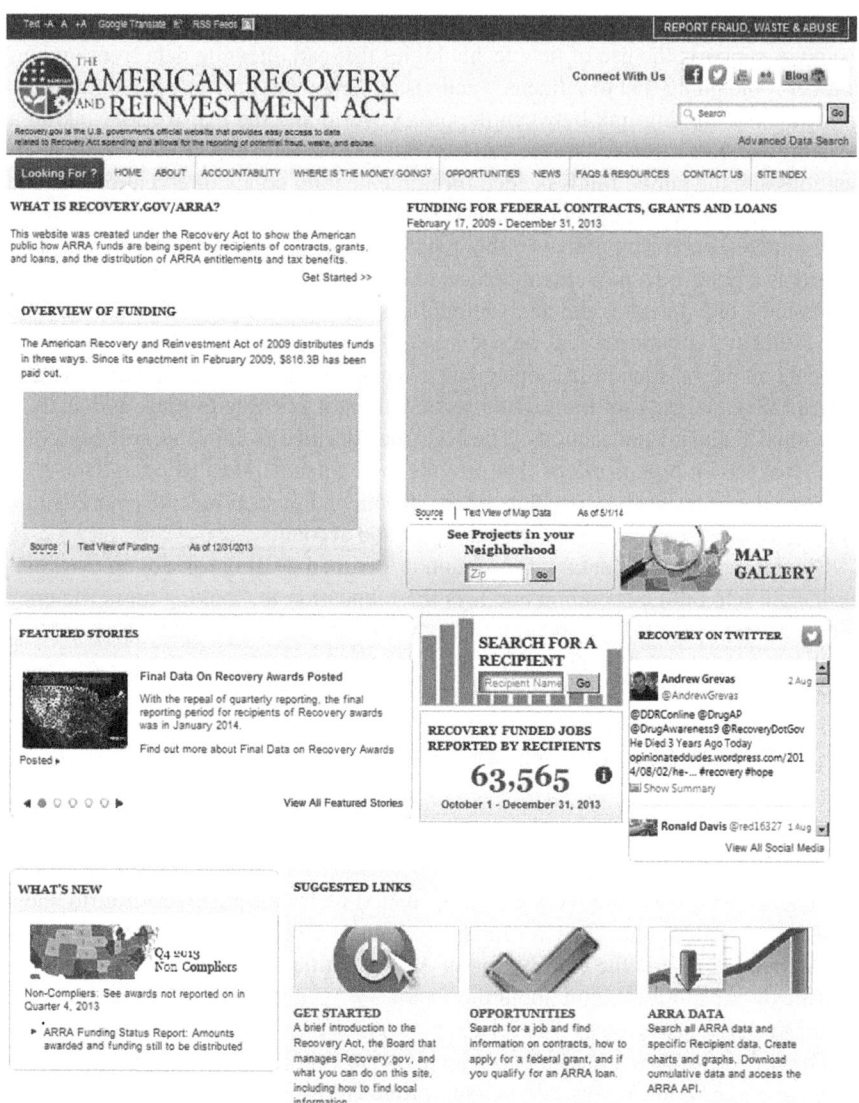

Fig. 22.5 Screen shots of the user interface of ARRA (www.recovery.gov)

22.5 Geodesigning Real Estate and Social Domain

It is the aim of the NCA to stimulate the design of policy information that connects the policy world with the real world and includes an audit trail. This means that the NCA has to be prepared to take a more pro-active role in contrast with its traditional role as an ex-post auditor. A more proactive role doesn't mean that the NCA will take responsibility for designing government policy or policy information. But that the NCA actively wants to bring in its expertise in the design process of policy and policy information, adopting the multidisciplinary approach of designing as advised by the geodesign framework of Carl Steinitz.

The NCA decided to experiment with this new role starting with contributing to the accountability and audit framework of the Dutch aid for Haiti in 2010 (see Sect. 22.2). At present the NCA takes up this more proactive role regarding real estate and the welfare system, discussed below.

22.5.1 Stimulating an Integrated Overview of Public Real Estate

In the Netherlands various public organizations own real estate objects, which include office buildings, jails, educational facilities, military facilities, land and infrastructure. These real estate objects have a certain economic value, but also value related to their use in the public domain (societal value). Due to the financial crisis and efforts to create a more efficient public administration, many public entities have to downsize staff and make more efficient use of their assets such as real estate objects. This has led to a number of large reorganizations within the Dutch public sector, like the transition to a national police and the reorganization of the ministry of Defense, the Public Prosecutors' Office and the courts. These reorganizations include relocation of staff and downsizing of real estate.

The NCA audits these individual organizations by looking into the reorganization process and possible risks or negative effects. It is also aware that it should look into the portfolio management of real estate assets due the economic and social value aforementioned. A crucial part of managing the portfolio of real estate assets is a proper administration and registration of these assets including economic value and geospatial data (location, neighborhood characteristics, etc.). If public entities want to manage their real estate portfolio and design their real estate strategy or policy they need policy information about their real estate assets. Based on its experiences, the NCA has observed that public entities—in general—can and should improve their information regarding real estate assets.

What the NCA also has identified is that information regarding real estate assets is hardly shared among public entities. In a situation where many public entities at the same time have to downsize their real estate assets, this poses a risk where separate decisions of selling real estate objects (office buildings for instance) or relocating staff in a certain neighborhood have negative consequences for the neighborhood as a whole. Empty buildings can lead to lower market rates in that neighborhood, but also to a deterioration of the neighborhood (e.g. rise of criminality, feelings of unsafety and pollution). That is why the NCA is dedicated to stimulating public entities to share their information about their real estate assets and the plans they have with these with other public entities and with the general public. The NCA intends to develop a real estate application together with public entities that own a large number of real estate objects. The real estate application should visualize the real estate data and information of various public entities on a map, thus creating a better overview and a stronger basis for joint decision-making. When real estate information also is available to the wider public it can stimulate the involvement of citizens with regard to the use of public real estate objects in their neighborhood.

22.5.2 Stimulating Slender and Smart Policy Information for Welfare System

As part of the aim to create a more efficient and effective public administration, the Dutch central government has decided to transfer the responsibility for parts of the welfare system to the municipalities. The main objectives of this transfer of responsibility are better and cheaper welfare services to citizens. These objectives are based on the conviction that municipalities have a better understanding how to support their citizens in a tailor made manner. The transfer of responsibility is a major operation not only in terms of tasks and duties but also in terms of euros (16 billion € will be transferred from the State budget to the municipalities). It also requires a strong policy information infrastructure for municipalities, and for the central government to assess whether the desired objectives will be realized.

The NCA identified the transfer of responsibilities for the welfare system to municipalities as a unique opportunity to use the geodesign framework to generate policy information for a slender and smart information infrastructure. This infrastructure should be based—in the view of the NCA—on households and their neighborhood (natural habitat). Furthermore, it should be based on single information and multiple-use meaning that the information need of municipalities is the basis for gathering data. The central government should rely on the fact that these data are sufficient to obtain a national overview needed for monitoring the realization of national objectives. This means a completely different way of designing policy information for central government, because central government in general has the tendency to think top-down and from its own information needs. The NCA will cooperate with municipalities and ministries to see if it is possible to construct a slender and smart policy information structure that serves the need for accountability and transparency of local and central government. The information structure should also make it possible for citizens to assess if they are better off after the transfer of responsibilities to their municipalities by making (a part of) its data available as open data.

22.6 Conclusion

Opening up data makes it possible to benchmark and to learn which approach works best at the lowest cost. In this sense our efforts regarding opening data on real estate and welfare could serve as important experiments for gathering evidence on the impact of policy interventions in the real world and hopefully a strong impulse for evidence based policy.

References

Cameron, D. (May 29, 2010). Transcript of PM's podcast on transparency. http://webarchive.nationalarchives.gov.uk/20130109092234/http://number10.gov.uk/news/pms-podcast-on-transparency/.

INTOSAI Task Force on the Accountability for and Audit of Disaster-related Aid. (2008). Lessons on accountability, transparency and audit of Tsunami-related aid. http://www.courtofaudit.nl/english. Accessed 31 March 2013.

NCA. (2010). Lessons on accountability, transparency and audit of Tsunami-related aid. http://www.courtofaudit.nl/english. Accessed on 31 March 2013.

Recovery Accountability and Transparency Board. (2009). www.recovery.gov. Accessed on 31 March 2013.

Steinitz, C. (2012). *A framework for geodesign: Changing geography by design*. Redlands: ESRI.

United Nations Economic and Social Commission for Asia and the Pacific. (2009). What is good governance. http://www.unescap.org/sites/default/files/good-governance.pdf. Accessed 31 March 2013.

http://www.courtofaudit.nl/english/Latest_News/All_newsitems/2010/01/SHO_and_Court_of_Audit_issue_joint_press_release_on_Haiti.

Chapter 23
Towards Geodesign: Building New Education Programs and Audiences

John P. Wilson

23.1 Introduction

Many of the descriptions of geodesign invoked to date have cast it as an iterative design and planning method where emerging designs are shaped using spatial knowledge acquired from geospatial technologies (e.g. Dangermond 2012). This often leads to a relatively elaborate discussion of geodesign workflows (e.g. Steinitz 2012) and how these can be invoked for storytelling, collaboration and public participation, among others (e.g. Niemann et al. 2011). These kinds of descriptions may not serve us very well. An economist, for example, might struggle to think of it differently from one of their own methodologies, such as cost–benefit analysis. Here, I argue that this would be a dangerous outcome and that the aforementioned approaches to defining the field of geodesign pay too little attention to the special place we find ourselves both in the world as a whole and in education in particular. This special "place" can be traced to the precarious state of the world, to the need to invoke spatial thinking to help find solutions to many of our most serious and enduring problems, and to the tremendous opportunities afforded by the Web for learning and collaboration.

Many scholars and commentators have written passionately about the state of the world and the difficult choices we will likely face in the coming decades. Fisher (2012), for example, has written that "the only way we can avoid such a fate is to realign our relationship with the natural world, to reorganize our considerable knowledge about it to reveal the forces that lead to unsustainable practices, and to relearn how to steward what remains of the planet that we have so altered". These observations point to the tremendous gains that our disciplinary-based education systems have yielded during the past few centuries on the one hand and the need for change, given we know much more about ecosystem drivers and outcomes and yet have struggled to put this knowledge to work to create more sustainable ways of living, on the other hand. Fisher (2012) sees GIS as a way to spatialize all of the knowledge about a place and to see the relationships among disciplines and the con-

J. P. Wilson (✉)
Spatial Sciences Institute, University of Southern California, Los Angeles, CA, USA
e-mail: jpwilson@usc.edu

nections among data. Goodchild (2010), views geodesign as a way for GIS to make good on one of its early objectives (the use of GIS as a tool for creating designs as was popularized by the late Ian McHarg (1969), among others) and thereby promote futuristic collaborations among scientists and designers to help empower efforts to improve and sustain the surface and near-surface environments of the Earth. Finally, Steinitz (2012) in his seminal book summarizing where geodesign is now and where major research and education efforts should be focused in the future, explained how the latter requires the training of "conductors" as well as "soloists" and how this might be accomplished within a master's level curriculum in geodesign.

This chapter explores some of these ideas in more detail and highlights some of the challenges universities are likely to face as they work to create and sustain successful geodesign degree programs in an education setting in which disciplinary silos are still the norm and they are continually challenged to do more with less. The next section describes the importance of spatial thinking and four additional characteristics of geodesign that suggest it represents an important turning point for spatial scientists and practitioners. Section 23.3 uses several examples to show how this "geodesign" concept is not new and highlights some of the challenges that derailed early geodesign projects and lessons learned. Section 23.4 discusses the role of the Web and why this may be an ideal time to accomplish meaningful change with the help of three modern geodesign initiatives. Section 23.5 explores the implications of all of this for geodesign education using the University of Southern California's new B.S. in Geodesign degree as an example and the chapter closes with some conclusions and ideas for future work in the final section.

23.2 Five Characteristics of Geodesign

The most popular definitions of geodesign position it as a new field built on top of the spatial sciences, assuming the latter spans all the various ways in which spatial information can be acquired, represented, organized, analyzed, modeled, and visualized (Wilson and Goodchild 2012). The successful pursuit of each of these activities involves spatial thinking at its core and spatial thinking, it turns out, is used in many (most?) occupations and many facets of everyday life to structure problems, organize knowledge, find answers, solve problems, and communicate solutions (Sinton and Lund 2007; Sinton 2012). However, spatial thinking is not explicitly taught in K-12 and college settings in the US and many other parts of the world and a large part of my own interest and fascination with geodesign is the special opportunity it provides to introduce formal training in the spatial sciences to a larger and more diverse audience compared to past years.

Geodesign provides new opportunities to use the spatial sciences to promote and guide design across a variety of spatial scales ranging from specific sites to neighborhoods, watersheds, regions, and the world as a whole. Given this setting, the best geodesign programs will incorporate spatial thinking and teaching that aims to develop student's capacities to conceptualize, visualize, and interpret location, distance, relationships, movement, and change across places and spaces (Goodchild 2010; Sinton 2012). Spatial thinking, viewed in this way, might serve as a kind of "glue" that: (1) provides the means to clarify and understand the role of different perspectives in describing the context for specific problems and/or opportunities; and (2) serves as a platform by which we could engage and use a series of diverse disciplinary backgrounds and perspectives for solving real-world problems. However, geodesign practice itself will need to evolve to encourage and enable the participation of non-spatial thinkers as well, because the growing number and popularity of map-based tools and scorecards threatens to leave these individuals behind on the sidelines.

The second distinguishing feature of geodesign is the use of a variety of geospatial technologies to help gather, organize, analyze, model, and visualize large volumes of both spatial and non-spatial data. These technologies, and the underlying science that they are built on top of, have grown enormously during the past four decades. Their scope and purpose are perhaps best described in the Geographic Information Science & Technology Body of Knowledge (BoK) that was published by the Association of American Geographers and University Consortium for Geographic Information Science in 2006 (DiBiase et al. 2006, 2007) and in the Geospatial Technical Competency Model (GTCM) that was published by the US Department of Labor a few years later (DiBiase et al. 2010). The latest geospatial software platforms provide an increasingly data-rich environment for all with access and endeavor to cover every feature or aspect of interest—including the less tangible attributes that are discussed later in this chapter and contribute to the sense or meaning of place. However, they have traditionally focused on current conditions and how they came to be this way (Fisher 2012). Geodesign then affords new opportunities for utilizing these platforms in ways that encourage spatial thinking and its use in decision making and problem solving.

This last observation brings us to the next two distinguishing characteristics of geodesign: its future orientation and focus on design, which in the most general sense, involves imagining and one hopes, doing something positive, to change conditions on or near the Earth's surface to improve the everyday lives of residents. This pair of characteristics acknowledges that geodesign has emerged at a uniquely important moment in history, given the rapid growth of the human population during the past two centuries, the emergence of cities as home to more than 50% of the human population, the growing numbers of educated people in the world coupled with our steadily expanding capacity to alter conditions at or near the Earth's surface (for better or worse), and the spread of the Web as a platform to share knowledge and aspirations of one kind or another (Worldwatch Institute 2013). The role of the Web is rapidly evolving and it is likely to provide many new opportunities for performing spatial analysis, building spatially explicit models, and visualizing potential solutions to problems across a range of scales in the next few years (Wang et al. 2013). It is perhaps not surprising then that many commentators have written about geodesign as a force for good—for helping individuals and societies to build more sustainable, livable and healthy communities for both current and future generations (e.g. Niemann et al. 2011; Steinitz 2012). For educators, this pair of characteristics points to the need to teach visioning along with systems thinking and analysis.

The fifth and final feature of geodesign is its focus on collaboration. This is manifested in at least two ways. The first is the need for multi-disciplinarity. It takes many workers from multiple fields to build a 20-story residential building for example. However, the size and diversity of this cast would be expanded if we decided to build multiple high-rise residential towers and locate a subway line and station below them. The latter is increasingly likely and may provide new opportunities to build more sustainable and healthy communities. This last scenario also speaks to the increasing complexity of modern life and the growing need for teamwork. The second way in which collaboration is manifested has to do with the need to involve the people who would be affected by these designs and subsequent actions. People vary tremendously in terms of interests, goals and aspirations and the often-cited sentiment that people will come if we build sustainable urban forms (i.e. transit-friendly high density housing for example) may be nothing more than wishful thinking.

This is an especially opportune time because the emergence of the Web and the tremendous opportunities to implement geodesign on top of this increasingly ubiquitous platform provides a multitude of exciting new ways we can engage policy-makers, regulatory agencies, architects, scientists, engineers, and everyday citizens to build more sustainable and healthy communities. The role of the Web and its various manifestations (cloud computing architectures, big data, data analytics, etc.) warrants special emphasis and attention because it dramatically expands the possibilities and the ease of collaboration. However, saying this will be so and making it happen may well constitute two different outcomes unless we proceed carefully, for the reasons articulated in the next two sections. Geodesign educators, of course, will need to use the Web to promote and sustain collaboration in their classroom activities as well for students to grasp the full range of collaboration possibilities moving forward.

23.3 Geodesign is Not New!

The five characteristics of geodesign noted above—the focus on spatial thinking, geospatial technologies, the future, design as a force for good and multi-disciplinary collaboration—are not necessarily new and one can find plenty of examples of similar efforts that extend back many decades and even centuries. Three examples from southern California are used below to illustrate these kinds of efforts.

In the first example, the Los Angeles Chamber of Commerce commissioned well-known landscape architects Olmsted and Bartholomew to create a regional plan for "parks, playgrounds, and beaches" in the 1930s. The geospatial technologies of the time consisted of pen and paper and these were used to construct and disseminate a series of maps depicting current and proposed land use conditions—just as modern geospatial software platforms such as ArcGIS and Google Earth are often used today (see, for example, the map books published as a part of the Esri International User Conference each year). Though the plan was never implemented, in the mid-1990s, it resurfaced and became a beacon, simulating new calls for a revitalized city connected by green corridors and major parklands (e.g. Hise and Deverell 2000; Wolch et al. 2012).

This is but one of many plans from Los Angeles that were never implemented and today's land use patterns bear little resemblance to the plan envisaged by Olmsted and Bartholomew. The metropolitan region today stretches 210 km from Oxnard in the west to Redlands in the east and 140 km from Santa Clarita in the north to San Juan Capistrano in the south. Some 18 million residents now call this region home and it is characterized by a series of enduring challenges related to employment, traffic, crime, pollution, fragmented natural resources and environmental goals, among others, that go unmet.

This state-of-affairs provided the backdrop for the second example: the Reality Check Los Angeles event convened by the University of Southern California's Lusk Center for Real Estate and the Greater Los Angeles Area Office of the Urban Land Institute in 2002 (see http://http://www.youtube.com/watch?v=-aHgIh6m3ns for additional details). This event gathered 250 politicians, policy-makers, planners and professionals from various domains. They spent a morning working in small groups to allocate new residents and employment opportunities specified in a future growth forecast on a series of specially prepared maps showing the existing distribution of settlement, economic activity and various kinds of infrastructure. These maps were collected and analyzed by a team of 12 GIS faculty and students in ArcGIS over the lunch break and used throughout the afternoon to delve into what the various groups had tried to accomplish and what would need to happen for their plans to be realized. This effort would not have been possible without modern GIS tools (how else could one have captured and summarized the results from 20 separate tables and compiled the results in map form in 2 h), but in the end it suffered the same fate as Olmsted and Bartholomew's plan given that the designs did not lead (as far as one can tell) to tangible action(s) and there was little engagement with everyday residents.

The third and final example concerns the GreenVisions Plan for twenty-first Century Southern California project that was funded by a consortium of regional conservancies in 2004. This ambitious project, which focused on parks, open space, watershed health, biological conservation and restoration, was organized around the development and deployment of web mapping tools that local residents and citizen groups could use to help prepare funding proposals that would be submitted to one or more of these regional conservancies. The motivation of the funders, led by the San Gabriel and Lower Los Angeles Rivers and Mountains Conservancy, was to provide a platform that could be used by potential grant applicants to identify projects that would provide tangible benefits by adding additional parkland and open space in park-poor areas, improving watershed health, and/or providing new opportunities for the conservation and restoration of important flora and fauna. A web mapping platform was built—see Ghaemi et al. (2009) and Sister et al. (2010) for additional details—and used for one or two rounds of grant funding but ultimately this effort failed because the funding to support the mapping infrastructure disappeared as general economic conditions deteriorated and technology advances rendered the ArcIMS platform on which the application was built more or less obsolete.

The challenges highlighted by these examples point to the need to gather the support of those making decisions as well as experts and those whose everyday lives will be affected on the one hand, and the need to choose the geospatial technologies that are to be deployed carefully on the other hand. Fortunately, the emergence of the Web as a ubiquitous platform for analysis and communication offers new opportunities and ways of meeting the aforementioned needs, as indicated to varying extents by the three recent initiatives described below.

23.4 Recent Examples: Getting It Right!

Several authors have spent considerable time clarifying the roles of experts and affected publics in participatory geodesign projects in the past few years. Goodchild (2010), for example, distinguished design and Design in which the former contemplates design as a simple optimization problem and the latter sees the process complicated by varying goals among stakeholders, feedback loops that modify objectives, constraints and data as the process proceeds, and uncertainties about implementation. Spatial optimization models and spatial decision support systems provide just two of many possible approaches for solving geodesign problems given the first view of the world (e.g. Ghosh and Rushton 1987; Malczewski 1999; Jankowski and Nyerges 2001; Faiz and Krichen 2013). Steinitz (2012, pp. 198–201), on the other hand, spent much of the final chapter in his influential book about geodesign describing how future geodesign projects will have greater involvement from the people of the place and will need larger numbers of more technically competent people (i.e. experts) who will be forced to take more active roles, due to changes in political attitudes and information technologies. Figure 23.1 illustrates this particular view graphically and shows how: (1) each of the six stages in a typical "regional" scale geodesign project might be formalized as a model and supported by computational tools; and (2) the ways the people of the place, experts and conductors might be engaged at each stage. Goodchild (2010) offered a similar commentary and noted the need for new sketch and simulation tools to support geodesign as it is conceived here. The beauty of the second approach is that spatial thinking becomes second nature for many of the participants in geodesign.

Three initiatives can be used to show some of the progress that has been made in the past few years. The first example is the Trust for Public Land's Greenprinting GIS-based service which provides a platform to help communities prioritize their park and conservation goals. The service utilizes GIS in a transparent mapping and modeling process and engages local residents and other stakeholders in a place-based planning exercise. These tools have been employed many times to delineate the lands with the highest conservation value and to meet the diverse goals identi-

23 Towards Geodesign: Building New Education Programs and Audiences

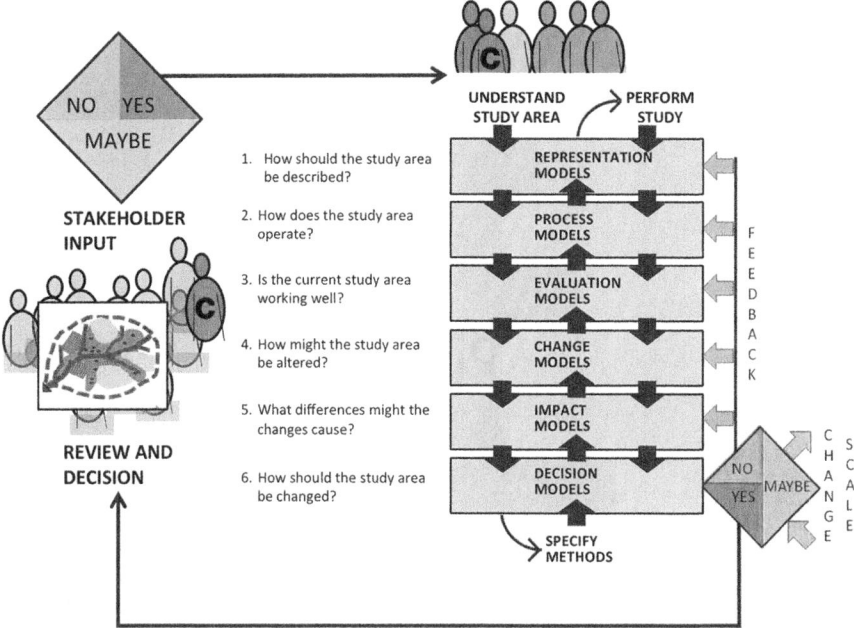

Fig. 23.1 Schematic showing how future, larger geodesign projects will also have greater involvement from the people of the place and more technically competent people who will have to take a more active role. (Source: Steinitz 2012, p. 200)

fied by stakeholders, such as preserving ranchlands, protecting water sources and creating new parks. These services have been deployed by the Trust for Public Land in a series of collaborative greenprinting projects across the US that have brought more and better geographic information technologies to the task at hand and encouraged some additional public involvement in the planning and design processes.

The second example is SeaSketch (http://www.seasketch.org), a collaborative geodesign software service that is being used for marine spatial planning around the world. Using this service, individuals can: (1) specify their own geographic area(s) of interest; (2) upload existing map services from ArcGIS Online; (3) create and invite users and groups to participate in their project(s); (4) define "sketch" classes for marine management zones; (5) create map-based discussion forums; and (6) create simple surveys to collect data on human uses of the ocean. The goal of marine spatial planning has long been to achieve an optimal balance of marine resource use by reducing conflicts between users and maintaining ecological processes and the ecosystem services they support (Beck et al. 2009). SeaSketch can be thought of as the successor to MarineMap, a web-based platform for collaborative marine protected area planning (Merrifield et al. 2013) that was used to engage a large number and variety of stakeholders and delineate a series of new marine protected areas along the north-central California coast (Gleason et al. 2010). SeaSketch brings similar

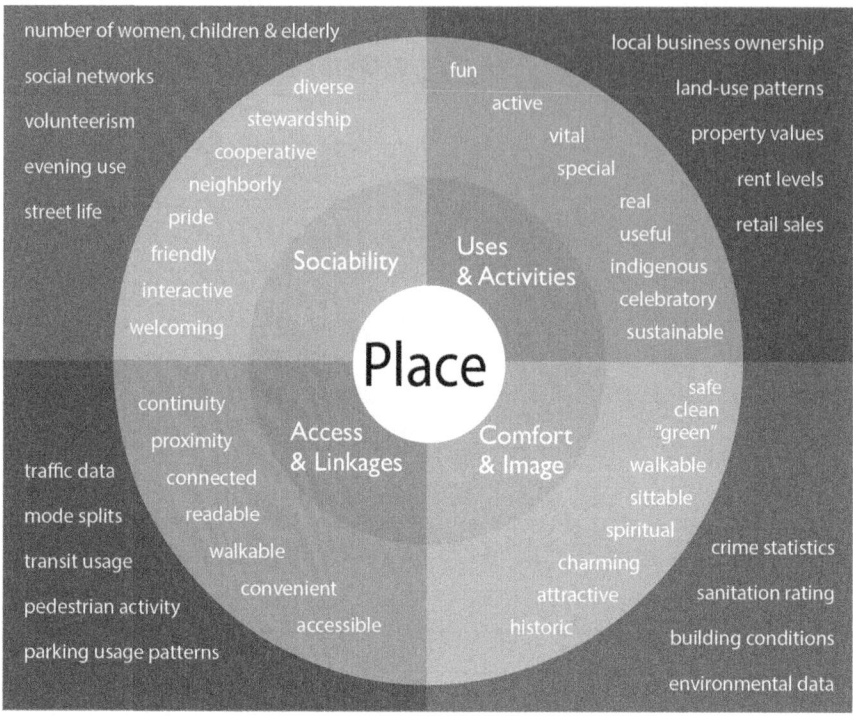

Fig. 23.2 Schematic showing key attributes, intangible qualities and measurable data included in the Project for Public Space's placemaking concept. (Reprinted with permission of Project for Public Spaces, Inc. © 2014. All rights reserved)

benefits as the first example and is especially noteworthy because it offers greatly expanded opportunities for Web-based stakeholder participation.

The third example is the "placemaking" concept promoted by the Project for Public Spaces (PPS; http://www.pps.org). PPS is a nonprofit planning, design and educational organization dedicated to helping people create and sustain public spaces that build stronger communities. Their pioneering placemaking approach is both an overarching idea and a "hands-on" tool for improving a neighborhood, city or region. It incorporates 11 key elements—(1) the community is the expert; (2) create a place, not a design; (3) look for partners; (4) you can see a lot just by observing; (5) have a vision; (6) start with the petunias: lighter, quicker, cheaper; (7) triangulate; (8) they always say "it can't be done"; (9) form supports function; (10) money is not the issue; and (11) you are never finished—and combines both intangible qualities and measurable data in helping residents imagine new public spaces (Fig. 23.2). Founded in 1975 to expand on the work of William (Holly) White, the author of "The Social Life of Small Urban Places" (Whyte 1980), PPS has completed more than 2,500 community projects in 40 countries and in all 50 US states and trained tens of thousands of people per year so they can transform their own public spaces into vital places that showcase local assets, spur investment and rejuvenation, and better serve local needs. This particular example speaks to the previously noted idea of invoking design as a force for good and the kinds of benefits that would ensue if these approaches could be modified and used with Web-based map services and tools to imagine new places across a range of geographic scales (i.e. extents).

The three initiatives, taken as a whole, are valuable because they indicate the new kinds of collaborations among experts, stakeholders, and everyday residents that are now possible and point to some of the gains that we have made in developing and deploying the kinds of sketch and simulation tools that Michael Goodchild wrote about in his landmark article on geodesign (Goodchild 2010).

23.5 Implications for Geodesign Education

The immediate challenge for educators is to plan and build programs that provide the spatial and systems thinking, collaboration, problem solving, experiential learning, and technical skills and experiences that geodesign professionals will need in the years ahead. Responding to this challenge, we have launched a new B.S. in GeoDesign degree at the University of Southern California. Viewed as a multidisciplinary program from the start, the collaboration is led by the Spatial Sciences Institute, housed in the Dana and David Dornsife College of Letters, Arts and Sciences, the planning faculty in the Sol Price School of Public Policy, and the architecture and landscape architecture faculty in the School of Architecture. The development of six learning outcomes supported by this new B.S. degree program follow more or less directly from the commentary and examples offered in the preceding sections:

1. Learn about the myriad ways in which places can be constructed, interpreted and experienced in different ways by different people (e.g. migrants, people of color, the elderly, the poor, teenagers, toddlers and working adults, among others).
2. Learn about the principles of design and how these can be used as a force for good in building healthy, livable and sustainable communities.
3. Learn how urban and regional planning provides a framework for promoting civic engagement and collective action.
4. Learn how geographically referenced data can be gathered and organized to support a large number and variety of collaborative projects.
5. Learn how geospatial data can be analyzed, modeled and visualized to inform design and planning and by doing so, support public participation and urban development.
6. Learn how form and function co-exist and evolve in urban settings and how globalization connects near and faraway places and actions.

Given these learning objectives, the program starts with a series of spatial classes that use geospatial technologies to build spatial thinking competency and then gradually integrates design and planning classes in the mix so our students can focus their time and energy on future challenges and see all that they do as a force for good in the world (Fig. 23.3). Some of the design classes place students in the field

PRE-MAJOR COURSES (8 UNITS)

ECON 203 – Principles of Microeconomics
MATH 116 – Mathematics for the Social Sciences

CORE COURSES (40 UNITS)

SSCI 301 – Maps and Spatial Reasoning
ARCH 203 – Visualizing and Experiencing the Built Environment
PPD 227 – Urban Planning and Development
SSCI 382 – Principles of Geographic Information Science
ARCH 303 – Principles of Spatial Design I
SOCI 314 – Analyzing Social Statistics
PPD 417 – History of Planning and Development
SSCI 401 – Spatial Science Practicum
ARCH 403 – Principles of Spatial Design II
PPD 425 – Designing Livable Communities

MAJOR ELECTIVES (20 UNITS FROM GROUPS A & B)
GROUP A – BUILT ENVIRONMENT (8-16 UNITS)

ARCH 361L – Ecological Factors in Design
ARCH 432 – People, Places and Culture: Architecture in the Public Realm
HIST 347 – Urbanization in the American Experience
POSC 363 – Cities and Regions in World Politics
PPD 410 – Comparative Urban Development
PPD 420 – Environmental Impact Assessment
PPD 461 – Sustainability Planning
SOCI 331 – Cities

GROUP B – DESIGN, ANALYSIS & COMPUTATION (8-16 UNITS)

ANTH 481 – GIS for Archaeology
ARCH 307 – Digital Tools for Architecture
ARCH 370 – Architectural Studies, Expanding the Field
FADN 102 – Design Fundamentals
HIST 493 – Quantitative Historical Analysis
PPD 306 – Visual Methods in Policy, Management, Planning and Development
PPD 427L – Geographic Information Systems and Planning Applications
SOCI 365 – Visual Sociology of the Urban City and Its Residents

CAPSTONE (4 UNITS)

SSCI / ARCH / PPD 412 – GeoDesign Practicum

Fig. 23.3 Schematic showing the preferred pathways students would follow to complete the University of Southern California's new B.S. in Geodesign degree

so they can learn first-hand how "form" and "function" work together and both the GIS and design classes utilize a mix of labs and studios in specially designed and dedicated learning spaces to promote skill development, teamwork, and collaboration. The planning classes, in turn, provide a series of pathways and protocols for combining collective and individual action to accomplish measureable and lasting change in the world. The capstone studio will be taught by faculty from the three contributing schools and will involve students working with real-world clients to solve one or more real-world problems.

Table 23.1 List of characteristics that are part of and not part of placemaking as envisaged by the Project for Public Spaces (PPS; http://www.pps.org)

Placemaking Is …	Placemaking Isn't …
Community-driven	Imposed from above
Visionary	Reactive
Function before form	Design-driven
Adaptable	A blanket solution
Inclusive	Exclusionary
Focused on creating destinations	Monolithic development
Flexible of the car	Overly accommodating
Culturally aware	One-size-fits-all
Ever changing	Static
Multi-disciplinary	Discipline-driven
Transformative	Privatized
Context-sensitive	One-dimensional
Inspiring	Dependent on regulatory controls
Collaborative	A cost benefit analysis
Sociable	Project-focused
	A quick fix

The overarching focus of this new geodesign degree is on placemaking, as conceived and promoted by the Project for Public Places (Table 23.1). The role of scale is highlighted throughout the program and the need for more sophisticated analytical and modeling approaches is introduced in both the core courses and in a series of electives that introduce and teach new skills and perspectives (Fig. 23.3). We realize that we cannot simultaneously train these new geodesign majors to be experts in architecture, computation, engineering, environmental design, geospatial technology, mathematics, science, and urban planning, and that the new B.S. degree will be a stepping stone for many of our students. With this in mind, we anticipate that some of our graduates will go directly to careers in environmental planning and design firms, in various government departments, and that others will go on to complete master's degrees in environmental science, geographic information science & technology, landscape architecture and planning, among others, before embarking on geodesign careers spread across the public, private and non-profit sectors.

The benefits of studying geodesign at the undergraduate level follow from the broad and deep introduction to design, planning and spatial sciences provided by this path and the opportunity for students to utilize this new knowledge and the accompanying skills to help clarify their future career and educational aspirations and needs. The reverse pathway (i.e. taking one of many undergraduate degrees and then a master's degree program in geodesign) relegates spatial thinking and the accompanying geospatial technologies to the role of Band-Aid in which spatial thinking and the accompanying geospatial technologies are introduced immediately prior to graduation. A much broader and deeper engagement with spatial ways of thinking are enabled by the B.S. degree in Geodesign.

No matter what the path chosen by our graduates, our hope is that some will grow into "conductors" and others will serve as "soloists" (as described in Steinitz 2012) and that all will help us to discover and implement future ways of living that are both rewarding and sustainable following some of the examples published in Yu and Padua (2006), Hou (2010), and McElvaney (2012).

23.6 Conclusions

This chapter has painted geodesign as an interdisciplinary field with five important and distinguishing characteristics: a focus on spatial thinking, geospatial technologies, the future, design as a force for good, and multi-disciplinary collaboration. Several examples were introduced to trace the evolution of the geodesign concept and to highlight some of the challenges that derailed early "geodesign" projects. The important role of the Web as a global platform and why this may be an ideal time to accomplish meaningful and lasting change were explained with the help of three recent geodesign initiatives. The Web provides a ubiquitous analysis and communication platform with the potential to transcend scale (i.e. move seamlessly across multiple geographic extents) and incorporate a multitude of voices and viewpoints in planning and decision-making workflows. The ways in which these aforementioned characteristics and a series of both early and recent examples have been utilized to guide the development of a new B.S. in Geodesign degree at the University of Southern California were briefly introduced and used to highlight some the challenges universities are likely to face as they work to create and sustain successful geodesign degree programs in an education setting in which disciplinary silos are still the norm and we are continually challenged to do more with less. My own hope is that this new degree will offer a vehicle to teach how spatial thinking can help to build vibrant and sustainable communities and lifestyles in the years ahead.

References

Beck, M. W., Ferdanìa, Z., Kachmar, J., et al. (2009). *Best practices for marine spatial planning.* Arlington: The Nature Conservancy.

Dangermond, J. (2012). Foreword. In C. Steinitz (Ed.), *A framework for geodesign: Changing geography by design* (p. vii). Redlands: Esri Press.

DiBiase, D. W., DeMers, M., Johnson, A., et al. (2006). *Geographic information science and technology body of knowledge.* Washington, DC: University Consortium for Geographic Information Science and Association of American Geographers.

DiBiase, D. W., DeMers, M., Johnson, A., et al. (2007). Introducing the first edition of the GIS & T Body of Knowledge. *Cartography and Geographic Information Science, 34,* 113–120.

DiBiase, D. W., Corbin, T., Fox, T., et al. (2010). The new geospatial technology competency model: Bringing workforce needs into focus. *URISA Journal, 22*(2), 57–72.

Faiz, S., & Krichen, S. (2013). *Geographical information systems and spatial optimization.* Boca Raton: CRC Press.

Fisher, T. (2012). Place-based knowledge in the digital age. *ArcNews, 34*(3), 1, 4–6

Ghaemi, P., Swift, J. N., Sister, C. E., et al. (2009). Design and implementation of a web-based platform to support interactive environmental planning. *Computers, Environment and Urban Systems, 33,* 482–491

Ghosh, A., & Rushton, G. (Eds.). (1987). *Spatial analysis and location–allocation models.* New York: Van Nostrand Reinhold.

Gleason, M., McCreary, S., Miller-Henson, M., et al. (2010). Science-based and stakeholder-driven marine protected area network planning: A successful case study from north-central California. *Ocean and Coastal Management, 53*(2), 52–68.

Goodchild, M. F. (2010). Towards Geodesign: Repurposing cartography and GIS. *Cartographic Perspectives, 66,* 7–22.

Hise, G., & Deverell, W. (2000). *Eden by design: The 1930 Olmsted–Bartholomew plan for the Los Angeles region.* Berkeley: University of California Press.

Hou, J. (Ed.). (2010). *Insurgent public space: Guerrilla urbanism and the remaking of contemporary cities.* London: Routledge.

Jankowski, P., & Nyerges, T. L. (2001). *Geographic information systems for group decision making: Towards a participatory geographic information science.* London: Taylor and Francis.

Malczewski, J. (1999). *GIS and multicriteria decision analysis.* New York: Wiley.

McElvaney, S. (Ed.). (2012). *Geodesign: Case studies in regional and urban planning.* Redlands: Esri Press.

McHarg, I. L. (1969). *Design with nature.* Garden City: Natural History Press.

Merrifield, M. S., McClintock, W., Burt, C., et al. (2013). MarineMap: A web-based platform for collaborative marine protected area planning. *Ocean and Coastal Management, 74*(2), 67–76.

Niemann, B. J., Moyer, D. D., Ventura, S. J., et al. (2011). *Citizen planners: Shaping communities with spatial tools.* Redlands: Esri Press.

Sinton, D. S. (2012). Spatial thinking. In J. Stoltman (Ed.), *21st Century Geography: A Reference-Handbook* (p. 733–744). Thousand Oaks: Sage.

Sinton, D. S., & Lund, J. J. (Eds.). (2007). *Understanding place: GIS and mapping across the curriculum.* Redlands: Esri Press.

Sister, C. E., Wolch, J., & Wilson, J. P. (2010). Got green? Addressing environmental injustice in park provision. *GeoJournal, 75,* 229–48.

Steinitz, C. (2012). *A framework for GeoDesign: Changing geography by design.* Redlands: Esri Press.

Wang, S., Anselin, L., Bhaduri, B. et al (2013) CyberGIS software: A synthetic review and integration roadmap. *International Journal of Geographical Information Science, 27,* 1–14.

Whyte, W. H. (1980). *The social life of small urban places.* New York: Project for Public Space.

Wilson, J. P., & Goodchild, M. F. (2012). Rethinking spatial sciences education programs. In T. Jekel, A. Car, J. Strobl, & G. Griesebner (Eds.), *Geoinfomatics forum 2012: Geovisualization, society and learning conference proceedings* (pp. 242–245) Berlin: Wichmann.

Wolch, J., Longcore, T., & Wilson, J. P. (2012). Unpaving paradise: The green visions plan for southern California. In D. Sloane (Ed.), *Planning Los Angeles* (p. 230–239) Los Angeles: APA Planners Press.

Worldwatch Institute. (2013). *State of the World 2013: Is sustainability still possible?* Washington, DC: Island Press.

Yu, K., & Padua, M. G. (Eds.). (2006). *The art of survival: Recovering landscape architecture.* Mulgrave: Images Publishing Group.

The manufacturer's authorised representative in the EU is Springer Nature Customer Service Centre GmbH, Europaplatz 3, 69115 Heidelberg, Germany. If you have any concerns regarding our products, please contact ProductSafety@springernature.com

Printed and bound by CPI Group (UK) Ltd, Croydon, CR0 4YY

23/03/2026

02076685-0001